Advances in Volcanology

Advances in Volcanology

Edited by Anna White

MURPHY & MOORE
www.murphy-moorepublishing.com

Published by Murphy & Moore Publishing,
1 Rockefeller Plaza,
New York City, NY 10020, USA
www.murphy-moorepublishing.com

Advances in Volcanology
Edited by Anna White

International Standard Book Number: 978-1-63987-023-3 (Hardback)

Cataloging-in-Publication Data

Advances in volcanology / edited by Anna White.
p. cm.
Includes bibliographical references and index.
ISBN 978-1-63987-023-3
1. Volcanology. 2. Volcanism. 3. Volcanoes. 4. Geology. I. White, Anna.
QE522 .N49 2022
551.21--dc23

TABLE OF CONTENTS

PREFACE

Volcanology is a discipline of geology that deals with all aspects of volcanic phenomena. It deals with the formation, distribution and classification of volcanoes, as well as their structure and the types of materials emitted during an eruption. Volcanology also includes research on the relationships between volcanic eruptions and other large-scale geological processes such as plate tectonics, mountain formation, and earthquakes. One of the main objectives of this research is to determine the nature and causes of volcanic eruptions with the aim of forecasting their occurrence. The topics included in this book on volcanology are of utmost significance and bound to provide incredible insights to readers. It aims to shed light on some of the unexplored aspects and recent researches in this field. This book includes contributions of experts and scientists which will provide innovative insights into the field of volcanology.

Significant researches are present in this book. Intensive efforts have been employed by authors to make this book an outstanding discourse. This book contains the enlightening chapters which have been written on the basis of significant researches done by the experts.

Finally, I would also like to thank all the members involved in this book for being a team and meeting all the deadlines for the submission of their respective works. I would also like to thank my friends and family for being supportive in my efforts.

Editor

1

Pyroclastic Density Current Hazards at the Baekdusan Volcano, Korea: Analysis of Several Scenarios from a Small-Case to the Worst-Case Colossal Eruption

Angelo Paone and Sung-Hyo Yun

Abstract

The Baekdusan volcano was formed through three stages of activity: (a) a basalt shield (aging between 22.6 and 1.48 Ma), (b) a trachytic comendite stratocone (aging between 1.19 and 0.02 Ma), (c) a trachyte-comendite ignimbrite deposits (aging from 20 ka till date). Volcanic seismicity, ground deformation, and volcanic gas geochemistry yield new evidence for magmatic unrest of the volcano between 2002 and 2006. The monitoring data suggest that Mt. Baekdusan is a potentially active volcano and that its close monitoring is needed. One of the possible volcanic hazards from this volcano is the pyroclastic density currents. In order to evaluate the small-scale pyroclastic flow emplacement scenario of the 1903 AD eruption, Titan2d mass-flow model is used. The 1668–1702 AD and the Millennium eruption are characterized by 4–5 and ~7 VEI, respectively. The Millennium eruption can be considered as the last colossal super-eruption like Tambora, and so Baekdusan volcano could have even a global effect. The parameters used are as follows: volume ($5–10 \times 10^7$, 1×10^9, 2×10^{10} m^3), the vents of the 1903, 1668–1702, and Millennium eruptions; the pyroclastic flow runout calculated in the field are small (3,000 m : 1903 eruption), intermediate (5,000 m :1668–1702 eruptions), and large (7–80,000 m, Millennium eruption). The initial velocities (m/s) range from 50 (1903 eruption) to as high as 300 (Millennium eruption). The input parameters have constructed three scenarios (1903, 1668–1702, and Millennium eruptions) following the recent volcanic history of Baekdusan volcano. These eruptions embrace all the possibly explosive eruptive scenarios that can occur at Baekdusan volcano in the future. The 1903 type scenario has been performed; according to the vent location, the flow moves in diverse directions (NE, SE) with a thickness of 3 m, and if the vent is the center of the caldera the flow fills the caldera with a thickness of 5 m. The emplacement of the 1668–1702 AD scenarios will involve at least two populated cities nearest to Baekdusan volcano. The scenario of the Millennium eruption will be much more severe and it will hit all the cities and towns within a range of 80 km. The impact will be from regional to global. This, however, is an underestimation of the runout for the diluted pyroclastic density currents.

Keywords: Baekdusan volcano, pyroclastic density currents, hazard, eruptive scenario, super-eruption, Titan2d

1. Introduction

When active volcanoes are located in strategic and controversial territory with a history of political and military conflicts [1, 2], their studies and the better understanding of how those volcanoes work are much more difficult, especially when the earth scientists have to deal with mitigation measures to protect the population living around [3–6]. This is surely the case of an active volcano known by a few until 15 years ago, situated between the China and North Korea borders, called by different names by the countries that surround it. The name of the volcanic field surrounding Tianchi(=Cheonji) caldera lake is 白頭山 in Chinese characters which means "White Head Mountain" These characters are Romanized as Baegdu-san, Baekdu-san, or Paektu-san in Korea, Hakuto-san in Japan, and Changbaishan(長白山) in China [7, and references therein]. From now on, we will use the Korean name, Baekdusan, to avoid confusion. Baekdusan volcano has recently become of greater interest to volcanologists after the sign of unrest during the period 2002–2009, with several volcanic precursors such as volcanic earthquake, surface inflation, specific volcanic gas emission, higher temperatures of hot springs [8–10], which have prompted the Eastern Asian Volcanological Community to pay major attention to this volcano by establishing a Volcanological Observatory (the Chinese Tianchi Volcanological Observatory has collected data since 1999 [11]) and by setting up an improved monitoring network around the volcano [12, 13]. Since 2012, the South Korean scientists have also taken steps in this direction, and the National Emergency Management Agency of Korea (Ministry of Public Safety and Security of Korea) has sponsored a vast research project on Baekdusan volcano (NEMA-BAEKDUSAN-2012-1-2), granted from the Volcanic Disaster Preparedness Centre (MPSS-NH-2015-81), and through the Natural Hazard Mitigation Research Group. Under the auspices of these groups, we analyzed the hazards from pyroclastic density currents and their impact on the population and infrastructures around Baekdusan volcano [14]. Further, we present arguments as to why this volcano has many characteristics as a potential site for the development of the conditions that can cause a super-eruption. Thus, we must also consider this extreme event if we do have to prepare for the next eruption, as it would have a global impact [15].

2. Regional tectonic of Baekdusan volcano and an outline of the plumbing system

Baekdusan volcano is an intraplate stratovolcano included in the larger Changbaishain area, where we can list a series of intraplate volcanoes (Wangtiane volcano, Bukpotaesan volcano, Zhenfengshan volcano, [7], **Figure 1**). Cenozoic intraplate basalts are widely distributed around the Changbaishan volcanic area. Beneath the Baekdusan volcano, a prominent low-velocity anomaly with a plume-like shape has been imaged in the upper mantle by P-wave tomography [16, 17]. This suggests an upwelling of hot and wet material from the mantle transition zone (410–670 km [18]), where a stagnant subducted Pacific slab produces fluids and deep earthquakes through faults preserved in the stagnant slab. Hence, fluids and magmas may be supplied to the upper mantle under Baekdusan volcano, unlike beneath other volcanic and nonvolcanic areas in NE Asia, because large deep earthquakes occur frequently in the vicinity (a distance of about 200 km) of the Baekdusan volcano. At least some of these earthquakes are caused by the reactivation of faults which were produced within the plate, even after plate subduction. Further, the fluids could be released to the overlying mantle wedge from the preserved faults in the slab when large deep earthquakes occur. As a result, the Baekdusan volcano has to be considered much more active and prominent than other intraplate volcanoes in NE Asia [9, 10], because more magmas are produced from the fluids released from the slab by the deep earthquakes [19]. Further, Kuritani et al. [20] use the trace element and Pb isotope data to demonstrate the intense hydration of the big mantle wedge beneath the Baekdusan volcano, through two different dehydration events of an ancient stagnant slab (~1.5 Gyr) and the most recent subducted Pacific slab (**Figure 2A** and **B** show the most recent setting). The wet mantle wedge is likely to result in more frequent rise of diapiric wet melts, as shown from the tomography [17, references therein], and develop a deep hot zone intruded by mafic sills repeatedly at the MOHO depth or/and scattered in the lower crust [21, and references therein]. Enrichment in fluid-mobile elements (e.g. Cs, Ba, K, Sr, U, Pb), a lack of enrichment in nonfluid-mobile elements (e.g. Nb, Ta, Ti, etc), together with specific isotopic signatures (e.g. $^{87}Sr/^{86}Sr$, $^{206}Pb/^{204}Pb$) in island-arc basalt or subduction- related magmatism are interpreted as a result of metasomatism of the magma source by slab-derived fluids or melts. Their detailed

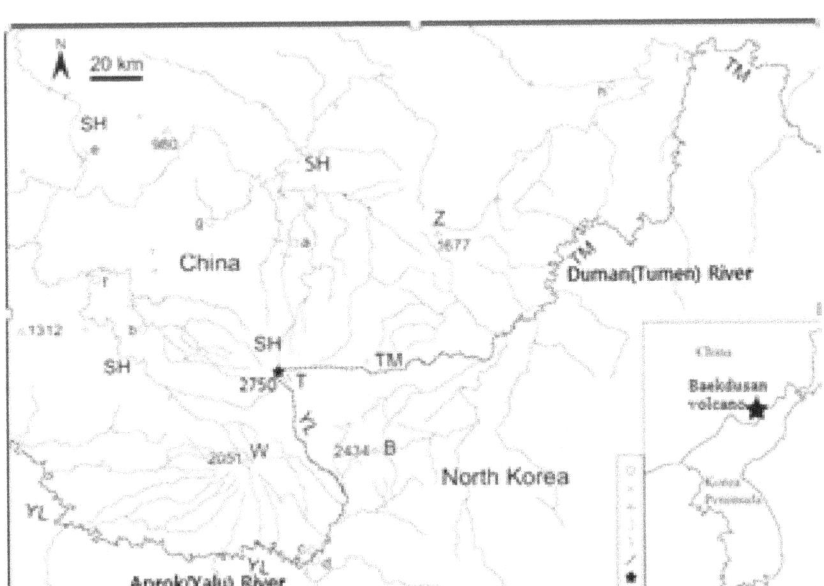

Figure 1. Regional setting of Baekdusan (Tianchi) volcano. Symbols mean the following. ○ Densely populated city or town: (a) Erdaobaihe; (b) Songjianghe; (c) Changbai; (d) Hyesan; (e) Baishan; (f) Fusong; (g) Lushuihe; (h) Yanji, and (i) Tumen. ° Populated village. Δ Mountain tops with the elevation asl. ★ Baekdusan (Tianchi) volcano. Capital letters indicate places referred to in text: T, Baekdusan (Tianchi) volcano; W, Wangtiane volcano; B, Bukpotaesan volcano; Z, Zhenfengshan volcano; SH, Songhua River; YL, Yalu(Aprok) River; and TM, Tumen (Duman) River.

investigation has helped to establish the current, rather empirical, model of slab dehydration and subduction-related melting processes. But, according to petrochemical characteristics and magma genesis of Holocene volcanic rocks at Mt. Baekdusan [22], trace elements from the Baekdusan volcano do not indicate significant contributions from a subduction slab of the Pacific plate. Volcanic activity in Mt. Baekdusan summit during Holocene resulted from intraplate magmatism [after 23, 24], caused by upwelling of asthenospheric mantle under extensional tectonic conditions (**Figure 3** [24]). Hot zone melts are H_2O-rich as in Baekdusan volcano (e.g., Millennium eruption [25]). Consequently, they have low viscosity and density, and can readily detach from their source and ascend rapidly. Crystallization begins only when the ascending magma intersects its H_2O-saturated liquidus at shallow depths. Decompression and degassing are the driving forces behind crystallization, which take place at shallow depths

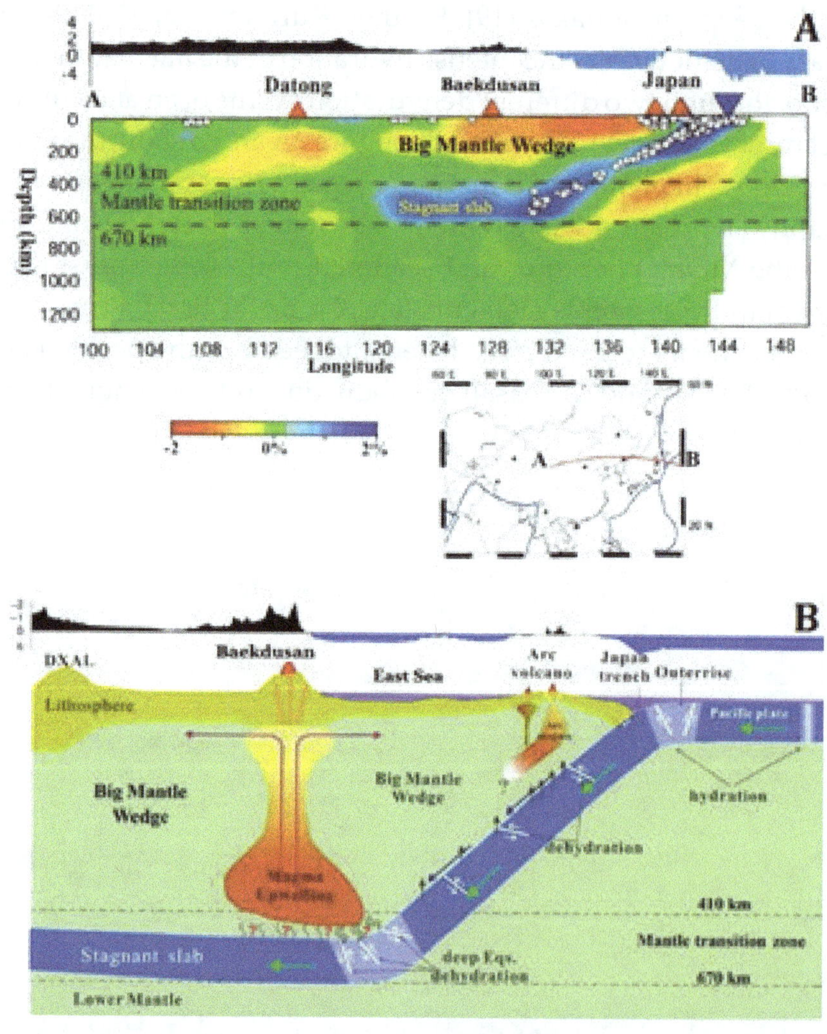

Figure 2. A. Regional tomography from the surface down to 1,300 km depth along a cross-section line shown on the inset map. The velocity perturbation scale is shown below the cross section. White dots denote earthquakes. The two dashed lines show the 410 and 670 km discontinuities. The red and black triangles show the active volcanoes. The surface topography along the profile is shown at the top of the tomographic image. The blue lines on the inset map denote the major plate boundaries (modified from [17]). **B.** A sketch showing the main features of the upper-mantle structure around Baekdusan volcano in East Asia, emphasizing the possible relationship between the intraplate volcanism and deep earthquakes in the Pacific slab [17], DXAL, Daxing-Anling (modified from [17]).

on timescales of decades or less. Degassing and crystallization at shallow depths lead to large increases in viscosity and stalling of the magma to form volcano-feeding magma chambers [26, 27]. The above model is well confirmed from the study of Wei [24] at the Baekdusan volcano. The heating of the deep crust was inefficient in early times, because much of the basalts erupted out directly, and much of the heat from magma was not used to heat the wall rocks. Annan and Sparks [27] state that there is an initial incubation period in which the basalt intrusions solidify. Generation of silicic melts initiates when the solidus temperature of either the basalt magma or the surrounding crust is reached. At an intrusion rate of 50 m per 10,000 years, incubation periods in the range 10^5–10^6 years are estimated, consistent with geochronology and stratigraphy at Baekdusan volcano, with its evolution from mafic to silicic volcanism (the stratigraphy and geochronology will be treated in the next section). Once a zone of residual partial melt develops in the deep hot zone and starts to prevent the basalt from eruption, the temperature rises more efficiently, and the production of residual melt is accelerated. Large amounts of evolved residual silicic melt have been formed in the last 10,000 years and then intruded into the upper crust to form magma chambers large enough to result in the caldera-forming explosive eruptions. The repose time between eruptions of size similar to the Millennium eruption was about thousands of years, one order greater than less-explosive parasitic eruptions of hundreds of years [7].

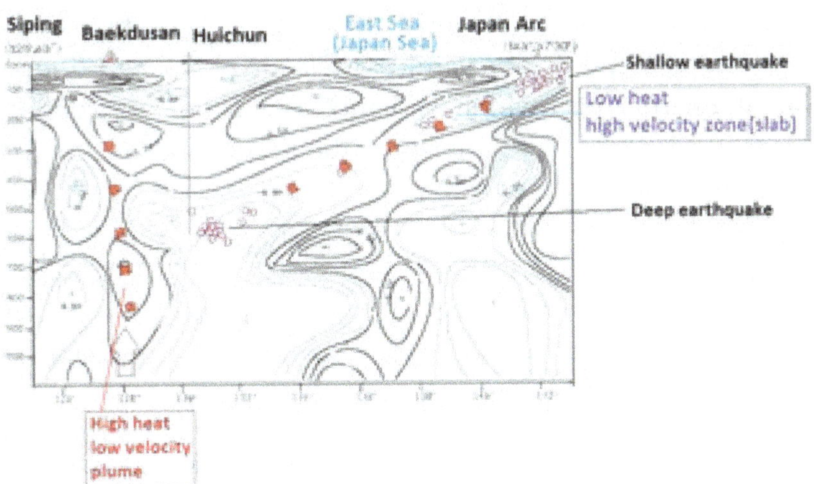

Figure 3. Siping–Changbaishan–Japan seismic profile of the structure of Changbaishan area [24].

3. Summary of stratigraphy and geochronology of Baekdusan volcano

The volcanic activity at the Baekdusan volcano goes back as far as nearly 25 Ma B.P [7, 28, 29]. Wei et al. [7] describe the volcanic activity in three stages: early shield-forming; middle cone-construction; and late ignimbrite-forming eruptions. The volcanic rocks were alkaline and tholeiitic basalts during the shield stage, trachyte and comendite lavas, and their pyroclastic equivalents during the cone-construction stage, and mostly comenditic pyroclastic rocks in the final stage (**Table 1**, **Figure 4**). However, Yang et al. [30] affirm on the basis of a large amount of age and stratigraphic data ($^{40}Ar/^{39}Ar$, U-series disequilibria) that Baekdusan volcano has been more active in the last 20 ka (~19, 16, 11 ka) than that was previously thought. From historical records, several authors [31, 32] affirm the high frequency of eruptive events, and since the Millennium eruption, more than 31 eruptive events have been documented, most of which are the Plinian eruptions with volcanic ash that dispersed into the regions in the vicinity of the volcano. Many authors concord that the last explosive eruptions are in chronological order, 1668 AD (**Figure 5**), 1702 AD (**Figure 6**), and 1903 AD (**Figure 7**) [32–34]. In **Figure 4** is shown the Baekdusan volcano's geological map with all the terrains outcropping (see the legend of **Figure 4** for more major details). The shield-forming stage lasted from the Miocene into the early Pleistocene. Alkali olivine and tholeiite basalt form the main part of a shield-like lava plateau that covers an area of 7,200 km^2 centered at Cheonji. In addition to the eruptions from central vents, there also were fissure eruptions, as evidenced by feeder dikes along the headwaters of the Tumen (Duman) and Heishihe Rivers. The youngest shield basalts are located near Cheonji cone and are overlain by the alkaline trachytes of the cone, except in the northwestern part of the shield. The shield basalts are ~150 m thick near the Baishan Forestry Center in the headwaters of the Yalu River (Aprok River). The cone-construction stage lasted from ~1.0 Ma to perhaps as young as ~20 ka. The latter is about the age of the youngest trachyte lava on the caldera rim near Tianwenfeng (K–Ar, 19 ± 5 ka) and one of the youngest comendite lavas, Qixiangzhan, which was dated recently at 17 ± 1 ka. This stage was characterized by eruption of trachyte and comendite lavas and pyroclastic rocks, and dominated by trachyte lava effusion from the center of Cheonji volcano. The cone-construction stage has been subdivided into four stratigraphic formations for the main cone and the Laofangzixiaoshan

Stages	Formation and age	Lithologies
	Liuhaojiie Tuff Ring (1903 AD)	Comendite phreatomagmatic eruption
Late:	Wuhaojie (1702 AD)	White gray comentite fine glass
Igninbrite-forming	Baguamiao (1668 AD)	Dark gray trachyte ignimbrite and pumice
	Millennium eruption (~936–1200 AD)	White gray comentite ignimbrite and Air fall pumice with minor trachyte Ignimbrite and air fall pumice
	Qixiangzhan (17 ka)	Comentite lava and pyroclasts
Middle: Cone	Baitoushan III (0.02–0.22 Ma)	Trachyte and comendite lava
construction	Baitoushan II (0.25–0.44 Ma)	Trachyte with Laohudong basalt
	Baitoushan I (0.53–0.61 Ma)	Trachyte
	Laofangzixiaoshan (0.75–1.17 Ma)	Basalt
	Xiaobaishan (0.75–1.17 Ma)	Trachyandesite and trachyte
	Baishan (1.48–1.66 Ma)	Basalt
Early: Shield-	Toudao (2.35–5.02 Ma)	Basalt
forming	Naitoushan (15.6–22.6 Ma)	Basanite, Basalt

Table 1. Baekdusan volcano stratigraphy and petrography of the volcanic products simplified after Wei et al. [7, 28, 29].

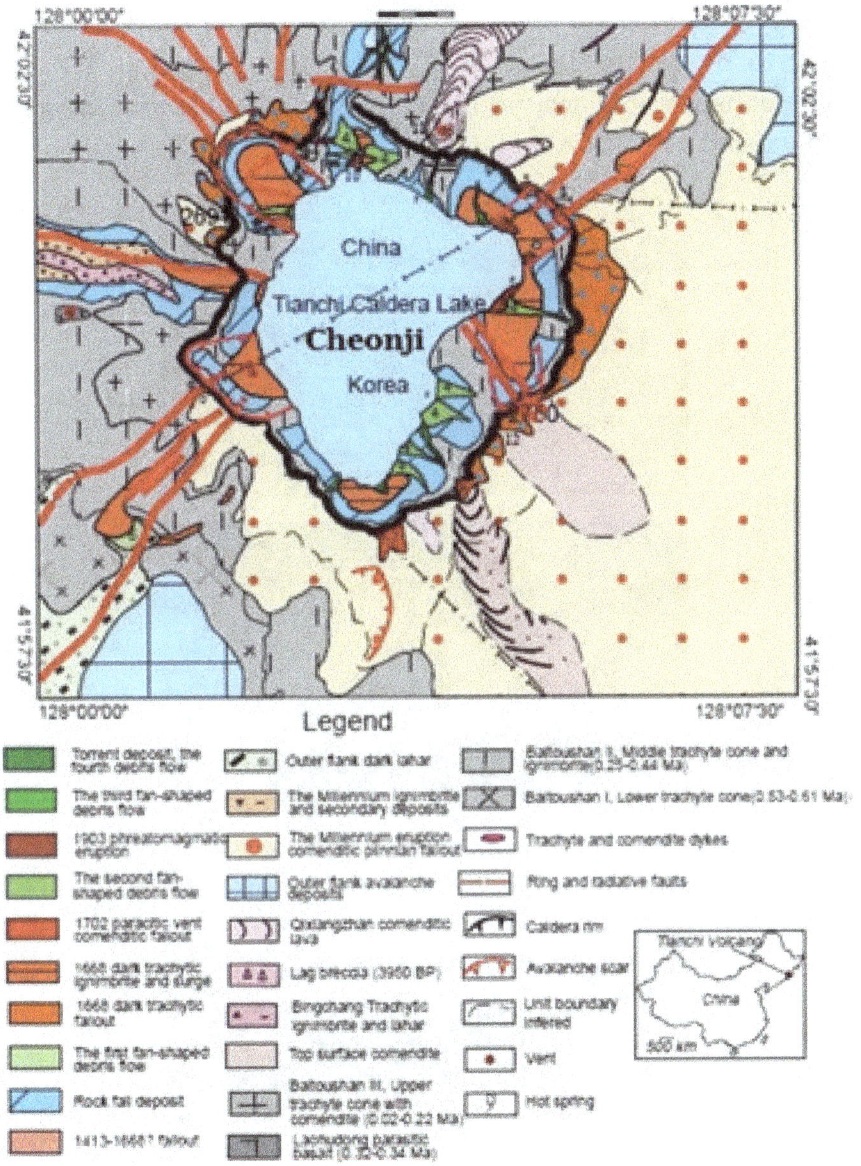

Figure 4. Geological map of the summit area of Baekdusan volcano. Northeast distribution of the basaltic shield, trachytic composite cone, and the comenditic ignimbrite sheet of the Baekdusan volcano. 1, Torrent deposit, the fourth debris flow (1994); 2, The third fan-shaped debris flow; 3, 1903 phreatomagmatic eruption; 4, The second fan-shaped debris flow; 5, Dark trachytic ignimbrite and surge (1668); 6, Dark trachytic fallout (1668); 7, The first fan-shaped debris flow; 8, Rock fall deposit; 9, Fallout (?1413–1668); 10, Outer flank dark lahar; 11, The Millennium ignimbrite and secondary deposits; 12, The Millennium valley and pond ignimbrite; 13, The Millennium eruption comenditic Plinian fallout; 14, Outer flank avalanche deposits; 15, Qixiangzhan distal comenditic lava; 16, Qixiangzhan middle comenditic lava; 17, Qixiangzhan proximal comenditic lava; 18, Qixiangzhan calstogenic lava, comendite, and obsidian; 19, Ice ground brown and gray trachytic ignimbrite, lahar; 20, Heifengkou Lag breccias (3950 ±120 a); 21, Upper trachyte and comendite cone (0.04–0.2 Ma); 22, Middle trachyte cone (0.25–0.44 Ma); 23, Lower trachyte cone (0.53–0.61 Ma); 24, Xiaobaishan trachyandesite and trachyte (1–1.49 Ma); 25, Laohudong parasitic basalt and scoria (0.32–0.34 Ma); 26, Laofangzixiaoshan basalt (0.31–0.58 Ma); 27, Baishan basalt (0.93–1.39 Ma); 28, Toudao basalt (1.91–2.77 Ma); 29, Comendite dyke; 30, Trachytic welded tuff dyke; 31, Trachyte dyke; 32, Ring and radioactive faults; 33, Caldera rim; 34,Avalanche scar; 35,Vent; 36, Hot spring; 37, Inferred geologic boundary; 38, Flow units boundary; 39, Facies boundary; 40, National boundary. β, Basalt; τ, Trachyte; λ, Comendite; π, Porphyry dyke (modified from [7]; Yun, unpublished data).

part of the shield (**Table 1**). In general, outcrops of lava decrease image with elevation on the cone. In plane view, the eastern sector of the cone consists of younger trachytes, whereas older lavas are distributed more widely. However, all the rocks share similar geochemical compositions and appear to be genetically related. The young Laohudong basalts are sparsely distributed as monogenetic vents around the caldera. Its largest and most explosive Plinian eruption occurred around 946–947 AD and is known as the "Millennium eruption."

Figure 5. The Buguamiao welded ignimbrite of the 1668 AD pyroclastic flow eruption is shown in all the pictures. The outcrops are in the vicinity of the Tianchi caldera. In the Baguamiao ignimbrite is also apparent its welded character.

Figure 6. The 1702 AD pyroclastic flow eruption on the summit of the cliff.

Figure 7. The 1903 AD pyroclastic flow deposits. The outcrop is separated by levels. The 1° level belongs to the 1702 AD eruption, and the others belong to the 1903 pyroclastic flow eruption. It is clear that the thickness of the 1903 deposit is near the caldera.

This eruption has a volcanic explosivity index (VEI) of 7, comparable to the 1815 eruption of Tambora in Indonesia. Horn and Schmincke [25] have suggested that the Cheonji eruption column reached ca 25 km in height, injecting about 9×10^{10} kg of volatile halogens into the atmosphere, with an estimated volume of pyroclastic material on the order of ~100 km^3. Volcanic ash from the eruption has been found 1,000 km away from the site of the eruption. A 5 cm thick deposit was observed in southern Hokkaido, Japan [38]. Accordingly, the Millennium eruption of the Baekdusan Cheonji volcano has been considered as one of the greatest volcanic eruptions of the past 2000 years.

3.1. An outline of the Millennium eruption

One of the most largest destructive eruptions from the Baekdusan volcano occurred on Earth in the last millennium, about 1,000 years ago, and it is named the Millennium eruption [25, 29, 34, 35]. It involved more than ~100 km^3 of magma erupted [10], and the Plinian column was estimated to have had a height of 25 km. A widespread ash-fall deposit has been traced across the Sea of Japan to Hokkaido, and it is several centimeters thick, at more than 1,000 km from the source [35, 37]. Unwelded pyroclastic flows associated with collapse events of the Plinian eruption column extended to more than 70 km from the crater rim (**Figure 8**). The Millennium eruption formed a large caldera with a diameter of about 5 km and an area of 20 km^2, forming the Cheonji (Tianchi = Sky) lake. Currently, the lake has a maximum depth of 374 m and an area of ~ 9.8 km^2, with a water surface elevation of 2,189 m. It has a volume of 2 km^3. A composite stratigraphic section of the Millennium eruption is given by Machida et al. ([35]; **Figure 9**). According to Machida et al. [35], the section is broadly composed of five levels (A, B, C, E, F; for major details, see **Figure 9**). **Figure 10** shows some unwelded Millennium pyroclastic flow outcrops with relative thicknesses and sites. However, there is a controversy if the complete volcanological section of the Millennium eruption has to be interpreted as a single continuous eruption rather than one with two distinct pulses, with the initial phase anticipating the major phases of tens or hundreds of years [39]. The scientific literature reports a large range of ages for the Millennium eruption [40]. The results show a variety of periods ranging from approximately AD eighth to fourteenth centuries, which are the dates of the Balhae(= Bohai) and Goryeo dynasties [40, and references therein]. The last geochronological data (radiocarbon wiggle-match dating) on the Millennium eruption show a much more restricted range, respectively, of 938–939 AD [41], 946–947 AD [42], and 1024 AD [28].

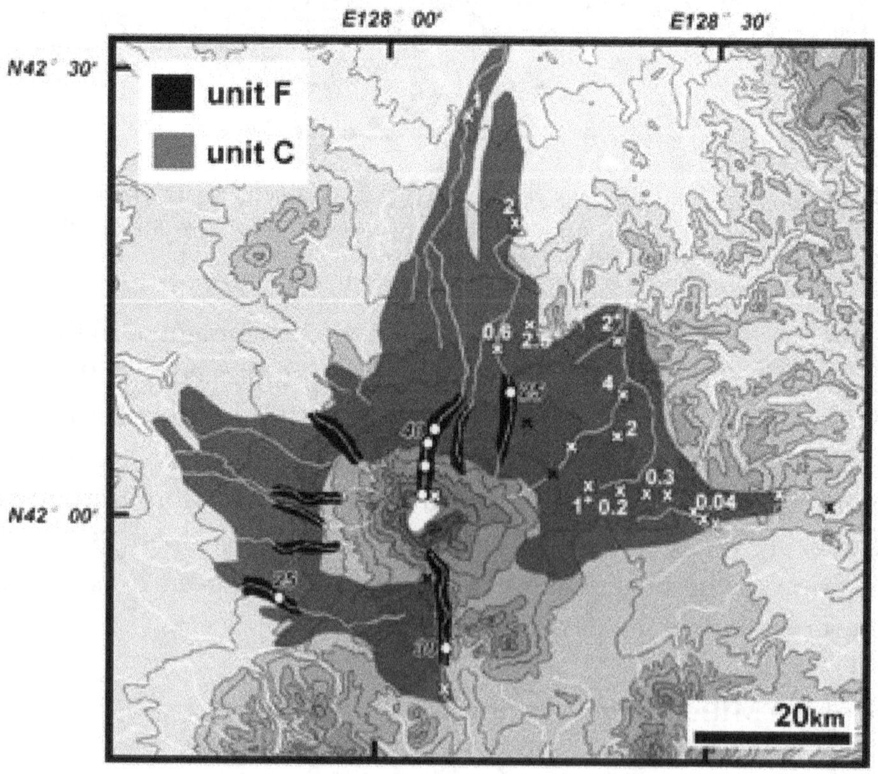

Figure 8. Distribution of pyroclastic density currents of the Millennium eruption in China side. Numbers are thickness (m). Units C and F are the two pyroclastic flow levels.

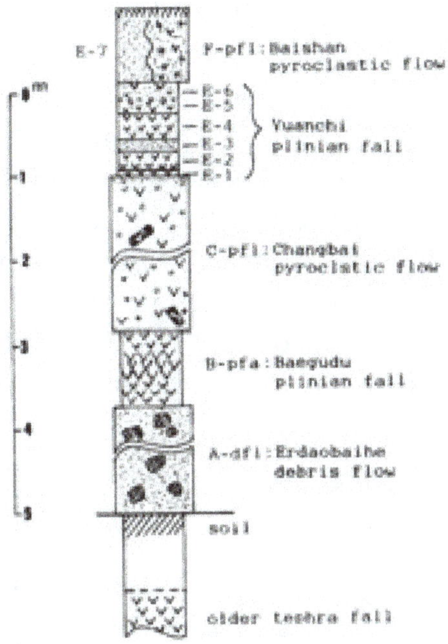

Figure 9. Stratigraphy of the Millennium eruption on the eastern slope of Baekdusan volcano (modified from [35]). pfa, pyroclastyc fall deposit; pfl, pyroclastic flow deposits; dfl, debris flow deposits.

Figure 10. Millennium unwelded pyroclastic flow outcrops at the north and south flanks of the caldera with the respective thicknesses. (**A**) Millennium eruption pyroclastic flow deposits, northern slope: 20 m thickness. (**B**) Millennium eruption pyroclastic flow deposits, southern slope: 60 m thickness.

4. Scenarios of the 1903 AD, 1668–1702 AD, and the Millennium eruptions

The scenarios envisaged here are based on the large amount of data that have been collected by the Korean research project, CATER2011-5210 [36], and by the new literature produced [25, 32–34, among others] in order to assign a selected VEI (Volcanic Explosivity Index [43]). The main parameters are summarized in **Table 2** (VEI, maximum volume erupted, and column height) for each eruption. The volumes considered for just the pyroclastic density currents are obviously less than the total volume and are listed in **Table 3**. **Figure 11** shows that between several volcanic-related disasters, the pyroclastic density currents have to be considered the most dangerous, given the number of fatalites in the last 2,000 years.

Eruption date	VEI	Volume (m^3)	Column height (km)
1903 AD eruption	2	10^6–10^8	1–5
	3		3–15
1668–1702 AD eruption	4	10^8–10^{10}	10–25
	5		>25
Millennium eruption	7	10^{11}–10^{12}	>25

Table 2. Summary of the main parameters for the 1093 AD, 1702 AD, 1668 AD, and the Millennium eruptions. Volumes are taken from Newhall and Self [43] compared with the VEI for each eruption.

Eruption type	Runout length (m)	Volume (m^3)	Initial velocity (m/s)	Time duration (s)
1903-type	~3000	10^7–10^8	30	3600
1668–1702 type	~4000–5000	10^9	50	>3600
Millennium eruption	7–80,000	10^{10}	80–300	>3600

Table 3. Main parameters used to analyze the impact of the pyroclastic density currents and to show small-scale simulations [34, 61].

Branney and Kokelaar [44] illustrate in detail how different types of pyroclastic density currents originate (**Figure 12**). From **Figure 12**, the pyroclastic density currents most likely to occur at Beakdusan volcano are the following: A, B, C, depending on the size and the amount of interaction with external water. For laterally moving systems, the pyroclastic density currents can be identified as two end-member types: (A) Concentrated currents or pyroclastic flows, (B) Dilute currents or pyroclastic surges (Note that these are often a spectrum, with gradations in between). The end-member A has the following main characteristics:

1. It has solids in concentrations of tens of volume percentage, and so they have higher densities than surges.
2. They have a free surface, above which solids' concentration decreases sharply.
3. Transport of material is by fluidization, and most flows are considered to be laminar.

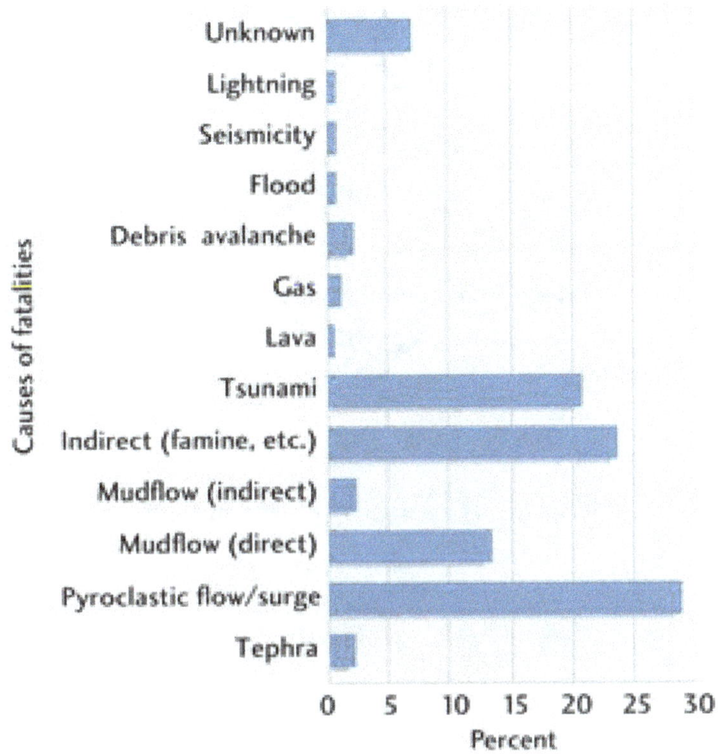

Figure 11. Statistical percentage of fatalities for volcanic-related phenomena in last 2000 years. Pyroclastic density current is the most deadly phenomena in the last 2,000 years (Figure taken from the web and modified).

4. Velocities vary, typically by 10s of meters per second up to several hundred meters per second, inferred from the heights of obstacles overcome by flows.

The end-member B has the following main characteristics:

1. It contains less than 0.1–1.0% by volume of solids, even near ground surface, and so they are relatively of low density.
2. They are density-stratified, with the highest particle concentration near the ground surface.
3. Material is transported primarily by turbulent suspension.
4. Transport system is modeled as one that loses particles by sedimentation, which depletes the system of mass. Eventually, the system may become buoyant, in which case it becomes a plume.
5. Velocities vary, typically by 10s of meters per second, but they can also be much faster [44–48].

In fluid dynamics, the equations that govern the concentrated current are well known [49, 50], and the mathematics based on this volcanic phenomena is broadly used to model pyroclastic flows, along with other gravity currents: avalanches, seafloor turbidity currents, lahars, and lava flows. Two numerical codes apply these equations with small differences: Titan2D [51] and Volcflow [52, 53], but so far they have not produced any modeling for diluted currents.

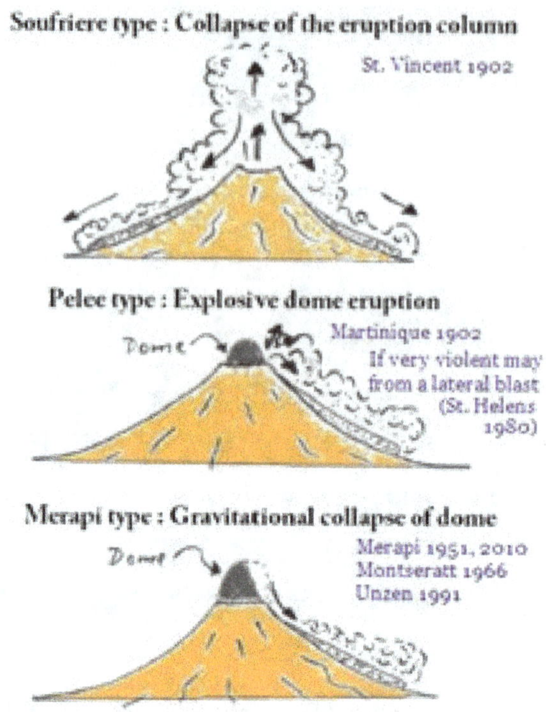

Figure 12. Origins of pyroclastic density currents. (A) Short single-surge current derived by momentary collapse from a Plinian eruption column. (B) Sustained current derived from prolonged pyroclastic fountaining. The height of the jet (gas thrust) that feeds the current may vary and is transitional into (C). (C) A sustained current derived from a prolonged low pyroclastic fountaining explosive eruption. (D) Current with a single (or multiple) surge derived from lateral blasts initiated by catastrophic decompression of a magmatic and/or hydrothermal system. (E) Single-surge current derived from a collapsing lava dome or flow front. Hot rock avalanches generate turbulent density currents. (F) Deposit-derived pyroclastic density current caused by gravitational collapse and avalanching of unstable loose ignimbrite. The current may be a single-surge or more sustained where the collapse is retrogressive. Most large-volume ignimbrites derive from current types (B) and (C), which may involve periods of quasi-steady flow. Many may include significant components derived from currents of type (F) [44, 45]. For Baekdusan, we chose the following three types: Soufriere, Pelee, and Merapi types (modified from [62]).

The only numerical code that models the diluted currents is called PDAC [54]. An application is shown for Somma-Vesuvius volcano, where pyroclastic density currents are modeled on the data of the 1631 AD sub-Plinian eruption ([55]; **Figure 13**). Kelfourn [53] tried to model two-fluid model block-ash flows and cloud-ash surge at Merapi volcano, developing new equations, but there is not yet an agreement (Kelfourn, personal communication). The importance of modeling diluted currents together with dense currents for the hazard of an explosive active volcano is commonly witnessed from the volcanological history, as seen in the work of Mastrolorenzo [56] who demonstrate the causes of mortality in PDCs at Pompeii and surroundings, on the basis of a multidisciplinary volcanological and bioanthropological study. Field and laboratory studies of the eruption products and victims, merged with numerical simulations and experiments, indicate that heat was the main cause of death. Their results show that exposure to at least 250°C hot surges at a distance of 10 km from the vent was sufficient to cause instant death, even if people were sheltered within buildings. Despite the fact that impact force and exposure time to dusty gas declined toward PDCs' periphery up to the survival conditions, lethal temperatures were maintained up to the PDCs' extreme depositional limits. This study suggests that diluted currents are very dangerous, and if the runout of the pyroclastic flow must be chosen, it is always underestimated when considering diluted currents, especially for large eruption.

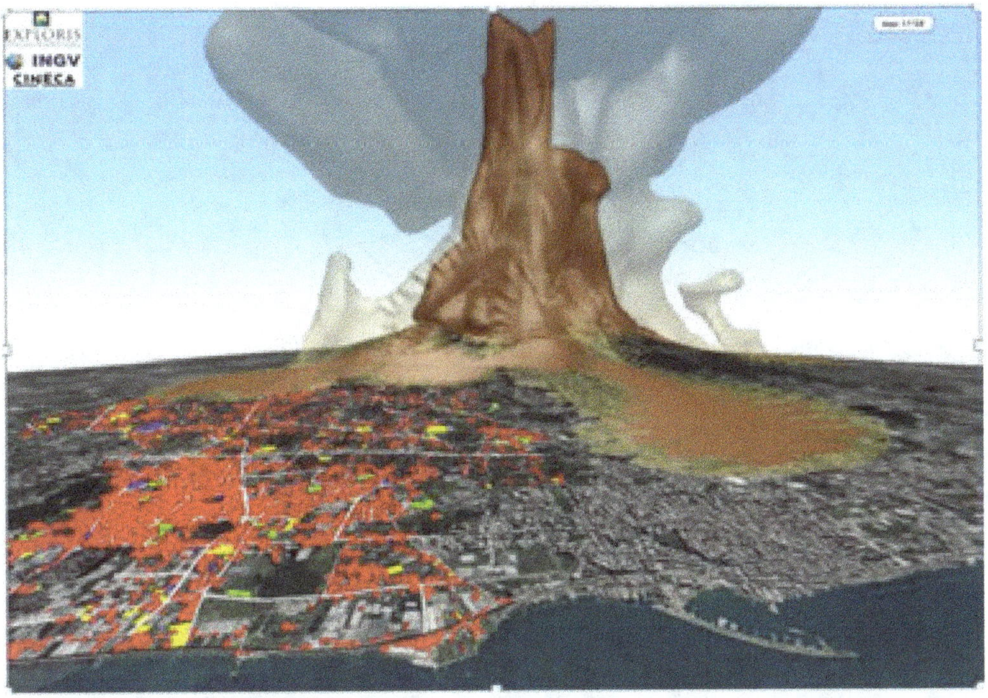

Figure 13. 4D immersive visualization of the simulated partial column collapse scenario at Vesuvius. The collapse of the volcanic eruption column and the propagation of pyroclastic density currents (PDCs), for selected medium-scale (sub-Plinian) eruptive scenarios at Vesuvius, Italy (modified from [55]).

5. Relationship between the Baekdusan eruptions: 1903 AD, 1668–1702 AD, and Millennium eruptions, and the most devastating eruptions in the recent history of the earth

As indicated in **Table 1**, the 1903 AD, 1668–1702 AD, and the Millennium eruptions are characterized by the following VEI, respectively: 2–3, 4–5, and ~7. From the classification of Newhall and Self [43], the magnitude of each eruption with all the volcanological parameters can be deduced. If we compare the total volume ejected from Baekdusan volcano in the last Millennium activity with the most devastating eruptions in the recent history of volcanology, it appears that the two most recent eruptions shown, from Mount St. Helens and Mount Pinatubo, each caused extensive damage and received intense media interest. Ejected volumes, however, were only about 1/10,000 (St. Helens) and 1/500 (Pinatubo) of the volumes associated with either the 74 ka Toba or the 2 Ma Yellowstone super-eruptions. Even the 1883 Krakatau eruption, which accounted for over 35,000 deaths, was smaller than the Toba and Yellowstone eruptions by two orders of magnitude. The 1815 Tambora eruption, which had an ejected volume of less than 5% of those of Toba or Yellowstone and was distinctly less than "super," affected global climate and was responsible for over 50,000 deaths (**Figure 14A** and **B** [57]).

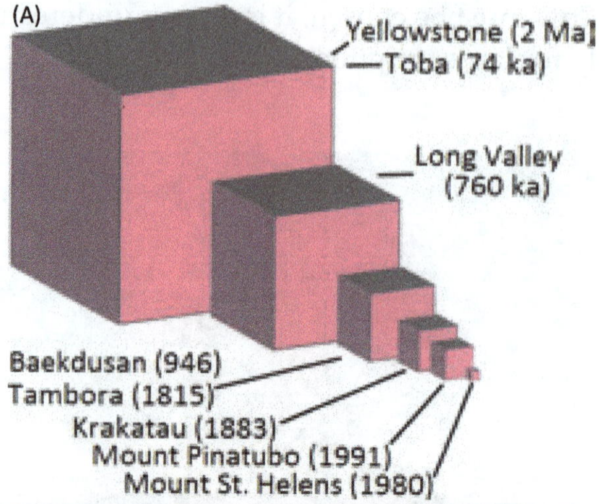

(B) VEI	Plume height (km)	Ejected volume (km³)	Frequency on Earth	Example
0	<0.1	>~10^{-6}	daily	Kilauea, Hawai'i
1	0.1–1	>~10^{-5}	daily	Stromboli, Italy
2	1–5	>~10^{-3}	weekly	Galeras, Colombia, 1993
3	3–15	>~10^{-2}	yearly	Nevado del Ruiz, Colombia, 1985
4	10–25	>~10^{-1}	~every 10 y	Soufrière Hills, West Indies, 1995
5	>25	>~1	~every 50 y	Mount St. Helens, USA, 1980
6	>25	>~10	~every 100 y	Pinatubo, Philippines, 1991
7	>25	>~100	~every 1000 y	Tambora, Indonesia, 1815
8	>25	>~1000	~every 10,000–100,000 y	*Supereruptions:* Toba, 74 ka

Figure 14. A. Relative volumes of pyroclastic material erupted from the large eruption in the 2 Ma [57]. **B**. Classification of the VEI modified after [4] (modified from [57]).

The 1903 AD eruption can be compared to two recent eruptions: (A) Nevado del Ruiz, Colombia, 1985, (B) Soufrière Hills, West Indies, 1995 (**Figure 14B**). The two eruptions of 1668–1702 AD of similar size can be compared to the following eruptions: (A) Soufrière Hills, West Indies, 1995, (B) Mount St. Helens, USA, 1980 (**Figure 14B**). The Millennium eruption may have been slightly larger than the 1815 Tambora eruption, Indonesia. Millennium eruption can be considered a super-eruption alike other eruptions (e.g., Campanian Ignimbrite, Campi Flegrei, Napoli [58]; **Figure 15**). This assumption is confirmed from some South Korean scientists that during the 2nd International Workshop for Volcanic Disaster Preparedness at Chungbuk National University, Korea affirmed that the Millennium eruption could have been as high as 7 VEI. Further, Xu et al. [42, 59] suggest that the Millennium eruption was a Toba-like "ash giant/sulfur dwarf" and had much smaller global climatic impacts. Sun et al. [60] confirm that the eruption most probably occurred around AD 940s, 7 years after the Eldgjá eruption on Iceland, and examining the eruption's potential for climate-forcing using the sulfate records from the ice cores, they conclude that it was unlikely to have had a global impact. Zircons U–Th data of Millennium eruption indicate that magma resided only ~8 ka prior to eruption. Therefore, Baekdusan volcano can produce catastrophic, explosive eruptions in the foreseeable future [36]. Thus, Baekdusan is the most active volcano in China [14], and it is a high-risk volcano because more than 100,000 people live on or near the slopes, with the addition of many tourists.

Eruption magnitude or VEI (Volcanic Explosivity Index)	Minimum erupted mass (kg)	Minimum volume of magma erupted (km^3)	Minimum volume of ash (km^3)	Example of typical eruption	Frequency (average number of eruptions per 100 years)	Minimum probability of one or more eruptions of this size during 21st century
7 (low)	1×10^{14}	40	100	An event a little larger than Tambora, 1815	0.1 – 0.5	10 – 50%
7 (moderate)	2.5×10^{14}	100	250	Possibly Kikai, Japan, 6000 years ago	0.01 – 0.06	1 – 6%
7 (high)	8×10^{14}	300	750	Campanian, Italy, 35,000 years ago	0.001 – 0.01	0.1 – 1%
8 (low)	1×10^{15}	400	1000	Taupo caldera, New Zealand, 26,000 years ago	<0.001	0.1%
8 (high)	8×10^{15}	1200	>5000	A Toba-size event	0.0001	Approximately 0%

Figure 15. The Volcanic Explosivity Index (VEI) is calculated with a logarithmic-based scale of erupted mass, as well as past examples, estimated frequency, and probability. It includes large eruptions from 40 km^3 in magma volume up to the largest super-eruptions known (modified from [15]).

5.1. An example of Titan2D simulation with the size of 1903 AD eruption

The size of the 1903 AD eruption has been evaluated, and consequently some Titan2D simulations are presented, taking into account where the vents of the 1903 AD, 1668 AD, and 1702 AD were located. Five vent locations are considered: three inside the caldera lake and two outside the caldera, as illustrated in **Figure 16**. The simulations start with an initial column (or pile) that collapses. The column is a vertical cylinder with a diameter 250 m and the height Hcol (**Table 4**). In **Figure 17** we report the thickness of the deposit when the flow is stopped. We consider 1 h (3,600s) after the initial collapse of the eruption column which can be formed by the Plinian eruption. The scenario performed shows, according to the vent locations, that the flow moves in diverse directions (NE, SE) with a thickness of 3 m, and that if the vent is located inside the caldera the flow fills the caldera with a thickness of 5 m. If the vent is on the border of the caldera, the flow will deposit thickly, mostly in the northern valley, the upper stream region of Erdaobaihe (**Figure 17** [34, 61]).

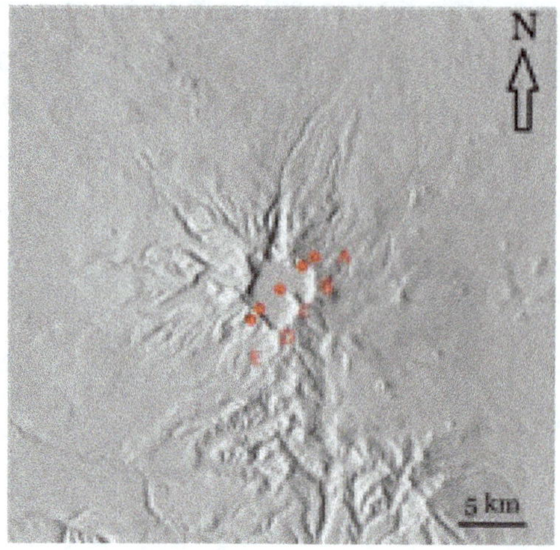

Figure 16. Location of the considered vents for the Titan2D simulation of the small-scale 1903 AD eruption.

Number	Vent	Int. frict.	Bed. Frict	Hcol (m)	Diameter (m)	Volume (m^3)
1	A	30	25	1000	250	4.9×10^7
2	A	30	16	1000	250	4.9×10^7
3	A	25	16	1000	250	4.9×10^7
4	A	25	16	2000	250	9.8×10^7
5	B	25	16	2000	250	9.8×10^7
6	C	25	16	2000	250	9.8×10^7
7	D	25	16	2000	250	9.8×10^7
8	E	25	16	2000	250	9.8×10^7

Table 4. Parameters used to simulate the pyroclastic flow of 1903 AD eruption needed by Titan2D code.

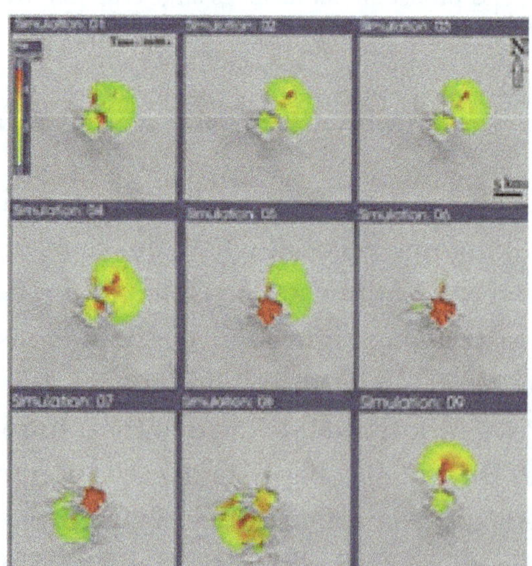

Figure 17. Results of the Titan2D simulation for a small-scale scenario (VEI 3) similar to the 1903 AD eruption [34, 61].

6. The impact of 1668–1702 AD and Millennium eruption around Baekdusan volcano

From the pyroclastic density currents scenario of the 1668–1702 AD eruptions with a runout around 5,000 m (**Figure 18**), two densely populated cities which are nearest to Baekdusan volcano would suffer significant damage: Erdaobaihe and Songjianghe. Possibly, the pyroclastic density currents could also affect two other cities located slightly further away, such as Fusong and Lushuihe. In the case of the Millennium eruption, considering only the direct impact, without considering the indirect effect which in the case of such eruption could be severe, and considering an underestimation of the runout, the following cities will be hit: Erdaobaihe, Songjianghe, Changbai, Huishan, Baishan, Fusong, Lushuihe, and Tumer (**Figure 19**).

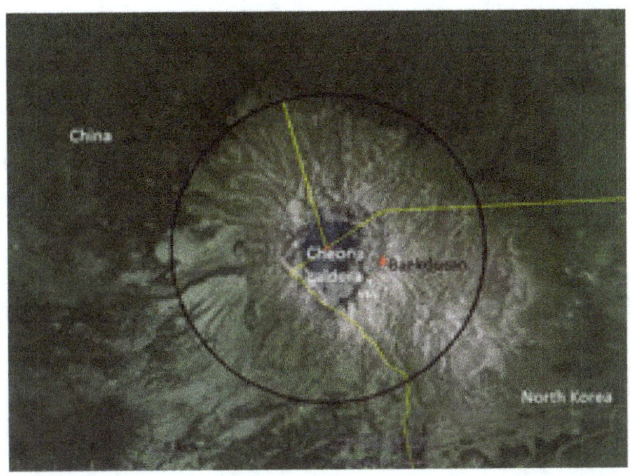

Figure 18. Possible area covered by a pyroclastic density current of the size of the 1668–1702 AD eruption. The runout chosen could have been underestimated so that the impact would be more severe than expected.

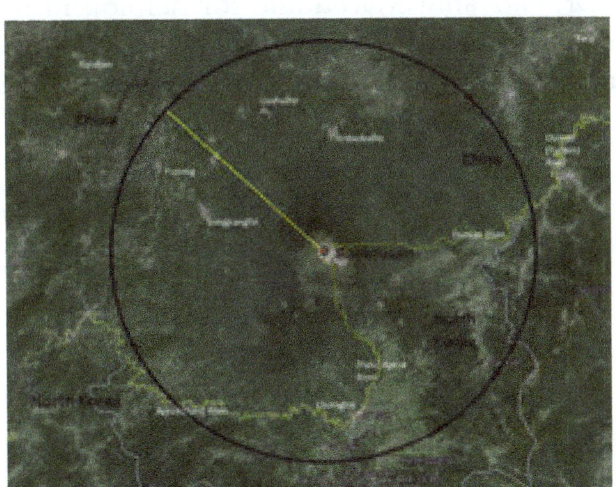

Figure 19. Possible area covered by a pyroclastic density current of the size of the Millennium eruption. The runout chosen could have been underestimated so that the impact would be more severe than expected and surely of regional scale, if not global.

7. Concluding remarks on the worst-case scenario

The worst-case scenario is surely an eruption of a colossal size (Millennium eruption), but it has to be excluded from the scenario of the 1668–1702 AD eruption. In this case, the hazard and the risk must be evaluated carefully, and the population around Baekdusan volcano must be informed in detail. We think in this case, the population at risk is not aware of the possible hazard of the volcano. In all sorts of circumstances, the population has to be alerted. Every sign of unrest has to be monitored and communicated to everybody involved in the decision-making to the people. As far as we are concerned there is not a robust monitoring system around Baekdusan volcano, especially on North Korea side. An emergency plan must be created for Baekdusan volcano.

8. Future research prospectives

In the near future, there are several points in research direction that the Asian scientists could undertake, if they have not done yet: (A) Evolution of the Baekdusan volcano plumbing system in the least 20,000 years. (B) Developing an open source numerical code to model diluted density currents (surge) for the worst-case scenario, (C) Making an event tree study of Baekdusan volcano like that of Somma-Vesuvius volcano, Napoli, Italy.

Acknowledgements

This research was supported by a grant [MPSS-NH-2015-81] through the Natural Hazard Mitigation Research Group funded by Ministry of Public Safety and Security of Korean government. Thanks to Dr. Giovanni Macedonio for helping in making the simulation of **Figure 17** at Osservatorio Vesuviano, INGV, Napoli, Italy. Thanks to Chris Hawkesworth for an early review of the manuscript. Thanks to Steve Sparks to critically looking at the manuscript and addressing the journal for the content of the manuscript.

Author details

Angelo Paone[1,2] and Sung-Hyo Yun[1*]

*Address all correspondence to: yunsh@pusan.ac.kr

1 Pusan National University, Department of Earth Science Education, Busan, Republic of Korea

2 Liceo Scientifico Meucci, Aprilia, Latina, Italy

References

[1] Gomà Daniel, 2004. Border Disputes between China and North Korea. China Perspectives, 52 Online since 23 April 2007. URL: http://chinaperspectives.revues.org/806

[2] Gomà Daniel, 2006. The Chinese-Korean border issue: an analysis of a contested frontier. Asian Survey, Vol. 46, No. 6, pp. 867–880. URL: http://www.jstor.org/stable/10.1525/as.2006.46.6.867

[3] Cashman K. V., Sparks R. S. J., 2013.How volcanoes work: a 25 year perspective. Geological Society of America Bulletin, Vol. 125, No. 5–6, pp. 664–690.

[4] Sparks R. S. J., Biggs J., Neuberg J. W., 2012. Monitoring volcanoes. Science, Vol. 335, pp. 1310–1311.

[5] Sparks R.S.J., 2003. Forecasting volcanic eruptions. Earth and Planetary Science Letters, Vol. 210, pp. 1–15.

[6] Newhall C., Andreastuti S., Lara L., Rosi M., Woo G., 2014. (3.I.F) Managing Worst-Case Scenarios (with Reference to Merapi 2010, St. Helens 1980, Pinatubo 1991, and The Next VEI 7 Eruption) Cities on Volcanoes 8, Living in Harmony with Volcano: Bridging the will of nature to society, Yogyakarta, 9–13 September 2014

[7] Wei H., Liu G., Gill J., 2013. Review of eruptive activity at Tianchi volcano, Changbaishan, northeast China: implications for possible future eruptions. Bulletin of Volcanology, Vol. 75, pp. 1–14.

[8] Yun S. H., Lee J. H., 2012. Analysis of unrest signs of activity at the Baegdusan volcano. Journal of Petrological Society of Korea, Vol. 21, pp. 1–12.

[9] Yun S. H., Lee J. H., Chang C., 2013. A study on the change of magma activity from 2002 to 2009 at Mt. Baekdusan using surface displacement. Journal of Korean Earth Science Society, Vol. 34, No. 6, pp. 470–478 (in Korean with English abstract).

[10] Xu J., Liu G., Wu J., Ming Y., Wang Q., 2012. Recent unrest of Changbaishan volcano, northeast China: a precursor of a future eruption? Geophysical Research Letters, Vol. 39, p. L16305, doi: 10.1029/2012GL052600

[11] Liu G., Liu G. R., Zhang H. R., Kong Q. J., Wu C. Z., Guo F., Zhang C. F., 2004. Elementary analysis of data from Tianchi volcano. Journal of Geoscience Research NE Asia, Vol. 7, No. 2, pp. 156–16.

[12] Gun U. Y., Ok J. U., Song K. M., Song R. G., Nam R. K., Hammond J. O., Oppenheimer C., Whaler K., Park S., Dawes G., Iacovino K., 2013.The Mt. Paektu geoscientific experiment. IAVCEI 2013 Scientific Assembly, July 20–24, Kagoshima, Japan.

[13] Liu G., 2013.Changbaishan seismic mornitoring network and recent unrest of the volcano. IAVCEI 2013 Scientific Assembly, July 20–24, Kagoshima, Japan.

[14] Wei H., Sparks R. S. J., Liu R., Fan Q., Wang Y., Hong H., Zhang H., Chen H., Jiang C., Dong J., Zheng Y., Pan Y., 2003. Three active volcanoes in China and their hazards, Journal of Asian Earth Sciences, Vol. 21, pp. 515–526.

[15] Sparks S., Self S., Grattan J., Oppenheimer C., Pyle D., Rymer H., 2005. Super-eruptions: Global Effects and Future Threats. Report of a Geological Society of London Working Group (2nd (print) Edn.). Original (Web) edition available at www.geolsoc.org.uk/supereruptions.

[16] Lei J., Zhao D., 2005. P-wave tomography and origin of the Changbai intraplate volcano in Northeast Asia. Tectonophysics, Vol. 397, pp. 281–295.

[17] Zhao D., 2012.Tomography and dynamics of western-Pacific subduction zones. Monographs on Environment, Earth and Planets, Vol. 1, No. 1, 1–70.

[18] Zhao D., Liu L., 2010. Deep structure and origin of active volcanoes in China. Geoscience Frontiers, Vol. 1, pp. 31–44.

[19] Zhao D., Tian Y., 2013. Changbai intraplate volcanism and deep earthquakes in East Asia: a possible link? Geophysics Journal International, Vol. 195, pp. 706–724.

[20] Kuritani T., Ohtani E., Kimura J.-I., 2011.Intensive hydration of the mantle transition zone beneath China caused by ancient slab stagnation. Nature Geoscience, Vol. 4, pp. 713–716.

[21] Annen C., Blundy J.D., Sparks R. S. J., 2006. The genesis of intermediate and silicic magmas in deep crustal hot zones. Journal of Petrology, Vol. 47, pp. 505–539.

[22] Yun S. H., Koh J. S., Chang C. W., 2015. Magma genesis of Holocene volcanism at Mt. Baekdusan. Abstract Book, 2015 Fall Joint Conference of Geological Science of Korea (2015. 10. 28-31), Jeju, Korea, pp. 518, 39.

[23] Yun S. H., Won C. K., Lee M. W., 1993. Cenozoic volcanic activity and petrochemistry of volcanic rocks in the Mt. Paektu area. Journal of Geological Society of Korea, Vol. 29, pp. 291–307 (in Korean with English abstract).

[24] Wei H., 2010. Magma up-moving process within the magma prism beneath the Changbaishan volcanoes. Earth Science Frontiers, Vol. 17, No. 1, pp. 11–23 (in Chinese with English abstract).

[25] Horn S., Schmincke H.U., 2000. Volatile emission during the eruption of Baitoushan Volcano (China/North Korea) ca. 969 AD. Bulletin of Volcanology, Vol. 61, pp. 537–555.

[26] Annen C., Blundy J. D., Sparks R. S. J., 2006. The sources of granitic melt in deep hot zones. Transactions of the Royal Society of Edinburgh: Earth Sciences, Vol. 97, pp. 297–309, doi: 10.1017/S0263593300001462

[27] Annen C., Sparks R. S. J., 2002. Effects of repetitive emplacement of basaltic intrusions on thermal evolution and melt generation in the deep crust. Earth and Planetary Science Letters, Vol. 203, pp. 937–955.

[28] Wei H., Wang Y., Jin J., Gao L., Yun S.H., Jin B., 2007.Timescale and evolution of the intracontinental Tianchi volcanic shield and ignimbrite-formation eruption, Changbaishan, Northeast China. Lithos, Vol. 96, pp. 315–324.

[29] Wei H., Taniguchi H., Myamoto T., Jin B., 2007. Stratigraphic sequences and magmatic cycles of the Tianchi volcano, Changbaishan. Northeast Asian studies Vol. 11, pp. 173–193.

[30] Yang L., Wang F., Feng H., Wu L., Shi W., 2014.$^{40}Ar/^{39}Ar$ geochronology of Holocene volcanic activity at Changbaishan Tianchi volcano, Northeast China. Quaternary Geochronology, Vol. 21, pp. 106–114.

[31] Yun S. H., 2013.Volcanological interpretation of historical eruptions of Mt. Baekdusan volcano. Journal Korean Earth Science Society, Vol. 34, No. 6, pp. 456–469.

[32] Yun S. H., Cui Z. X., 1996. Historical eruption records on the Cheonji caldera volcano in the Mt. Peaktu. Journal of Korean Earth Science Society, Vol. 17, No. 5, pp. 376–382 (in Korean with English abstract).

[33] Yun S. H., Lee J. H., 2011. Volcanological interpretation of historic record of 1702 fallout-ash from the Mt. Baekdusan. Journal of Petrological Society of Korea, Vol. 20, No. 4, pp. 243–250 (in Korean with English abstract).

[34] Yun S. H., Lee J. H., Kim S., Chang C., Cho E., Yang I., Kim Y., Kim S., Lee K., Kim S., Macedonio G., 2013. TITAN2D simulations of pyroclastic flows from small scale eruption at Mt. Baekdusan. Journal of Korean Earth Science Society, Vol. 34, No. 7, pp. 615–625 (in Korean with English abstract).

[35] Machida H., Hiroshi M., Zhao D. C., 1990, The recent major eruption of Changbai eruption and its environmental effects. Geophysical Reports of Tokyo Metropolitan University, Vol. 25, pp. 1–20.

[36] Yun et al., 2012. A study on the cooperative diagnosis on eruption possibility of Baegdusan volcano. Pusan National University Project. CATER 2011–5210 pp. 1–131 (in Korean with the references in English).

[37] Zou H., Fan Q., Zhang H., 2010. Rapid development of the great Millennium eruption of Changbaishan (Tianchi) volcano, China/North Korea: evidence from U–Th zircon dating. Lithos, Vol. 119, pp. 289–296.

[38] Machida H., Arai F., 1992, Atlas of Tephra in and around Japan. University of Tokyo Press, Tokyo. 276 p (in Japanese).

[39] Nakagawa M., Nishimoto J., Miyamoto T., Taniguchi H., 2013. Magma system and its eruption processes of the caldera-forming 10th century eruption of Changbaishan (Baitoushan) Volcano: inferred from petrological and geochemical characteristics. IAVCEI 2013 Scientific Assembly, July 20–24, Kagoshima, Japan.

[40] Chu K. S., Tsuji Y., Baag C. E., Kang T. S., 2011.Volcanic eruptions of Mt. Baekdu Changbai occurring in historical times. Bulletin of Earthquake Research Institute, University of Tokyo, Vol. 20, pp. 11–27 (in Japanese with English abstract).

[41] Yin J., Jull T., Burr G. S., Zheng Y., 2012. A wiggle-match age for the Millennium eruption of Tianchi volcano at Changbaishan, Northeastern China. Quaternary Science Reviews, Vol. 47, pp. 150–159.

[42] Xu J., Pan B., Liu T., Hajdas I., Zhao B., Yu H., Liu R., Zhao P., 2013. Climatic impact of the Millennium eruption of Changbaishan volcano in China: new insights from high-precision radiocarbon wiggle-match dating. Geophysical Research Letters, Vol. 40, pp. 1–6, doi: 10.1029/2012GL054246

[43] Newhall C., Self S., 1982.The volcanic explosivity index (VEI): an estimate of explosive magnitude for historical volcanism. Journal of Geophysical Research, Vol. 87, pp. 1231–1238.

[44] Branney M. J., Kokelaar P., 2002. *Pyroclastic Density Currents and the Sedimentation of Ignimbrites.* Geological Society, London, Memoirs, 27, pp. 152.

[45] Sparks R. S. J., Bursik M. I., Carey S. M., Gilbert J. S., Glaze L. S. Sigurdsson H., Woods A. W., 1997.Volcanic Plumes. Wiley, Chichester.

[46] Gilbert J. S., Sparks R. S. J. (eds), 1998. The Physics of Explosive Volcanic Eruptions. Geological Society, London, Special Publications, Vol. 145, pp. 193.

[47] Parfitt E. A., Wilson L., 2008. Fundamentals of Physical Volcanology. Blackwell Publishing Ltd., pp. 256.

[48] Fagents S. A., Gregg T. K. P., Lopes R. M. C., 2012. Modeling Volcanic Processes. The Physics and Mathematics of Volcanism. Cambridge University Press, pp. 448.

[49] Savage S. B., Hutter H., 1989. The motion of a finite mass of granular material down a rough incline. Journal of Fluid Mechanics, Vol. 199, pp. 177-215

[50] Iverson Denlinger, 2001. Flow of variably fluidized granular masses across three-dimensional terrain 1. Coulomb mixture theory. Journal of Geophysical Research, Vol. 106, p. 537.

[51] Patra A. K., Bauer A. C., Nichita C., Pitman E. B., Sheridan M. F., Bursik M. I., 2005.Parallel adaptive simulation of dry avalanches over natural terrain. Journal of Volcanology and Geothermal Research, Vol. 139, pp. 1–21.

[52] Kelfoun K., 2011.Suitability of simple rheological laws for the numerical simulation of dense pyroclastic flows and long-runout volcanic avalanches. Journal of Geophisical Research, Vol. 116, p. B08209, doi: 10.1029/2010JB007622

[53] Kelfourn K., 2011. A two-fluid model for block-and-ash flows and ash-cloud surge confronted with the 1994 eruption of Merapi volcano (Indonesia). European Geosciences Union. General Assembly, Vienna, Austria, Poster.

[54] Esposti O. T., Cavazzoni C., Erbacci G., Neri A., Salvetti M. V., 2007. A parallel multiphase flow code for the 3D simulation of explosive volcanic eruptions. Parallel Computer. doi: 10.1016/j.parco.2007.04.003

[55] Neri A., Esposti Ongaro T., Menconi G., De' Michieli Vitturi M., Cavazzoni C., Erbacci G., Baxter P.J., 2007. 4D simulation of explosive eruption dynamics at Vesuvius. Geophysical Research Letters, Vol. 34, p. L04309.

[56] Mastrolorenzo G., Petrone P., Pappalardo L., Guarino F.M., 2010. Lethal thermal impact at periphery of pyroclastic surges: evidences at Pompeii. PLoS ONE, Vol. 5, No. 6, pp. 1–12.

[57] Miller C. F., Wark D. A., 2008. Supervolcanoes and their explosive supereruptions. Elements, Vol. 4, pp. 11–16.

[58] Costa A., Folch A., Macedonio G., Giaccio B., Isaia R., Smith V. C., 2012. Quantifying volcanic ash dispersal and impact of the Campanian Ignimbrite super-eruption. Geophysical Research Letters, Vol. 39, p. L10310, doi: 10.1029/2012GL051605

[59] Xu J., Pan B., Liu T., Hajdas I., Zhao B., Yu H., Liu R., 2013.The millennium eruption of Changbaishan volcano in northeast China: high-precision wiggle-match radiocarbon chronology and implications. IAVCEI 2013 Scientific Assembly, July 20–24, Kagoshima, Japan.

[60] Sun C., Plunkett G., Liu J., Zhao H., Sigl M., McConnell J. R., Pilcher J. R., Vinther B., Steffensen J. P., and ValerieHall, 2014. Ash from Changbaishan Millennium eruption recorded in Greenland ice: implications for determining the eruption's timing and impact. Geophysical Research Letters, Vol. 41, pp. 694–701, doi: 10.1002/2013GL058642.

[61] Yun S. H., Paone A., Macedonio G., Jeong Lee H., Kim S. K., Cho E., Kim Y. J., 2013. A preliminary evaluation of the impact of pyroclastic flows using Titan 2D at the Baegdusan volcano according to three different scenarios. 2nd International Workshop for Volcanic Disaster Preparedness. Chungbuk National University. Abstract and Oral communication.

[62] Francis P., 1993, Volcanoes: A planetary perspective. Oxford University Press, Walton Street, New York, 249, 443 p.

2

Fumarolic Minerals: An Overview of Active European Volcanoes

Tonči Balić-Žunić, Anna Garavelli,
Sveinn Peter Jakobsson, Kristjan Jonasson,
Athanasios Katerinopoulos,
Konstantinos Kyriakopoulos and
Pasquale Acquafredda

Abstract

The fumarolic mineralogy of the Icelandic active volcanoes, the Tyrrhenian volcanic belt (Italy) and the Aegean active arc (Greece) is investigated, and literature data surveyed in order to define the characteristics of the European fumarolic systems. They show broad diversity of mineral associations, with Vesuvius and Vulcano being also among the world localities richest in mineral species. Volcanic systems, which show recession over a longer period, show fumarolic development from the high-temperature alkaline halide/sulphate, calcic sulphate or sulphidic parageneses, synchronous with or immediately following the eruptions, through medium-temperature ammonium minerals, metal chlorides, or fluoride associations to the late low-temperature paragenesis dominated by sulphur, gypsum, alunogen, and other hydrous sulphates. The situation can be different in the systems that are not recessing but show fluctuations in activity, illustrated by the example of Vulcano where the high-temperature association appears intermittently. A full survey of the mineral groups and species is given in respect to their importance and appearance in fumarolic associations.

Keywords: fumarolic minerals, Iceland fumaroles, Tyrrhenian volcanic belt fumaroles, Aegean active arc fumaroles, mineral sublimates

1. Introduction

At present, there are three active volcanic provinces in Europe (**Figure 1**). The first one is situated in NW corner, on the spreading Mid-Atlantic Ridge, nurtured by the Iceland hot spot and causing intermittent activation of numerous volcanoes. The second is the Tyrrhenian volcanic belt with a very active volcanism. It comprises the Roman comagmatic province, which is related to the extension of the Tyrrhenian basin and the rollback of the Apennine chain, and the Aeolian Arc, the genesis of which is generally ascribed to the subduction of the Ionian microplate under Calabria. The third one is the Aegean Active Volcanic Arc, related to the eastern Mediterranean lithosphere subduction under the Aegean-Anatolian microplate. It is characterized by a dormant or recessing volcanism. Is this remarkable variety in causes of volcanism and its activity reflected in the products of volcanic emanations, and in which way? This chapter tries to summarize the mineralogical evidence and give the answer to this question.

Figure 1. Europe from the space with locations of the three volcanic provinces described in this chapter.

There is no process in geology that, on closer inspection, can be called simple. Compared to the processes on the similar scale, the mineralization of a fumarole or a fumarolic field might well be the most complex one. It happens in an open system with high kinetic energy, high mass transport rate, and subject to constantly fluctuating conditions. It involves reactions of gases, fluids, and solids at the boundary between the atmosphere, volcanic gases, meteoric water, hydrothermal solutions, lava, and country rock. Minerals in fumaroles form either as direct sublimates due to changes in temperature or composition of gases, as accumulations of aerosols carried by gas, through gas-solid reactions either by the assimilation of gas components or by leaching of solid, through temperature-induced reactions (mineral reactions, phase transitions, dehydration), through the action of water or vapour (hydration or dissolution) or by crystallization from solution. Inspection of the literature about fumaroles reveals that authors use the following two distinct attributions: sublimates and encrustations. The term sublimate is used for the minerals formed by deposition from gases, and encrustations are formed from fumarolic fluids, but the study reveals many inconsistencies. The reason is that the processes mentioned earlier happen simultaneously at the same place and are entangled. Moreover, most of the minerals can be formed through more than one process. Here, we will consider all minerals in fumaroles that are formed under whatsoever influence of volcanic gases as fumarolic minerals and avoid making classifications that would have too many exemptions and in many cases be misleading. Another question is the definition of a fumarole itself. In their work on Icelandic fumaroles, Jakobsson et al. [1] distinguished the following two types of fumarolic associations: volcanogenic, which are formed by short-lived, shallow-rooted thermal systems and characterized by no discharge of water and by encrustations that are primarily product of magmatic degassing; and solfataric hydrothermal systems, which are long-lived with a deep-rooted source and surface exposures of high-temperature hydrothermal activity with extensive water-rock interaction. We keep our choice of the European localities as close as possible to the former type (volcanogenic), although it, in the context broader than specific Icelandic conditions, becomes more arbitrary because the fumaroles of a quiescent volcano closely resemble solfataras in their process of formation and mineralogy. The fumaroles on the Solfatara crater, which gave the name to the second category, are not identical in all their characteristics to those encountered on Iceland and illustrate how difficult it is to draw a border line. We are aware that our choice of what to include and what to exclude among the European places with gas emanations can be disputed, but we think that it illustrates well all stages of surface pneumatic processes connected to volcanism from its birth to its old stage and the last breaths of a volcano.

Measured with geological scales, fumaroles are short-lived phenomena. They are a surface or close to the surface feature of a volcano and much more sensitive to weathering than surrounding lava. The formation and dissolution and erosion of fumarolic minerals are contemporary processes, and when the gas emanations ultimately stop, the mineral content in the vents of fumaroles disappears in a short time and just the roots of a fumarolic system can be found in a fossilized form. Besides high chemical reactivity and instability, the fumarolic minerals often appear in microcrystalline aggregates and in the mixtures of many phases. The methods of research therefore require specific approaches and strategies.

2. Methods

2.1. Sample collection and preparation

In this work, the samples from 12 European volcanoes and other fumarolic systems have been investigated (Krafla, Askja, Hekla, Fimmvörðuháls, Eldfell, Surtsey, Vulcano, Etna, Soussaki, Milos, Santorini, and Nisyros). The data for Campi Flegrei and Vesuvius, plus additional data for investigated volcanoes, have been taken from the literature.

The samples have been left to cool down at atmospheric conditions after extraction. The fragile samples were mounted in plastic boxes fixed to avoid the physical damage. The less sensitive ones or those that were extracted in crushed form were sealed in plastic bags. After separating the chosen parts under the microscope, the samples for the X-ray diffraction (XRD) were crushed in agate mortar, and the samples for the scanning electron microscope with energy-dispersive spectrometer (SEM-EDS) analyses were sputtered with a 30-nm-thick carbon film using an Edwards Auto 306 thermal evaporator. The fumarolic samples are mostly composed of intimate mixtures of several minerals with micrometer-sized crystals. Even the most painstaking separation work is many times not in stand to produce a pure one-phase sample for powder X-ray diffraction (PXRD), and hence, the separation of multiple seemingly different portions of a sample was almost always performed in order to get the best overview of the phase composition.

2.2. X-ray diffraction

PXRD is the method of choice for the identification of minerals, alone or in mixtures, and for the quantitative phase analysis. Essential for handling complex mixtures is a diffractometer with high resolution, and therefore, the D8ADVANCE Bruker-AXS powder diffractometer with the primary Ge_{111} monochromator and the LinxEye silicon strip detector in reflection geometry was used for the measurement of bulk samples (at the Department of Geosciences, University of Copenhagen). The wavelength of X-ray used for measurement was 1.54059 Å. The separated portions of samples for PXRD were usually in very small quantities (≤ 20 mm^3). They were therefore mounted as thin layers on specially cut single-crystal quartz plates that produce no scattering inside the instrument's measuring range that was 5° to 70° or 90° 2θ, with measurement steps of 0.02° and sufficient measuring time for producing good diffraction patterns (usually 4 s per step). The identification of minerals was done with the help of the JCPDS powder diffraction database (ICDD product) and the own set of calculated patterns from the ICSD database (FIZ Karlsruhe product) plus use of the Rietveld method (Topas 4.1 program, Bruker-AXS product). Modern single-crystal diffractometers enable fast analyses of small grains (only several tens of micrometer in diameter). The full crystal structure characterisation they offer is essential in the definition of new mineral species encountered in fumaroles and builds the basis for many of results reported here.

2.3. SEM-EDS

The use of a SEM equipped with EDS (Si(Li), Ge or silicon drift) gives a possibility to obtain very good images by secondary and backscattered electrons together with a chemical micro-analysis. The ED detector has the advantage of providing a full spectrum in a very short time that can be very helpful for quick, preliminary identification of minerals. The high counting efficiency of the solid-state ED detectors allows a complete analysis of a mineral in very short time (50 s or less) and with very low probe currents of the beam (about 500 pA). The low probe current involves the generation of a small X-ray escape volume and therefore accurate analysis of minerals even in the case of crystals of extremely small size and in the co-presence of other mineral phases, without the risk of evaporating the sample. Moreover, an ED detector that does not require critical positioning of the sample [2, 3] can give good quantitative chemical data analysing directly the natural faces of the crystals without a need of polished surfaces required for microprobe, but mostly impossible for fumarolic samples, which as a rule consist of loose aggregates of tiny crystals. The last generation of solid-state silicon drift detectors (SSD), normally equipped with a very thin polymeric SuperAtmosphere Thin Window©, gives an output signal with much higher count rates that guarantee also a better sensitivity for light elements even at very low probe currents. Their use in conjunction with the newest software for the correction of the matrix effects [4] is highly recommended for the samples of fumarolic minerals.

Our investigations were carried out with two SEM instruments: the Stereoscan 360 of Cambridge Instruments and the EVO-50XVP of Zeiss-Cambridge. Microanalyses were carried out with Oxford-Link Ge ISIS energy dispersive spectrometer or Oxford X-max (80 mm^2) silicon drift detector, both equipped with a SuperAtmosphere Thin Window©. The operating conditions were as follows: 15 kV accelerating potential, 500 pA probe current, counting time 100 s. The software used was Z AF4/FLS and eXtended Pouchou and Pichoir (XPP) correction scheme, respectively, both Oxford-Link Analytical products.

3. The minerals

More than 200 different minerals have been described from European fumaroles. With the rest of the world, the number might be over 300. Most of them are rare or extremely rare but can be important as indicators of specific conditions in the fumarole where they have been found. **Table 1** gives the chemical and crystallographic parameters of the scientifically confirmed European fumarolic minerals. In the following text, we describe the main features of the mineral groups and individual minerals in fumaroles. The exposition of species follows the mineral classification of Strunz [5] with small modifications. Jakobsson et al. [1] listed in their work a number of supposed new minerals and labelled them each with a two-letter symbol. For those that still do not have mineral names, we used the same notation. The labels for the localities are as follows: A (Askja), CF (Campi Flegrei), El (Eldfell), Et (Etna), F (Fimmvörðuháls), H (Hekla), Kr (Krafla), M (Milos), N (Nisyros), Sa (Santorini), So (Soussaki), Su (Surtsey), Ve (Vesuvius), and Vu (Vulcano).

Minerals	**Formula**	**Sp.gr.**	**Unit cell parameters (Å, °)**	**ICSD**
Sulphur-α*	S_8	Fddd	10.465 12.866 24.49	63082
Sulphur-β	S_8	$P2_1/c$	10.926 10.855 10.790 95.9	870
rosickýite	S_8	P2/c	8.455 13.052 9.267 124.9	66517
sulphurite*	S+As,Se	amorph		amorph
Selenium-α	Se_8	$P2_1/n$	9.054 9.083 11.601 90.8	2718

Minerals	Formula	Sp.gr.	Unit cell parameters (Å, °)	ICSD
siderazot	Fe_3N	P6$_3$22	4.698 4.379	79983
Gold	Au	Fm-3m	4.079	52700
tellurium	Te	P3$_1$2	4.456 5.921	65692
covellite	CuS	P6$_3$/mmc	3.796 16.36	41975
sphalerite	ZnS	F-43m	5.409	77082
wurtzite	ZnS	P6$_3$mc	3.823 6.261	67453
chalcopyrite	$CuFeS_2$	I-42d	5.289 10.423	2518
pyrrhotite	$Fe_{11}S_{12}$	Cc	6.897 11.954 17.702 101.3	166063
millerite	NiS	R3m	9.611 3.151	151599
galena*	PbS	Fm-3m	5.934	38293
pyrite*	FeS_2	Pa-3	5.416	109377
realgar*	As_4S_4	P2$_1$/n	9.325 13.571 6.587 106.4	15238
pararealgar	As_4S_4	P2$_1$/c	9.909 9.655 8.502 97.29	80125
alacranite1	As_8S_9	P2/c	9.942 9.601 9.178 101.9	98792
alacranite2	As_4S_4	C2/c	9.943 9.366 8.908 102.0	95290
dimorphite-α (HT)	As_4S_3	Pnma	9.158 8.033 10.200	188058
dimorphite-β (LT)	As_4S_3	Pnma	11.217 9.922 6.607	188059
demicheleite	BiS(Br,Cl,I)	Pnam	8.042 9.851 4.033	161637
bismuthinite*	Bi_2S_3	Pnma	11.269 3.972 11.129	153946
mozgovaite	$PbBi_4(S,Se)_7$	Bbmm?	13.18 37.4 4.05?	[28]
galenobismutite*	$PbBi_2S_4$	Pnam	11.802 14.569 4.076	158392
cannizzarite*	$Pb_{48}Bi_{56}(S_{1-x}Se_x)_{132}$	P2$_1$/m	38.87 4.090 39.84 102.3	169960
cosalite	$Pb_2Bi_2S_5$	Pnma	23.89 4.062 19.143	169944
lillianite	$Pb_3Bi_2S_6$	Bbmm	13.540 20.64 4.110	246062
heyrovskyite	$Pb_6Bi_2S_9$	Bbmm	13.719 31.39 4.132	180078
kirkiite	$Pb_{10}Bi_3As_3S_{19}$	P2$_1$/m	8.700 26.24 8.774 119.7	156249
vurroite	$Pb_8(Pb,Bi)_2(Sn,Bi)(As,Bi,Pb)(Bi,As)_{10}S_{27}Cl_3$	C2/c	8.371 45.50 27.27 98.8	160401
sylvite*	KCl	Fm-3m	6.285	165593
halite*	NaCl	Fm-3m	5.620	18189
salammoniac*	NH_4Cl	Pm-3m	3.877	22141
lafossaite	TlCl	Pm-3m	3.901	5253
sellaite	MgF_2	P4$_2$/mnm	4.621 3.052	394
chloromagnesite	$MgCl_2$	R-3m	3.636 17.666	86439

Minerals	Formula	Sp.gr.	Unit cell parameters (Å, °)	ICSD
scacchite	$MnCl_2$	R-3m	3.711 17.59	33752
fluorite*	CaF_2	Fm-3m	5.463	41413
oskarssonite*	$Al(F,OH)_3$	R-3c	4.982 12.387	5243
parascandolaite	$KMgF_3$	Pm-3m	4.003	192409
molysite*	$FeCl_3$	R-3	6.056 17.407	39764
carnallite	$KMgCl_3(H_2O)_6$	Pnna	16.119 22.47 9.551	64691
eriochalcite	$CuCl_2(H_2O)_2$	Pbmn	7.414 8.089 3.746	40290
ammineite	$CuCl_2(NH_3)_2$	Cmcm	7.688 10.645 5.736	180189
chloraluminite	$AlCl_3(H_2O)_6$	R-3c	11.827 11.895	22071
ferruccite	$NaBF_4$	Cmcm	6.837 6.262 6.792	36067
avogadrite	KBF_4	Pnma	8.659 5.480 7.030	9875
barberiite	NH_4BF_4	Pnma	9.077 5.679 7.279	9918
rosenbergite	$AlF_3(H_2O)_3$	P4/n	7.715 3.648	66691
HI	$FeF_3(H_2O)_3$	P4/n	7.846 3.877	14134
HU	$FeF_3(H_2O)_3$	R-3	9.253 4.675	unp
pachnolite	$NaCaAlF_6H_2O$	Fd	12.117 10.414 15.680 90.4	40132
gearksutite	$CaAlF_4(OH)H_2O$	P-1	4.94 6.81 6.978 101.1 94.9 110.1	89800
jakobssonite*	$CaAlF_5$	C2/c	8.601 6.290 7.219 114.6	188924
leonardsenite*	$MgAlF_5(H_2O)_2$	Imma	7.064 10.131 6.774	411650
HG	$Na_2Ca_3Al_2F_{14}$	$I2_13$	10.257	202657
coulsellite	$CaNa_3AlMg_3F_{14}$	R-3m	7.161 17.594	168054
thermessaite	$K_2AlF_3SO_4$	Pbcn	10.810 8.336 6.822	161272
thermessaite-(NH_4)	$(NH_4)_2AlF_3SO_4$	Pbcn	11.301 8.612 6.850	[45]
ralstonite*	$Na_xMg_xAl_{2-x}(F,OH)_6(H_2O)_y$	Fd-3m	9.91	31345
HH	$Ca_3Al_2F_{10}(OH)_2(H_2O)_3$?	C2/m?	6.257 22.19 6.311 115.5 (?)	unp
HB	Na-Ca-Al-F?	?	?	[1]
HD*	$(NH_4)_xFe_2F_6$	Fd-3m	10.325	202047
meniaylovite	$Ca_4AlSi(SO_4)F_{13}(H_2O)_{12}$	Fd-3	16.71	100728
cossaite	$Mg_{0.5}Al_6H(SO_4)_7F_6(H_2O)_{33}$	R-3	22.01 9.238	188928
malladrite*	Na_2SiF_6	P1	8.859 8.859 5.038 90 90 120	40917
hieratite	K_2SiF_6	Fm-3m	8.134	29407
demartinite	K_2SiF_6	$P6_3mc$	5.646 9.232	158483
heklaite*	$KNaSiF_6$	Pnma	9.339 5.503 9.796	183232

Minerals	Formula	Sp.gr.	Unit cell parameters (Å, °)	ICSD
cryptohalite*	$(NH_4)_2SiF_6$	Fm-3m	8.395	18033
bararite	$(NH_4)_2SiF_6$	P-3m	5.784 4.796	18027
HT	$FeSiF_6(H_2O)_6$	R-3m	9.616 9.675	41535
knasibfite	$K_3Na_4(SiF_6)_3BF_4$	Imm2	5.522 17.106 9.175	160430
chlormanganokalite	K_4MnCl_6	R-3c	11.926 14.787	24475
erythrosiderite	$K_2FeCl_5H_2O$	Pnma	13.75 9.92 6.93	30321
kremersite	$(NH_4)_2FeCl_5H_2O$	Pnma	13.706 9.924 7.024	200322
mitscherlichite	$K_2CuCl_4(H_2O)_2$	$P4_2/mnm$	7.477 7.935	16052
melanothallite	Cu_2OCl_2	Fddd	7.469 9.597 9.700	96610
atacamite*	$Cu_2Cl(OH)_3$	Pnma	6.030 6.865 9.120	61252
paratacamite*	$Cu_2Cl(OH)_3$	R-3	13.654 14.041	1810
connellite	$Cu_{36}Cl_{7.86}(SO_4)_{0.67}(NO_3)_{0.5}(OH)_{62.3}(H_2O)_{5.6}$	$P6_3/mmc$	15.787 9.101	415129
cumengeite	$Pb_{21}Cu_{20}Cl_{42}(OH)_{40}(H_2O)_6$	I4/mmm	15.101 24.49	157067
bismoclite	BiOCl	P4/nmm	3.887 7.354	74502
lucabindiite	$(K,NH_4)As_4O_6((Cl,Br)$	P6/mmm	5.239 9.014	289996
cotunnite	$PbCl_2$	Pnam	7.619 9.043 4.534	202130
pseudocotunnite	K_2PbCl_4	?	?	[10]
challacolloite*	KPb_2Cl_5	$P2_1/c$	8.849 7.918 12.472 90.1	416430
hephaistosite	$TlPb_2Cl_5$	$P2_1/c$	9.003 7.972 12.569 90.05	166293
brontesite	$(NH_4)_3PbCl_5$	Pnma	8.435 15.773 8.445	166092
panichiite	$[NH_4)_2SnCl_6$	Fm-3m	10.064	163660
steropesite	Tl_3BiCl_6	Cc	26.69 15.127 13.014 108.1	163661
argesite	$[NH_4)_7Bi_3Cl_{16}$	R-3c	13.093 102.68	5283
tenorite	CuO	C2/c	4.684 3.423 5.129 99.5	16025
spinel	$MgAl_2O_4$	Fd-3m	8.089	40030
magnesioferrite	$MgFe_2O_4$	Fd-3m	8.394	158402
magnetite	Fe_3O_4	Fd-3m	8.396	30860
hausmannite	Mn_3O_4	$I4_1/amd$	5.765 9.442	68174
minium	Pb_3O_4	$P4_2/mbc$	8.811 6.563	4106
corundum	Al_2O_3	R-3c	4.761 12.995	9770
hematite*	Fe_2O_3	R-3c	5.035 13.749	82904
ilmenite	$FeTiO_3$	R-3	5.088 14.085	30664
pseudobrookite	Fe_2TiO_5	Ccmm	9.793 3.730 9.976	51225

Minerals	Formula	Sp.gr.	Unit cell parameters (Å, °)	ICSD
opal*	$SiO_2(H_2O)_x$	amorph		amorph
cristobalite*	SiO_2	$P4_12_12$	4.978 6.948	9327
tridymite	SiO_2	Cc	18.494 4.991 23.76 105.8	1109
quartz*	SiO_2	$P3_22$	4.913 5.405	174
anatase	TiO_2	$I4_1/amd$	3.784 9.515	9852
rutile	TiO_2	$P4_2/mnm$	4.592 2.957	33837
akaganeite	$FeOOHCl_x$	I2/m	10.600 3.034 10.513 90.2	69606
portlandite	$Ca(OH)_2$	P-3m	3.585 4.895	15471
gibbsite	$Al(OH)_3$	$P2_1/n$	8.684 5.078 9.736 94.5	6162
doyleite	$Al(OH)_3$	P-1	5.00 5.168 4.983 97.4 118.7 104.7	50581
calcite*	$CaCO_3$	R-3c	4.988 17.061	40107
magnesite	$MgCO_3$	R-3c	4.635 15.019	40117
dolomite*	$CaMg(CO_3)_2$	R-3	4.806 16.006	66333
hydromagnesite	$Mg_5(CO_3)_4(OH)_2(H_2O)_4$	$P2_1/c$	10.105 8.954 8.378 114.4	920
aragonite	$CaCO_3$	Pmcn	4.961 7.970 5.742	166085
cerussite	$PbCO_3$	Pmcn	5.180 8.492 6.134	6178
azurite	$Cu_3(CO_3)_2(OH)_2$	$P2_1/c$	5.011 5.850 10.353 92.4	158577
thermonatrite	$Na_2CO_3H_2O$	$P2_1ab$	6.472 10.724 5.259	1852
natron	$Na_2CO_3(H_2O)_{10+x}$	Cc	12.750 9.001 12.590 115.8	97924
trona	$Na_3H(CO_3)_2(H_2O)_2$	C2/c	20.41 3.493 10.333 106.5	192710
sassolite*	H_3BO_3	P-1	7.039 7.053 6.578 92.6 101.2 119.8	24711
metaborite	HBO_2 (HT>140°C)	P-43n	8.886	34639
clinometaborite	HBO_2 (LT)	$P2_1/a$	7.127 8.842 6.773 93.2	183581
ameghinite	$NaB_3O_3(OH)_4$	C2/c	18.428 9.882 6.326 104.4	4219
chalcocyanite	$CuSO_4$	Pnma	8.409 6.709 4.833	71017
vanthoffite	$Na_6Mg(SO_4)_4$	$P2_1/c$	9.797 9.217 8.199 113.5	16607
EN	$Na_3Al(SO_4)_3$	R-3	13.376 8.933	unp
EA	$NaMgAl(SO_4)_3$	R-3	8.303 21.87	unp
EI	$Na_2Mg_2(SO_4)_3$	?	?	unp
langbeinite*	$K_2Mg_2(SO_4)_3$	$P2_13$	9.921	40986
manganolangbeinite	$K_2Mn_2(SO_4)_3$	$P2_13$	10.114	200897
steklite	$KAl(SO_4)_2$	P321	4.728 8.001	[75]
pyracmonite	$(NH_4)_3Fe(SO_4)_3$	R3c	15.217 8.932	169964

Minerals	Formula	Sp.gr.	Unit cell parameters (Å, °)	ICSD
aluminopyracmonite	$(NH_4)_3Al(SO_4)_3$	R-3	15.032 8.878	190666
eldfellite	$NaFe(SO_4)_2$	C2/m	8.043 5.139 7.115 92.1	166768
yavapaiite	$KFe(SO_4)_2$	C2/m	8.150 5.162 7.855 94.9	26004
thenardite*	Na_2SO_4	Fddd	5.860 12.304 9.817	2895
metathenardite*	Na_2SO_4 (HT)	$P6_3/mmc$	5.326 7.126	63077
aphthitalite*	$KNaSO_4$	P3m	5.607 7.177	26014
glaserite	$K_3Na(SO_4)_2$	P-3m	5.680 7.309	26018
palmierite	$K_2Pb(SO_4)_2$	R-3m	5.497 20.86	94234
mascagnite	$(NH_4)_2SO_4$	Pnam	7.782 10.636 5.993	34257
mercallite	$KHSO_4$	Pbca	8.415 9.796 18.967	249738
FB	$Na_3H(SO_4)_2$	$P2_1/c$	8.644 9.641 9.139 108.8	249553
therasiaite	$(NH_4)_3KNa_2Fe_2(SO_4)_3Cl_5$	Cc	18.284 12.073 9.535 108.1	5296
anhydrite*	$CaSO_4$	Amma	6.991 6.996 6.238	15876
gypsum*	$CaSO_4(H_2O)_2$	C2/c	6.277 15.181 5.672 114.1	161622
bassanite	$CaSO_4(H_2O)_{0.5}$	C2	17.559 6.962 12.070 133.4	380286
omongwaite	$Na_2Ca_5(SO_4)_6(H_2O)_3$	C2	12.089 6.903 6.354 90.1	88942
glauberite	$CaNa_2(SO_4)_2$	C2/c	10.129 8.306 8.533 112.2	16901
barite	$BaSO_4$	Pbnm	7.154 8.879 5.454	76926
celestite	$SrSO_4$	Pbnm	6.867 8.354 5.346	92608
anglesite	$PbSO_4$	Pbnm	6.955 8.472 5.397	92609
dolerophanite	Cu_2OSO_4	C2/m	9.370 6.319 7.639 122.3	61513
antlerite	$Cu_3SO_4(OH)_4$	Pnma	8.289 6.079 12.057	96348
d'ansite	$Na_{21}Mg(SO_4)_{10}Cl_3$	I-43d	15.90	[86]
d'ansite-(Mn)	$Na_{21}Mn(SO_4)_{10}Cl_3$	I-43d	15.929	190535
d'ansite-(Fe)	$Na_{21}Fe(SO_4)_{10}Cl_3$	I-43d	15.882	190536
aiolosite	$Na_4Bi(SO_4)_3Cl$	$P6_3/m$	9.626 6.880	166960
natroalunite*	$(Na,K)Al_3(SO_4)_2(OH)_6$	R-3m	6.973 16.877	166305
alunite*	$KAl_3(SO_4)_2(OH)_6$	R-3m	7.020 17.223	12106
natrojarosite	$NaFe_3(SO_4)_2(OH)_6$	R-3m	/.315 16.587	160409
jarosite*	$KFe_3(SO_4)_2(OH)_6$	R-3m	7.315 17.224	12107
adranosite	$(NH_4)_4NaAl_2(SO_4)_4Cl(OH)_2$	$I4_1/acd$	18.118 11.320	169963
adranosite-(Fe)	$(NH_4)_4NaFe_2(SO_4)_4Cl(OH)_2$	$I4_1/acd$	18.261 11.562	[90]
euchlorine*	$KNaCu_3O(SO_4)_3$	C2/c	18.41 9.43 14.21 113.7	69451

Minerals	Formula	Sp.gr.	Unit cell parameters (Å, °)	ICSD
chlorothionite	$K_2CuSO_4Cl_2$	Pnma	7.732 6.078 16.292	22364
linarite	$PbCuSO_4(OH)_2$	P2$_1$/m	9.701 5.650 4.690 102.6	68173
baliczunicite	$Bi_2O(SO_4)_2$	P-1	6.739 11.184 14.175 80.1 88.5 89.5	[13]
leguernite	$Bi_{38}O_{42}(SO_4)_{15}$	P2	11.249 5.657 11.914 99.2	[91]
kieserite	$MgSO_4H_2O$	C2/c	6.891 7.624 7.645 117.7	68345
rozenite*	$FeSO_4(H_2O)_4$	P2$_1$/n	5.979 13.648 7.977 90.4	23912
chalcanthite	$CuSO_4(H_2O)_5$	P-1	6.116 10.716 5.961 82.4 107.3 102.6	20657
pentahydrite	$MgSO_4(H_2O)_5$	P-1	6.314 10.565 6.030 81.1 109.8 105.1	2776
hexahydrite*	$MgSO_4(H_2O)_6$	C2/c	10.110 7.212 24.41 98.3	16546
nickelhexahydrite	$NiSO_4(H_2O)_6$	C2/c	9.880 7.228 24.13 98.4	65018
römerite*	$Fe_3(SO_4)_4(H_2O)_{14}$	P-1	6.463 15.309 6.341 90.5 101.1 85.7	15207
coquimbite	$(Fe,Al)_2(SO_4)_3(H_2O)_9$	P-31c	10.937 17.081	169956
halotrichite*	$FeAl_2(SO_4)_4(H_2O)_{22}$	P2$_1$/c	6.195 24.26 21.26 100.3	96598
pickeringite	$MgAl_2(SO_4)_4(H_2O)_{22}$	P2$_1$/c	6.184 24.27 21.23 100.3	90028
meta-alunogen	$Al_2(SO_4)_3(H_2O)_{14}$	?	?	[43]
alunogen*	$Al_2(SO_4)_3(H_2O)_{17}$	P-1	7.42 26.97 6.062 89.6 97.3 91.5	12129
mirabilite	$Na_2SO_4(H_2O)_{10}$	P2$_1$/c	11.474 10.356 12.788 107.8	411348
campostriniite	$(NH_4,K)_2Bi_{2.5}Na_{2.5}(SO_4)_6 \cdot H_2O$	C2/c	17.748 6.982 18.221 114.0	[92]
rhomboclase*	$FeH(SO_4)_2(H_2O)_4$	Pnma	9.742 18.333 5.421	183662
voltaite*	$K_2Fe_8Al(SO_4)_{12}(H_2O)_{18}$	Fd-3c	27.25	9254
löweite	$Na_{12}Mg_7(SO_4)_{13}(H_2O)_{15}$	R-3	18.924 13.538	15209
eugsterite	$Na_4Ca(SO_4)_3(H_2O)_2$	?	?	[93]
hydroglauberite	$Ca_3Na_{10}(SO_4)_8(H_2O)_6$	?	?	[94]
syngenite	$K_2Ca(SO_4)_2H_2O$	P2$_1$/m	9.771 7.145 6.247 104.0	157072
kröhnkite	$Na_2Cu(SO_4)_2(H_2O)_2$	P2$_1$/c	5.518 12.666 5.808 108.4	422593
ferrinatrite	$Na_3Fe(SO_4)_3(H_2O)_3$	P-3	15.560 8.666	14003
cyanochroite	$K_2Cu(SO_4)_2(H_2O)_6$	P2$_1$/a	9.066 12.13 6.149 104.4	2925
bloedite	$Na_2Mg(SO_4)_2(H_2O)_4$	P2$_1$/a	11.135 8.248 5.542 100.8	151453
polyhalite	$K_2Ca_2Mg(SO_4)_4(H_2O)_2$	F-1	11.690 16.330 7.600 91.6 90 91.9	6304
tamarugite*	$NaAl(SO_4)_2(H_2O)_6$	P2$_1$/a	7.353 25.22 6.097 95.2	15187
picromerite	$MgK_2(SO_4)_2(H_2O)_6$	P2$_1$/a	9.072 12.212 6.113 104.8	26772

Minerals	Formula	Sp.gr.	Unit cell parameters (Å, °)	ICSD
metavoltine	$K_2Na_6Fe^{2+}Fe_6^{3+}O_2(SO_4)_{12}(H_2O)_{18}$	P3	9.575 18.17	8270
kalinite	$KAl(SO_4)_2(H_2O)_{11}$	C2/c	19.92 9.27 8.304 98.8	[95]
alum-(K)	$KAl(SO_4)_2(H_2O)_{12}$	Pa-3	12.164	280548
kainite	$KMgSO_4Cl(H_2O)_{2.75}$	C2/m	19.72 16.23 9.53 94.9	26003
SH	$Na_2Mg_3(SO_4)_2(OH)_2(H_2O)_4$	$Cmc2_1$	19.735 7.223 10.028	425875
chessexite	$Na_4Ca_2Mg_3Al_8(SiO_4)_2(SO_4)_{10}(OH)_{10}(H_2O)_{40}$	?	?	[96]
chlorapatite	$Ca_5(PO_4)_3(Cl,OH)$	$P6_3/m$	9.564 6.816	1706
mimetite	$Pb_5(AsO_4)_3Cl$	$P6_3/m$	10.240 7.440	188080
phoenicochroite	Pb_2OCrO_4	C2/m	14.001 5.675 7.137 115.2	34831

Chemical formula, space group, unique unit cell parameters and the ICSD identification number (ICSD = Inorganic Crystal Structure Database, Fachinformationszentrum Karlsruhe, Germany). For the species that still lack ICSD entry, the reference is given as a citation, as unp = unpublished data, or amorph = amorphous compound. Minerals common in fumaroles are marked with a star.

Table 1. Minerals from European fumaroles.

3.1. Elements

Sulphur is the most abundant native element in fumaroles and generally one of the most common minerals in this environment. It appears in the following various forms: as thick yellow blocky crystalline masses, well-formed crystals (in cavities or on surfaces), granular crusts and acicular spear-like dendritic crystals (at vents of fumaroles, **Figure 2**) or as an admixture to earthy altered rock material, giving it grey colour. Sulphur forms through oxidation of H_2S. It is typically accompanied by alunite, gypsum or salammoniac, but it is also found in other associations. The molecules in the crystal structure consist of puckered eight-member rings of atoms in all three polymorphs found in nature. Native sulphur or α-sulphur is orthorhombic and at atmospheric pressures stable up to 95°C when it transforms reversibly to the monoclinic β-form with the melting temperature of 119°C. It can therefore be supposed that a significant part of sulphur found in fumaroles was formed as the high-temperature form and on cooling transformed to the alpha form observed typically on XRD diagrams. As solid unstable over 120°C, sulphur can be found only in the low-temperature fumaroles or at the outer rims of the hotter ones. It was found on all localities except F.

Rare: **Rosickýite or γ-sulphur**: observed once in a small transparent green-yellow crust made of platy crystals (Vu [6]). Metastable at all temperatures [7]. Conditions of formation still not completely understood. **Sulphurite**, amorphous S with As and Se, Vu [8]. **Selenium**: observed as red crusts (Ve [9], Vu [10]) and so it seems that it was the metastable form, which contains ring molecules similar to that of sulphur and not the stable polymorph with infinite spiral molecules, which is metallic in appearance. Melts at 217°C. **Siderazote**, Ve [9]: iridiscent cover on lavas. **Gold** and **tellurium**, Vu [11], observed as several micrometer-sized grains in the silicic alteration zone of the crater.

Figure 2. Spear-like crystals of sulphur. Fumarole on Nisyros.

3.2. Sulphides

Metallic sulphides are confined to deeper parts of a volcanic system with its high-temperature hydrothermal conditions and rarely appear as sublimates and then in small amounts on the surface of the fumaroles. Here, their formation depends on the persistence of reducing conditions in parts of these largely fluctuating systems. These were especially characteristic for Vesuvius in the periods immediately following the eruptions and for Vulcano during the thermal "crisis" (temperature increase in the fumaroles). Semimetals form stronger covalent bonds with sulphur and are more volatile, and therefore, they can appear in fumaroles in significant quantities either as simple sulphides or, when combining with metals, as sulphosalts (see below). Observed in the investigated localities are As and Bi, whereas Sb did not appear in quantities sufficient to form its minerals. Arsenic forms discrete molecules with sulphur and its simple sulphides are confined to low-temperature hydrothermal deposits and low-temperature fumaroles where they can appear in important quantities (CF). On Vulcano, however, it is mostly a constituent of sulphosalts together with Bi. The transport of the latter is supposed to occur as Bi-Cl complexes, so the abundance of its sulphides, sulphosalts and other minerals in the fumaroles of Vulcano is interpreted as a combined action of both elements [12].

Rare: **Covellite**, Ve [9]: deep-blue thin crusts. **Sphalerite**, Vu [13]: Cd-rich, in HT association of sulphides and sulphosalts. **Wurtzite**, Vu [13]: Metastable under 1020°C, accompanies sphalerite. **Chalcopyrite**, Ve [9]: metallic thin crusts composed of minute crystals. **Pyrrhotite**, Ve [14], Vu [13]: metallic hexagonal lamellae. **Millerite**, Ve [9]: observed once in capillary crystals. **Galena**, Ve [9]: with Fe-sulphides and oxides; Vu [12]: Se-rich, in company of Bi-Pb sulphosalts of the similar silver colour and metallic lustre. **Pyrite**, Vu [13], Ve [9]: with other sulphides; CF [15]: minute crystals dispersed in altered tuffs; So: in opalized rock; M: in microcrystals with rhomboclase and voltaite.

As-sulphides realgar, pararealgar, alacranite and **dimorphite** have molecular crystal structures with As_4S_n cage molecules. The commonest of them is **realgar (Figure 3)**. As_4S_4 melts at 309°C [16] and is therefore confined to low-to-intermediate *T* fumaroles. CF [17]: the main constituent in a fumarole on Solfatara in grainy or crusty aggregates of sub-mm prismatic light red lustrous crystals associated with salammoniac; Ve [9]: rare small red crystals associated with sulphur and selenium. Realgar transforms under the influence of visible or UV light into the orange-yellow **pararealgar** [18], earlier often mistaken for orpiment. CF: it is the alteration product of realgar. The synthetic HT polymorph β-As_4S_4 is stable over 267°C [16]. In natural occurrences, this type of molecule is partly substituted by As_4S_5 until the 1:1 proportion as in the type specimen of **alacranite (As_8S_9)** that has the ordered arrangement of the two types of molecules and different symmetry than β-As_4S_4 [19]; CF [20]: orange small crystals with realgar and dimorphite. **Dimorphite** has two polymorphs. LT β-form transforms to HT α-form at 130°C [21], which melts at 211°C [16]. CF, Ve [17]: tiny orange or yellow crystals with adamantine lustre. **Demicheleite** (Vu [22–24]) is a solid solution with three end-members containing different halogene elements (Cl, Br, I). Very rare prismatic red to black sub-mm crystals associated with other Bi minerals.

Figure 3. Realgar, Solfatara. Sample from the Natural History Museum in Copenhagen.

3.3. Sulphosalts

Sulphosalts, sulphides containing thioarsenide, thioantimonide, or thiobismuthite group(s), are relatively rare constituents of fumaroles. Among the fumaroles reported here, they appear only on Vulcano where they were abundant during the "thermal crises" when the fumarolic temperatures exceeded approximately 450°C. Recent analysis of the roots of paleofumarole at El Indio, Chile, revealed subsurface formation of sulphosalts crystallized from a melt condensing from hot fumarolic gases [25]. We could expect the same situation in the roots of the fumaroles on Vulcano during the periods of lower gas dynamics. When it increased during the thermal crisis, the hot front of the sulphosalt formation reached the surface and this established the new thermodynamical conditions where their crystals formed as sublimates at fumarole vents through a quenching process, producing generally very small crystals, homogeneous and lacking traces of decomposition [26]. They are silver-grey in colour, with metallic lustre and acicular (the latter with the exceptions of kirkiite and cannizzarite). A simple sulphide bismuthinite is here described together with sulphosalts because of its close structural and genetic relation to these minerals.

Figure 4. (a) A sample with sulphosalts, Vulcano; (b) SEM photograph of vurroite; (c) SEM photograph of bismuthinite and (d) SEM photograph of cannizzarite.

Sulphosalts belonging to the PbS-Bi_2S_3 system predominate on Vulcano due to high concentration of Pb and Bi in the gases, transported as volatile chlorides or chlorosulphide complexes [12]. **Bismuthinite** is relatively common and with galena mixed with Pb-Bi sulphosalts [27]. **Mozgovaite** [28] is known only from Vulcano and very rare. **Galenobismutite** is fairly common [12]. **Cannizzarite** [27] is a mineral with an incommensurate structure, which gives it a complex stoichiometry, large unit cell and specific crystal form (thin long leaves reaching sometimes mm diameters but only tens of microns thick). It is relatively abundant. **Cosalite** is rare [12] like **lillianite** [29]. The crystal structure of the latter typifies a homologous series [30] where also the rare **heyrovskyite** [31] with a higher Pb/Bi proportion belongs. Some new or very rare minerals were formed by the admixture of other elements to the Bi-Pb sulphosalts.

Kirkiite [32] and **vurroite** [33] are characterized by a partial substitution of Bi by As in the crystal structure, a feature rare in hydrothermal sulphosalt minerals. The former appears in short prismatic crystals and the latter in slender needle-like crystals, both among the smallest crystals in this sulphosalt association. The sulphosalts from Vulcano are specific in having small amounts of Se substituting for S. Another unusual feature is that some of them have also Cl substituting for S (galenobismutite, lillianite) [26] or playing a distinct role in the crystal structure (vurroite) [34] (**Figure 4**).

3.4. Halides

Halides are abundant in some fumaroles and some of them appear commonly in fumaroles (e.g. halite and salammoniac). Where the halides are abundant, they usually appear with a number of species many of which are unique for the fumarolic environment or have it as the main mode of occurrence. As compounds with predominately ionic bonding, they mostly are colourless or light colour and are relatively soft. Characteristic for fluorides in fumaroles is the partial substitution of F by OH. In chlorides, the substitution of OH for Cl is rare or non-existing due to the difference in ionic sizes. When present together, Cl, OH, and O play distinct structural roles and form oxy- or hydroxychlorides. Br and I often partly substitute Cl. Among complex fluorides, the groups of borofluorides, aluminofluorides, and silicofluorides can be defined. Finally, halides in combination with other anions, usually of complex type (like $[SO_4]^{2-}$) also exist in fumaroles. Degassing dynamics of Cl, F and other halogens is a complex function of physical and chemical magma conditions [35] and can vary largely between different volcanoes. Cl generally degasses faster than F [36] and the abundance of H_2O enhances it further [37], which can produce a temporal segregation of the major amounts of chlorides and fluorides in the fumaroles after volcanic eruptions. This, however, can be modified if other sources than magma contribute to gas production. Significant amounts of fluorides appear in most of the Icelandic volcanoes. They are less abundant, but still in important quantities on Vulcano, whereas they are rarer on Vesuvius and Etna. Alumino-fluorides are characteristic for all Icelandic volcanoes (except Askja that does not have registered fluorides). Fumaroles on Hekla contain also significant silicofluorides. They are found on Vulcano as well, but here more characteristic are borofluorides. Contrary to fluorides, chlorides are more important in Italian fumaroles (especially Vesuvius and Vulcano) than in Icelandic ones. On Icelandic volcanoes, they were present in important quantities in the first fumaroles after the eruptions, but a decrease with time is also observed while fumaroles still were producing significant amounts of fluorides (El, H). An important thermodynamic distinction exists between the most abundant chlorides in fumaroles, where those of alkali elements represent the high-temperature product, whereas salammoniac characterizes the low-temperature fumaroles. The border line is approximately at 300°C.

Sylvite and **halite** form a complete solid solution at temperatures higher than approximately 480°C and limited solid solution at lower temperatures with higher solubility of Na in sylvine than K in halite. This is often observed in fumaroles and can be deduced from the shift of the powder diffraction maxima. Halite is much more abundant than sylvite. Su [1]: in crusts and stalactites with Ca and Na sulphates. F: crusts or powdery covers of sub-mm cubic crystals with anhydrite, gypsum and other sulphates. Ve [9]: halite and sylvite are present in high-temperature "dry fumaroles" as earthy or dendritic crusts, rarely as crystals. Et [38]: Halite appears in stalactites and concretions in volcanic tubes accompanied by sylvite; El, H, A, Kr: halite is common in crusts on lava [39].

Figure 5. Salammoniac, Etna. Sample from the Natural History Museum in Copenhagen.

Salammoniac (CsCl structure type) (**Figure 5**) is one of the characteristic minerals in low-temperature fumaroles. Its sublimation temperature is 338°C. El [1]: the main constituent of white and yellowish crusts on lava shortly after eruption, but 15 years later no more present. H [1]: one of the main first fumarolic minerals, but already after one year disappeared from the fumaroles. Ve [14]: the representative main mineral of a type of fumaroles mostly on lower lava flows and rarely close to the crater with the temperature around 300°C. Vu [40]: the enrichment in Br follows the enrichment in the magmatic component of the gases. Crusts and aggregates of colourless soft isometric crystals sometimes reaching mm size. Also A, Kr, CF, Et. Typical accompanying minerals are cryptohalite (El, H) sassolite and sulphur (Vu), mascagnite and As sulphides (CF). Isostructural is **lafossaite** [41]; Vu: small grey-brown crystals (<100 μm) covering sulphosalts and pyrite or as opaque crusts; Ve [42]: <100 μm colourless cubic crystals, very rare.

Sellaite, El: in a brown crust with ralstonite and fluorite covering white microcrystalline anhydrite; Ve [43]: in ejected blocks with anhydrite; Et [43]: with fluorite. Very rare. **Chloromagnesite** and isostructural **scacchite**: very rare, with halite and sylvite (Ve [9]). **Fluorite** is common in fumaroles. Su [1]: the main fluoride in a lava tube and the only fluoride on the inner wall of a nearly closed crater, but not on the surface of lava where ralstonite is the dominating fluoride. H [1]: it is fairly common but subordinate to ralstonite and jakobssonite. Ve [44]: it is observed on pre-existing tenorite and rare. El: in a brownish crust with ralstonite and sellaite. Very rare. **Oskarssonite** [45] is the stable trigonal low-temperature form of AlF_3 with a perovskite-type framework of AlF_6 octahedra. El: in thick white microcrystalline crusts with other fluorides, opal, and anatase, more abundant than ralstonite in late fumaroles due to decreased amounts of Na and Mg in the gases and rock. Chemical analysis reveals partial replacement of F by OH, consistent with observations by Rosenberg on synthetic material [46]. **Parascandolaite**, Ve [47]: in colourless transparent sub-mm cubic crystals. Very rare.

Molysite Ve [9]: in yellow and red covers of the altered rocks near the crater. **Carnallite**, Su [1]: in a yellow-brown stalactite together with K-enriched halite. **Eriochalcite**, Ve [9]: rare blue wool-like aggregates together with other Cu minerals. **Ammineite**, Ve [48]: bluish crusts on altered tenorite together with opal and artroeite. Very rare. **Chloraluminite**, Ve [14]: small rhombohedral or prismatic crystals in high-temperature fumaroles. Very rare.

Ferruccite and **avogadrite** (Ve [43]) are very rare fumarolic borofluorides. **Barberiite**, Vu [49]: globular aggregates of colourless platy sub-mm crystals. With sulphur, salammoniac, sassolite, malladrite, and realgar.

Rosenbergite, H ([1]: mineral HM): in yellow to brown crusts with other fluorides. Rare. El: secondary hydration phase on oskarssonite. Very rare. **Mineral HI**, isostructural to rosenbergite. H [1]: in yellow to brown crusts together with its polymorph **mineral HU** [1] and other fluorides. Both rare. **Pachnolite**, H [1]: in a yellow/brown/green crust with ralstonite. **Gearksutite**, Vu [50]: in veins, impregnations, pockets, and concretions as alteration product of plagioclase under influence of F-rich solutions; Ve [48]: on post-1944 eruption fumaroles.

Jakobssonite [51] (mineral HA [1]) H, El, F, Kr: white crusts or thin overgrowths of sub-100 μm acicular crystals together with leonardsenite, ralstonite and other alumino- and silicofluorides. Common. Less abundant are **leonardsenite** [52] in orthorhombic-prismatic colourless crystals, and the presumed new mineral **HG** [1] in rhombic-dodecahedral crystals, both sub-100 μm. In similar associations as jakobssonite (H, El). The latter is identical to the synthetic compound of the same composition [53]. **Coulsellite**, Ve [47]: accompanies parascandolaite. **Thermessaite** [54] and **thermessaite-(NH_4)** [55] Vu: in medium temperature fumaroles (around 300°C) in sub-mm colourless prismatic crystals associated with alunite, sassolite, anhydrite, and metavoltine. Very rare.

Ralstonite has a crystal structure consisting of a pyrochlore-type framework of AlF_6 octahedra, with Al partly substituted by Mg, and Na and H_2O in framework cavities [56]. Common, dominating among fluorides in microcrystalline yellow crusts on Iceland (**Figure 6**) (H, El, Su [1]; F, Kr). The presumably new mineral HS [1] (H) shows the same diffraction characteristics as Rosenberg's AHF phase [46] with the composition $Al(F,OH)_3(H_2O)_n$ isostructural to ralstonite. The compositional range of ralstonite is still not sufficiently investigated and it is not clear if HS is a new mineral or a variant of ralstonite. The presumed new mineral **HH** (El, H [1]): thin platy colourless crystals of hexagonal habit up to over 100 μm in diameter in the surface part of the fumaroles. The composition and crystallography are preliminary determined. Rare. The presumed new mineral **HB** (H, El [1]): in white to yellowish crusts typically with ralstonite. Composition and crystallography unknown. Fairly common. The presumed new mineral **HD** (H, El, Su [1]) is probably an ammonium-free variant [57] of the zeolitic $NH_4Fe^{2+}Fe^{3+}F_6$ with a pyrochlore framework of FeF_6 octahedra [58]. In yellow to brownish crusts mixed with alumino- and silicofluorides. Fairly common. **Meniaylovite,** [59] is a minor constituent in yellow to brownish crusts in association with various other fluorides (Su, El, F, Kr). **Cossaite,** [60] Vu: only two 100 μm crystals found in a medium-T (~350°C) fumarole.

Figure 6. Fluoride crust from Eldfell. Mixture of ralstonite, jakobssonite, oskarssonite and other minor fluorides.

Malladrite, H [1]: fairly common in yellow crusts with ralstonite and jakobssonite; El [1], Kr: rare, in a yellow-brown crust with ralstonite; Vu [49]: rare, in association with barberiite. **Hieratite,** Vu [61]: in small colourless crystals with hexahedral habit associated with sulphur, realgar, sassolite, and several sulphates. Also Ve [43], H [1]. Rare. **Demartinite,** Vu [62] polymorph of hieratite, probably metastable. In colourless hexagonal pyramidal sub-mm crystals. **Heklaite,** H [63]: fairly common with malladrite and sometimes hieratite, because crystal structure differences prevent miscibility of Na and K silicofluorides. **Cryptohalite,** Ve [14]: in colourless cubic or octahedral crystals. Typically follows salammoniac, like on H and El in the first fumaroles [1]. **Bararite** is the low-temperature polymorph of the former, stable under 5°C. Found with salammoniac and cryptohalite (H). Probably a climatic influence. Also Ve [43]. Very rare. Presumed new mineral **HT** [1] is identical with synthetic $FeSiF_6(H_2O)_6$. H: very rare aggregates of micron-sized granular crystals overgrown with aggregates of HU crystals. **Knasibfite,** Vu [64]: with hieratite, avogradite and demartinite in low temperature fumaroles. Very rare.

Chlormanganokalite, Ve [14]: rare yellow crystals in high-temperature fumaroles. **Erythrosiderite**, Ve [14]: rare, in stalactites and crusts of tabular and pseudooctahedral red-orange deliquescent crystals in medium- and high-temperature fumaroles. **Kremersite**, Ve, Vu, Et [43]: rare small orange crusts. **Mitscherlichite**, Ve [43]: very rare in greenish-blue thin crusts. **Melanothallite**, Ve [9]: in rare thin black scales. **Atacamite**, Ve [14]: fairly common in thin green crusts or small prismatic crystals on lava. **Paratacamite**, Ve [9]: fairly common green alteration of tenorite. **Connellite**, Ve [65]: very rare, in post-1994 eruption fumaroles. **Cumengéite**, Ve [44]: very rare small transparent crystals of deep blue colour, adamantine lustre and pseudocuboctahedral habit, with tenorite, gypsum, and cotunnite. **Bismoclite**, Vu [22]: very rare sub-mm brilliant dipyramidal or tabular anatase-like orange crystals. **Lucabindiite**, Vu [8]: rare μm-size platy colourless transparent crystals with samammoniac, sulphur, and arsenolite. **Cotunnite**, Ve [9, 14]: rare colourless lustrous acicular crystals, scales, arborescent aggregates and fused masses, sometimes as pseudomorphs after galena in the high-temperature fumaroles. Vu [66]: variously coloured, fairly common in association with challacolloite. **Pseudocotunnite**, Ve [9]: very rare thin crusts and needles. Not fully crystallographically characterized species. **Challacolloite**, Ve [14]: rare colourless acicular crystals with adamantine lustre in HT fumaroles; Vu [66]: colourless to yellow, red, and brown, fairly common. **Hephaistosite**, Vu [67]: rare aggregates of pale yellow-green tabular crystals associated with Bi and Pb sulphides and chlorides. **Brontesite**, Vu [68]: very rare minute white tabular crystals. **Panichiite**, Vu [69]: very rare minute colourless to pale yellow octahedral crystals with adranosite, alunite, anhydrite. **Steropesite**, Vu [70]: very rare yellow-green minute crystals or crusts covering lafossaite. **Argesite**, [71] Vu: in sub-mm pale yellow very rare crystals in association with bismuthinite.

3.5. Oxides and hydroxides

Oxides are mostly products of reaction of gases with scoria, lava or other rocks in which fumaroles form and in lesser amount form as sublimates. The leaching of lava by aggressive fumarolic gases can lead to material consisting exclusively of oxides of Si, Al, Fe, and Ti, opal-cristobalite, corundum, hematite, and anatase, respectively, mixed with anhydrite (in Icelandic fumaroles) or alunite (in Greek fumaroles).

Tenorite, CuO, Ve [14]: lamellar, leaf-like or fan-like aggregates of crystals of a black metallic appearance in the HT fumaroles ($T > 400°C$ [72]); Et [38]: thin scales with hematite and polyhalite in volcanic tubes. Rare. **Spinel, $MgAl_2O_4$**, El: very rare, in a cellular brown crust consisting mainly of hematite, also accompanied by corundum, anhydrite and steklite. **Magnesioferrite, $MgFe_2O_4$**, and **magnetite, Mg_3O_4**, Ve [9]: minute iron-grey crystals. Rare. **Hausmannite, Mn_3O_4**, Ve [9]: rare thin brown coatings on lavas. **Minium** Ve [9]: very rare dusty red layers cementing volcanic conglomerate. **Corundum**, El: one of the main constituents of the wall rock (with cristobalite, hematite, and anhydrite) of a fumarole (**Figure 7**). **Hematite:** abundant (Su, El, H, F, Ve, Et, M, Sa) and the main reason for the red colour of scoria- and lava-bearing fumaroles. Mostly a decomposition product; also a sublimate in small black crystals with metallic lustre (El). **Ilmenite**, H [1]: in a brownish crust with hematite and fluorides. **Pseudobrookite**, El: very rare, admixture in a sample of white to grey granular to earthy crust of cristobalite and anhydrite on a substrate of hematite.

Figure 7. Hematite (red) and corundum (white) with opal in the altered lava on Eldfell.

Opal is common in all fumaroles. Primarily a product of decomposition of silicate minerals, but observed amorphous material could also be a primary glassy part of lava. In a sample from El, opal-CT forms the compact part of the leached lava, whereas opal-A fills holes as hydrothermal deposit. Opal is often associated with **cristobalite** (El, CF, Et, So, M, Sa, N) or **tridymite** (Et, Sa, N). They could also be products of fumarolic reactions or just unaltered remnant of lava. **Quartz** is absent in the majority of fumaroles but abundant in some (M, N, El). Mostly remnant of the primary rock, but appearance as granular crust suggests also secondary origin (M). **Anatase:** fairly common in late fumaroles with other oxides (El); rare in the altered rock with quartz, sericite and alunogen (M). Analyses of partly to fully leached lava on Eldfell show that TiO_2 remained as the only other component besides SiO_2 in the opalized rock, which shows it to be relatively immobile [1]. **Rutile**, high-temperature polymorph of TiO_2, stable over 600°C at atmospheric pressures, is rarely observed (M, N) probably primary from lava.

Portlandite, Et [38]: coralloid concretions with calcite in a volcanic tube. **Gibbsite**, Ve [9]: with basanite in hexagonal scales. Its polymorph **doyleite** was confirmed by XRD in one sample from Surtsey, in association with hematite [1].

3.6. Carbonates

Carbonates are rare constituents in fumaroles. It might look strange considering how abundant CO_2 is in volcanic gases, but the scarcity of solid products is due to the typical high acidity of the fumarolic environments making most carbonates unstable. Where they appear in significant quantities, specific conditions are present, for example, low acidic or even basic environments and high water activity, like in the case of occurrences of Na carbonates on Vesuvius or calcite on Surtsey and Etna.

Calcite is the most frequent carbonate in fumaroles. Common in caves and cavities of lava where steam emanations were vigorous (Su [1]). Fairly common in stalactites and concretions in volcanic tubes (Et [38]). Fairly common in altered rock with **magnesite** and **dolomite** (So [73]). **Hydromagnesite**, Su [1]: in a white botryoidal crust with calcite and fluorite in a lava tube. **Aragonite**, So [73]: rare, with anhydrite and Fe-Mg hydrous sulphates. **Cerussite** accompanies parascandolaite (Ve [47]). **Azurite**, Ve [9]: blue covers on older lavas. **Sodium carbonate hydrates** appear in the high-temperature fumaroles on Vesuvius dominated by potassium and sodium salts [9, 14]; **thermonatrite** in white crusts, sometimes stalactites, **natron** in transparent granules and small translucent crusts as efflorescence in the interior of lavas, and **trona** in small tabular pseudohexagonal crystals or whitish crusts. The latter was also found with halite in concretions in a lava cave on Etna [38].

3.7. Borates

Over 210 boron minerals are known. Among them only three borofluorides mentioned earlier and four borates are observed in fumaroles. The boric acid, **sassolite**, melts at 171°C and is therefore constrained to the low-temperature fumarole. Ve [9]: rare thin colourless scales together with gypsum or sulphur or realgar; Vu [10]: common in colourless plates, up to 5 mm or white covers with sulphur and salammoniac in low-temperature fumaroles. Exploited in the past as the raw material. **Metaborite** and **clinometaborite**, Vu [74]: with sassolite and andranosite in a medium-temperature fumarole (~250°C). **Ameghinite**, Et [38]: very rare, in macrocrystalline crusts with thenardite.

3.8. Sulphates

Together with halides sulphates are the group of minerals with the largest number of various species in fumaroles. Calcium sulphates, such as gypsum and in lesser grade anhydrite, are common and in many fumaroles very abundant. Sodium sulphate, thenardite, and K-Al sulphate, alunite, can also appear in large quantities in some types of fumaroles. Specific for sulphates is existence of often several hydrous forms related to the anhydrous ones. They possess different stability fields that depend on the humidity and temperature, and can, therefore, appear in various zones of the same fumarole. The anhydrous sulphates of fumarolic origin are generally instable and readily hydrate under atmospheric conditions (anhydrite and barite group minerals are exemptions).

Chalcocyanite, Ve [9]: fairly common in greenish to sky-blue crusts, rapidly changing to chalcanthite. **Vanthoffite**, F: main phase in a white "snow-like" compact crust on lava with thenardite, löweite, and glauberite; El: in white to greyish crusts with thenardite, anhydrite, glauberite, and other sulphates. Presumed new **mineral EN**, F: in orange spongy crust together with halite, gypsum, and a mineral from voltaite group; El: in paragenesis with presumed new

mineral EA, anhydrite, langbeinite, tamarugite, and hexahydrite. The composition and crystallography of **EN** and **EA** is confirmed by SEM-EDS and XRD. They form crusts of intermixed < 10 μm isometric grains and leaves, respectively. Presumed new **mineral EI**, El [1]: in saussage-like thin fragile hollow white or red (coloured by hematite) crusts, which easily hydrate to löweite and hexahydrite when exposed to the humid atmosphere. Composition confirmed, but crystal structure unknown. **Langbeinite**, El ([1]: mineral EB): in white crusts on scoria together with sulphates or with fluorides at temperatures from 80 to 230°C; F: With anhydrite in a glassy vesicular crust covering lava. **Manganolangbeinite**, Ve [9]: in microscopic pink tetrahedra and stalactites with thenardite and Na-K chlorides. **Steklite**, [75] El: accompanying hematite with corundum, spinel and anhydrite. Very rare. **Pyracmonite**, [76] Vu: aggregates of sub-mm colourless elongated hexagonal prismatic crystals with salammoniac and kremersite in a medium-temperature fumarole (~250°C). **Aluminopyracmonite**, Vu [77]: very rare globular aggregates of prismatic colourless crystals. **Eldfellite**, [78] El: in a frothy yellow to brown crust together with tamarugite, EN and anhydrite. Very rare. Isostructural **yavapaiite**, Vu [79]: very rare minute pink crystals in association with other Fe sulphates.

Figure 8. Dendritic aphthitalite crystals with hematite and bluish granular Cu-enriched metathenardite. Field of view 1 cm.

Thenardite is a frequent mineral in fumaroles. In Iceland, four different parageneses were observed: with gypsum, glauberite, and minor hydrous sulphates (Su, H); with metathenardite, aphthitalite, and glaserite (F, Kr, Su) with vanthoffite, anhydrite, and halite (El); with halite and minor other sulphates (A). In stalactites (Su, Ve, Et), white to lightly coloured friable microcrystalline crusts or in glass-like crusts and visible crystals when in association with aphthitalite and glaserite. The latter association includes its high-temperature polymorph **metathenardite**, stable over 271°C. Green due to Cu content, rare (F, Kr). **Aphthitalite (Figure 8)**, Ve [9], F, Kr: fairly common in millimetre size dendritic "snow flake" crystals on lava variously coloured from white to yellowish and greenish or bluish due to Cu content; Et [38]: in macrocrystalline crusts and stalactites in volcanic caves. **Glaserite**, F, Kr: in platy crystals mixed with aphthitalite. Rare. The crystal name was abandoned [80], although it is a distinct LT phase (<150°C) in the phase system [81] with a distinct crystal structure [82]. **Palmierite**, Ve [9]: in minute hexagonal scales together with aphthitalite, ferronatrite, jarosite, and euchlorine in the "dry" fumaroles. **Mascagnite**, Ve [9]: in white crystalline crusts mixed with salammoniac, halite, and sylvite in the high-temperature fumaroles; Vu [10]: rare sub-mm colourless platy crystals with adranosite and sassolite.

Mercallite, Et [83]: in a lava tunnel; F: in fine-grained colourless crystalline aggregates on the bluish dendritic thenardite with Kröhnkite. Rare. Presumed new **mineral FB** appears with the former (F). **Therasiaite**, Vu [84]: in very rare brown equant sub-mm crystals with salammoniac. Instable.

Figure 9. Gypsum crystals. Soussaki.

Anhydrite, Su, F, El, Vu, So: common, with hematite the main constituent of the altered wall-rock, in compact to earthy materials mixed with other minerals or as separate acicular crystals; Ve [14]: rare in tabular colourless or whitish crystals. **Gypsum (Figure 9)** is one of the main fumarolic minerals in general. It forms at lower temperatures because it dehydrates already at 120–130°C. In various forms: large (> cm) crystals (So), crusts of various thicknesses from thin colourless and glassy to white or lightly coloured massive, also as efflorescence and in various associations (Su, El, CF, Ve, Et, Sa, F, N, Vu). Rare on H and M, not observed on A. **Bassanite** is stable to about 200°C. Ve [14]: in altered volcanic bombs as acicular white crystals; Su [1]: in crusts with anhydrite and gypsum; El: mixed with oskarssonite in a yellowish earthy material. Rare. Isostructural **omongwaite** [85] Su ([1]: mineral SA): in a solid white to colourless crust on lava together with halite, anhydrite, glauberite, and other sulphates. **Glauberite**, Su,H [1] F: in small quantities accompanying thenardite and/or anhydrite. Very rare. **Barite**, M: mixed with alunite in a yellowish-white earthy crust that covers quartz crystals. Rare. **Celestite**, Et [38]: rare, in concretions in the lava caves. **Anglesite**, Ve [9]: rare, in small pale violet crystals with acicular and pseudooctahedral habit; Vu [13]: very rare in association with Bi oxy-sulphates.

Dolerophanite, Ve [9]: very rare dark brown crystals. **Antlerite**, Ve [9]: very rare, in grass-green aggregates of minute crystals as the alteration product of dolerophanite. **D'ansite** [86], A [1]: in a stalactite together with thenardite, halite, löweite, and blödite; Et [38]: in a macro-crystalline crust accompanying thenardite in a lava cave. Very rare. **D'ansite-(Mn)** [87] Ve: very rare colourless tris-tetrahedral crystals, with halite and aphthitalite. **D'ansite-(Fe)** [87] Vu: very rare aggregates of minute isometric colourless crystals with sassolite. **Aiolosite**, Vu [88]: very rare colourless acicular crystals in a medium-temperature (250°C) fumarole. **Natroalunite**, Su [1] Sa: in white-greyish crust on the inner wall of a crater, with opal, gypsum, and fluorite. **Alunite** is one of the main fumarolic minerals on Greek islands. In granular aggregates, dendritic crystals, flakes, white powdery to compact cover on lava, colourless, orange and red glassy crusts, white and yellowish botryoidal crusts, and grey cauliflower-like crusts accompanied by sulphur and alunogen and rarely by jarosite or barite, rhomboclase, and voltaite (M); constituent of the altered rock and around the fumaroles on the crater floor (N); in alteration with natroalunite accompanies sulphur in different fumaroles (Sa); in a yellow crust on lava block with halite, sylvite, gypsum, and erythrosiderite in the low-temperature fumarole (Ve [9, 14]); rare, in concretions in lava caves (Et [38]). **Natrojarosite**, El: rare in altered scoria with hematite, anhydrite, and opal with ralstonite, fluorite, and oskarssonite. **Jarosite**, El [1]: in yellow crusts together with gypsum, anhydrite, and fluorides; Ve [9]: in thin crusts with aphthitalite, euchlorine, and some chlorides; M: mixed with alunite. Rare. **Adranosite** and **adranosite-(Fe)**, Vu [89, 90]: very rare aggregates of colourless or pale yellow, respectively, acicular sub-mm crystals in medium-temperature (~250°C) fumaroles. **Euchlorine**, Ve [9]: fairly common in grass-green to emerald-green encrustations and tabular crystals together with chalcocyanite and with metavoltine. **Chlorothionite**, Ve [9]: rare greenish blue to green crusts. **Linarite**, Ve [9]: rare microscopic crystals in gypsum. **Balićžunićite** [13] and **leguernite** [91]: in cavities of a sample taken from the high-temperature fumarole ($T = 600°C$) in minute transparent prismatic crystals in association with anglesite and in the vicinity of sulphosalt formation. Very rare.

Kieserite, Su [1]: very rare, in association with halite, kainite, löweite, and pentahydrite in stalactites in a lava tube; Et [38]: very rare in stalactites in a lava cave. **Rozenite**, So [73]: the main mineral in white efflorescences. **Chalcanthite**, Ve [9, 14]: rare sky-blue granular masses or crusts, a product of alteration of primary fumarolic Cu minerals. **Pentahydrite**, F: very rare white sublimate on volcanic ash together with halite and **hexahydrite**: in a spongy yellowish crust with tamarugite and kainite (F); a hydration product of mineral EI and in a white crust with löweite and anhydrite (El); rare constituent of concretions in association with thenardite in lava caves (Et [38]); among sublimates at the exhausts of fumarolic vents (So). **Nickelhexahydrite**, So [73]: trace amount in a sample with römerite. **Römerite**, So [73]: the main mineral in green efflorescences. **Coquimbite**, So: rare in a grey earthy mass mixed with gypsum, anhydrite rhomboclase, and voltaite; Sa: with rhomboclase and tamarugite accompanying alunogen. **Halotrichite:** fairly common silky white needles (CF [PSCF], N). Characteristic is the "daisy" form: halotrichite "petals" growing on the core of globular yellow core of rhomboclase (N) or presumably coquimbite [15](**Figure 10**). **Pickeringite**, CF [15]: in thick wool-like aggregates of needle-like white crystals on the wall of a cave; Et (ferroan) [83]: in an eruptive fracture (*T* around 100°C) and on the rim of the NE crater (*T* 300–400°C). **Meta-alunogen**, Et [38]: with thenardite in stalactites in a lava cave; N: in colourless platy dendritic crystals together with alunogen, tamarugite, and voltaite on the floor of the crater. **Alunogen**, Ve [9, 14]: with kalinite and as silky fibrous masses in low-temperature fumaroles; M: accompanying alunite; Sa: alternating with alunite and natroalunite as the main mineral in crusts covered by sulphur crystals; N: around the fumaroles on the crater floor. **Mirabilite**, Su [1]: in a colourless crust with thenardite in a lava tube; Ve [9]: forming from solutions and by recrystallization of fumarolic salts; Et [38]: stalagmites in a lava cave. **Campostriniite**, [92] Vu: very rare white prismatic micrometre crystals, with adranosite, sassolite, and other Bi minerals.

Figure 10. Yellow globules of rhomboclase with acicular halotrichite, Nisyros.

Rhomboclase, So: rare constituent of the earthy mass with gypsum, anhydrite, coquimbite, and voltaite; M: accompanying alunite; Sa: accompanying alunogen together with coquimbite and tamarugite; N: microscopic yellow globular aggregates associated with silky needles and plates of halotrichite. **Voltaite** is of a dark-green or blue to black colour due to its content of mixed-valence iron. Ve [9, 14]: rare granular aggregates of cubic crystals in the low-temperature fumarole; CF [15]: rare black lustrous crystals mixed with halotrichite; So: rare part of a grey earthy mass; M: very rare together with alunite, alunogen, rhomboclase, and pyrite; N: in granular aggregates on the crater floor with tamarugite, alunogen, and meta-alunogen. A light-coloured mineral from the voltaite group is observed in H and F samples but not characterized in detail. **Löweite**, Su [1]: rare, in stalactites together with halite and mineral SH or Mg sulphates; A [11]: with other Na-Mg sulphates in a stalactite; El: a hydration product of mineral EI. **Eugsterite** [93] Su [1]: rare, in a soft white-to-yellow crust containing also glauberite, anhydrite, gypsum, thenardite, and Blödite. **Hydroglauberite** [94] H [1]: component of a yellow-whitish crust on lava together with thenardite and glauberite. **Syngenite**, Ve [14]: rare transparent minute tabular crystals; Et [38]: very rare, with thenardite in stalactites in a lava cave. **Kröhnkite**, F,Kr: rare, in bluish dendritic crystalline crusts with aphthitalite, glaserite, metathenardite, and thenardite. **Ferrinatrite**, Ve [9]: very rare thin fibres with adamantine lustre with aphthitalite and palmierite. **Cyanochroite**, Ve [9]: rare pale blue crystalline crusts.

Blödite, Su [1]: rare, mixed with Na-Ca sulphates and halite in stalactites and soft white-yellowish crusts; A [1]: rare, in a stalactite together with other Na-Mg sulphates; El: very rare hydration product of mineral EI; Et [38]: rare in concretions, stalactites, and macrocrystalline crusts, primarily with thenardite in lava caves. **Polyhalite**, Su [1]: very rare, in a solid white to colourless crust made of halite, anhydrite, and other sulphates; Et [38]: very rare in polycrystalline aggregates with hematite and tenorite in a lava cave. **Tamarugite**, El [1]: rare, accompanies minerals EN and EA, most probably as their hydration product and also accompanying eldfellite; F: very rare in a yellowish spongy crust with a smooth glassy surface together with hexahydrite and kainite; Sa: rare, together with alunite, alunogen, rhomboclase, and coquimbite; N: on the crater floor together with voltaite, alunogen, and meta-alunogen. **Picromerite**, Ve [9, 14]: in white to colourless crusts of elongated minute crystals; Et [38]: very rare, in concretional crusts in a lava cave. **Metavoltine**, Ve [9, 14]: in thin yellow crusts of tabular hexagonal crystals associated with euchlorine. **Kalinite** [95] Ve [9]: rare white crusts in the sulfurous fumaroles. **Alum-(K)**, Ve [14]: rare whitish crusts or octahedral to cuboctahedral crystals in medium-temperature fumarole. **Kainite**, Su [1]: rare white to colourless crusts and stalactites in mixture with other Mg, Ca, and Na sulphates; A [11]: rare, in a stalactite with thenardite and halite; F: rare, in a spongy crust with tamarugite and hexahydrite. Presumed new **mineral SH**, Su [1]: rare, in a colourless to white crust with halite, anhydrite, glauberite, kainite, omongwaite, and polyhalite and in stalactites with halite and löweite. **Chessexite** [96] El [1]: rare, in white crusts with gypsum and ralstonite at 80–100°C.

3.9. Other minerals

Chlorapatite and **mimetite** are isostructural. The former was identified in a stalactite accompanying thenardite in a lava cave (Et [38]). The latter and **phoenicochroite** accompany parascandolaite (Ve [47]).

4. European fumaroles

4.1. Icelandic volcanoes

Iceland is one of the most active terrestrial volcanic regions, with eruption frequencies of ≥20 events per century. It is also one of the most productive with magma output rates of ~8 km^3

per century in historic time [97]. Volcanism in Iceland is caused by the interaction of the Mid-Atlantic Ridge and the Iceland mantle plume. This volcanism can be traced back to the opening of the North Atlantic at 61 Ma as evidenced by massive volcanism in East Greenland, Ireland, Scotland, and the Faroe Islands. The oldest volcanic rocks exposed in Iceland are about 16 Ma [98]. During Late-Pleistocene and Holocene times, some 41 volcanic systems have been active in Iceland and its insular shelf. They are confined to volcanic zones, which are either rift zones or non-rifting flank zones. Volcanic systems in the rift zones produce rocks belonging to the tholeiitic series, while rocks of the mildly alkalic and transitional alkalic series are confined to the flank zones [99]. The majority of volcanic rocks are basalts, but significant amounts of silicic and intermediate rocks have also been produced. Encrustations have been studied from the products of seven eruptions in five volcanic systems in Iceland (**Figure 11**).

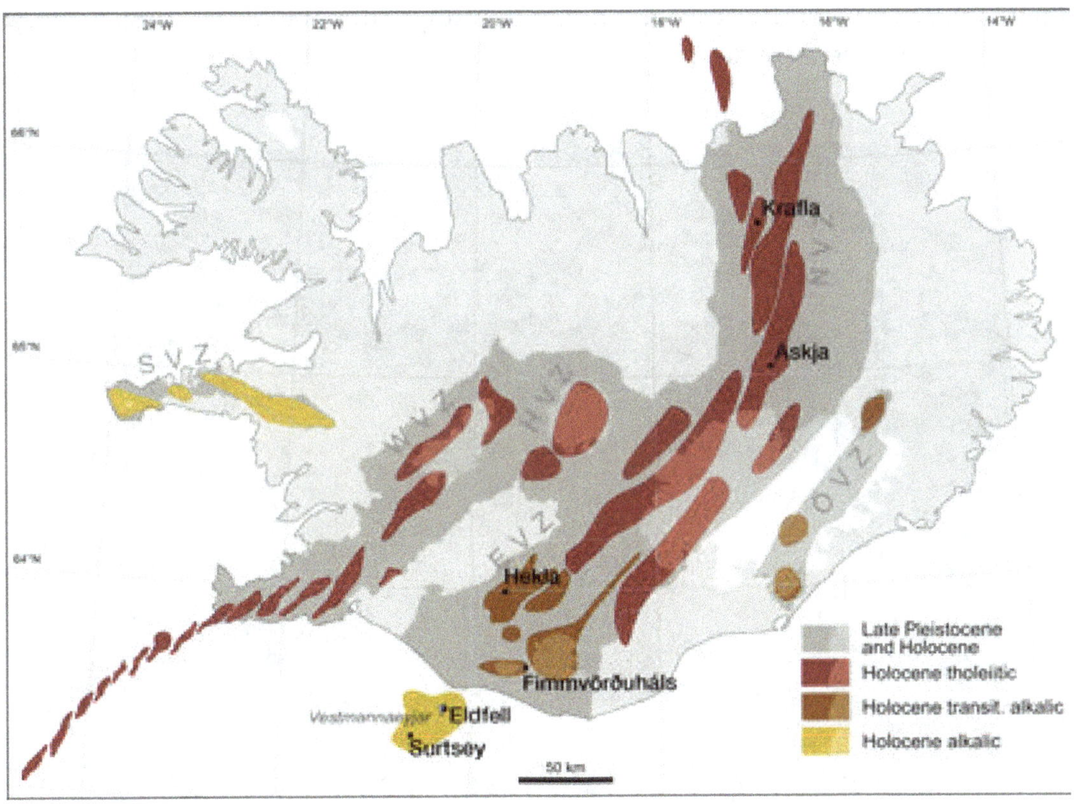

Figure 11. The volcanic systems of Iceland with the locations of investigated volcanoes.

4.1.1. The Vestmannaeyjar volcanic system

The Vestmannaeyjar archipelago represents the southernmost volcanic system in the eastern volcanic zone of Iceland. It is a partly submarine system-producing basaltic to intermediate rocks of the alkalic series. Encrustation samples from two separate eruptions were examined, the 1963–1967 Surtsey eruption and the 1973 Eldfell eruption.

The 1963–1967 Surtsey eruption. Surtsey is the south-westernmost island of the Vestmannaeyjar archipelago. It was constructed from the sea floor in a volcanic eruption occurring from

1963 to 1967 [100–102]. During the hydromagmatic explosive submarine phase of the eruption, from November 14, 1963 to April 4, 1964, alkali basalt tephra formed two crescent-shaped cones which merged. The eruption then evolved into an effusive alkali basalt lava phase which lasted until June 5, 1967. Two lava shields were formed with a thickness of 70–100 m. However, individual lava flow units are thin, usually only a few metres thick. The eruptive temperature of the lava was about 1140–1180°C. It is estimated that the total output of the 1963–1967 Surtsey eruption was 1.1 km^3 of basalt tephra and lava [102].

The surface encrustations on Surtsey were mainly deposited in two types of environments, as sublimates deposited directly from a gaseous state on lava and scoria at relatively high temperatures, and in a vapour-dominated system in lava craters and shallow lava caves, where steam emanation was vigorous [103]. Encrustation samples were collected in 13 expeditions from 1965 to 1998 [1]. One can recognize the following main mineral associations: 1. gypsum with thenardite, calcite, and fluorite; 2. halite with anhydrite, glauberite, thenardite, and Na-K-Mg hydrous sulphates; 3. ralstonite with other fluorides (**Figure 12**).

Figure 12. Surface encrustations on Surtsey in 1967. Photograph by H Bárðarson.

The 1973 Eldfell eruption. The Eldfell volcano is situated on Heimaey, the largest and the only inhabited island of the Vestmannaeyjar archipelago. The Eldfell eruption started on January

23, 1973, on a 1.5-km-long fissure, producing lava along its entire length. Eruptive activity quickly concentrated at the central part of the fissure and a scoria cone, Eldfell, was built up, reaching a height of 245 m asl. [104]. The Eldfell eruption ended on June 26, 1973. The magma was of hawaiite-mugearite composition [105]. At the end of the eruption, the lava covered 3.2 km^2, and the total output of lava and scoria was estimated to be 0.25 km^3 [106]. The Eldfell lava reaches a thickness of about 110 m at the eastern side of Eldfell, and large sections of the lava are 40–60 m thick. The eruptive temperature of the lava was 1030°C.

The Eldfell magma was more evolved than the magma erupted at Surtsey and as a result released a larger amount of volatiles. After the cessation of the Eldfell eruption, the extrusives and the feeder dikes have continued to release a considerable amount of gases, especially at the Eldfell scoria cone. The Eldfell lava is blocky, although large parts of it are covered with an apron of scoria. Early on, volcanic gases formed extensive encrustations on the surface of the lava and the first encrustations were already collected in February 1973 [39]. The lava encrustations on Eldfell were deposited directly as sublimates from a gaseous phase discharged from the cooling lava at a range of temperatures. Due to its thickness, the lava has been cooling down at a considerably slower rate than on Surtsey. The Eldfell crater is made up of coarse scoria mixed with volcanic bombs and lava fragments. When the crater was visited in June 2012, a narrow section of the crater rim was still very hot, reaching maximum temperatures of 290°C at 15 cm depth. Encrustation specimens were collected on the lava in 1973 and 1975, and at Eldfell crater between 1988 and 1995, and in 2009, all from the areas that were not affected by the cooling operations undertaken on the island during the eruption. The observed mineral associations are as follows: (1) (early) salammoniac, cryptohalite, gypsum, jarosite; (2) anhydrite, bassanite, gypsum; (3) anhydrite with ralstonite (later oskarssonite), jakobssonite and other fluorides; (4) anhydrite with Na-K-Al-Mg anhydrous and hydrous sulphates and (5) (late) anhydrite, hematite, corundum, anatase, opal/cristobalite.

4.1.2. The Eyjafjallajökull volcanic system

The Eyjafjallajökull volcanic system is located in the southern part of the eastern volcanic zone. It produces rocks belonging to the transitional alkalic series and consists of a ridge-shaped central volcano elongated E-W, reaching an altitude of 1651 m. It has a large summit crater flanked by mainly E-W-oriented eruptive fissures.

The 2010 Fimmvörðuháls eruption. Two connected eruptions occurred at Eyjafjallajökull in 2010. First, a basaltic fissure eruption occurred at Fimmvörðuháls on the eastern flank of the volcano. It started on March 20 and ended on April 12. Two days later, on April 14, an explosive eruption started in the summit crater of Eyjafjallajökull, which lasted until May 22nd [107]. The magma erupted at Fimmvörðuháls was a porphyritic transitional basalt, the majority of which solidified as aa-type lava, in addition to two scoria cones. The same type of basalt was involved in the explosive eruption at Eyjafjallajökull, where it was mixed with rhyolite, forming benmoreite and trachyte [108], much of which was dispersed as very fine-grained ash.

Encrustations were sampled at Fimmvörðuháls during the eruption, and at three separate times in the following months of 2010, and once in 2011. Very high gas temperatures (up to 800°C) were recorded during sampling. Volcanic ash from the Eyjafjallajökull eruption was cemented into a crust on the hot surfaces of craters and lava. This provided favourable conditions for the formation and preservation of volcanogenic encrustations beneath the crust. During the eruption, thin white coatings were observed on the fresh lava. Such coatings also appeared on samples collected while hot. The coatings were found to be thenardite, a readily soluble mineral that is quickly washed away [109]. The mineral associations determined through the analysis of samples are as follows: (1) halite and sulphates (Na-K-Mg-Ca-Al) and (2) ralstonite with other fluorides (**Figure 13**).

Figure 13. Formation of fumarolic minerals at 575°C in a fissure on the main scoria crater on Fimmvörðuháls about 2 months after the eruption.

4.1.3. The Hekla volcanic system

The Hekla central volcano is one of the most active volcanoes in Iceland, with more than 18 eruptions recorded in historical time. It is the production centre of the Hekla volcanic system in south Iceland. It produces basaltic to intermediate rocks of the transitional alkalic series and intermediate to silicic rocks belonging to the tholeiitic series.

The 1947–1948 Hekla eruption occurred solely within the top fissure (Heklugjá) of the volcano. The eruption started on March 29, 1947 and lasted until April 21, 1948. About 0.2 km^3 of tephra was produced during the plinian phase of the eruption. Lava covered 40 km^2 and its volume was about 0.8 km^3 [110]. The chemical composition of the extrusives changed from transitional mugearite at the beginning of the eruption, to transitional hawaiite at the end. A large amount of gas and vapour emanated from the craters, especially Axlargígur at the SW shoulder of the volcano. The encrustation samples from the 1947–1948 Hekla eruption were collected from five localities during 1947–1952. Encrustations have been observed in many of the eruptions of Hekla. For example, Schythe reports large deposits of encrustations at the craters which erupted in 1845–1846 [111].

The 1991 Hekla eruption started on January 17 with a short-lived plinian phase that was accompanied with an effusive lava phase [112]. After two days of eruption, the volcanic activity was mainly restricted to a single fissure trending east-southeast from the top of the mountain. The eruption came to an end on March 11, 1991. The 1991 Hekla extrusives are transitional mugearite [99]. On flat ground, the lava is of a block lava type; however, on the slopes of the mountain, it is mainly of an aa type. The lava has commonly a thickness of only 4-8 m on flat ground and covers about 23 km^2. The main crater is made up of coarse scoria mixed with volcanic bombs [112]. The total amount of tephra and lava produced is estimated to be 0.15 km^3. A considerable amount of volcanic gases and vapour was released during the eruption. Pollution of groundwater and rivers around the volcano was observed already a few days after the onset of the eruption, by the rise of concentration of fluorine, carbonate, sulphate, and other dissolved solids. The average content of soluble fluorine from tephra formed on January 17 was 1600 ppm [112]. Two lava surface localities and one lava cave were visited in January and March 1991. The most prolific area with respect to encrustations was on the eruption fissure above the main crater at 1105 m asl. The lavas, the main crater and the linear eruption fissure of the 1991 Hekla eruption cooled down to ambient temperatures at the surface in less than 10 years. The surface encrustations at Hekla were mainly deposited in two types of environments, as sublimates deposited directly from a gaseous state at relatively high temperatures on lava and scoria at the linear eruption fissure, and in a vapour-dominated system in the shallow lava cave where steam emanation was vigorous.

The observed mineral associations in fumaroles are as follows: (1) (early) salammoniac, cryptohalite, gypsum, sulphur; (2) ralstonite with jakobssonite and other alumino- and silicofluorides and (3) thenardite, anhydrite or gypsum, glauberite.

4.1.4. The Askja volcanic system

The Askja volcanic system is located in the Northern Volcanic Zone in the central highland of Iceland. It consists of a central volcano with nested calderas bisected by a volcanic fissure swarm. The youngest caldera was formed following a plinian eruption in 1875. The system produces mainly tholeiitic basalts, with minor rhyolite, dacite and hybrid intermediate rocks.

The 1961 Askja eruption occurred within the large Askja caldera. It was preceded by the formation of big solfataras on a N-S line along the eastern caldera wall of Askja, about 10 days before the eruption [113]. The volcanic eruption started along a fissure northwest of the solfatara area on October 26 and lasted until November 28 [114]. It was a small eruption, only producing about 0.1 km^3 of tholeiitic basalt lava. The lava, Vikrahraun, covered 11 km^2 and large parts of it are estimated to be only few meters in thickness.

Temperatures of 420°C were still measured at a depth of 20 cm at the crater in June 1962 [113]; however, in August 1962, all soluble salts had been washed away and, apparently, no new encrustations were being formed. Encrustations formed both at the lava surface and in caves but were apparently not widespread [115]. Óskarsson [39] studied samples collected in June 1962 and found encrustations of sulfur, sal ammoniac and an NH-Fe-Cl-compound at surface, and stalactites of metathenardite in lava caves. The 1961 Askja samples available to us, all of which were presumably collected during the summer of 1962, derive from two shallow lava caves. The observed mineral associations are as follows: (1) thenardite with halite and Na-Mg-K (chloro)sulphates and (2) salammoniac.

4.1.5. The Krafla volcanic system

The Krafla volcanic system is located in the northern part of the northern volcanic zone. It consists of a central volcano with rhyolite formations around the rim of a caldera. A volcanic fissure swarm bisects the central volcano. The rocks belong to the tholeiitic series ranging from olivine tholeiites to rhyolites.

The 1980 Krafla eruption. The Krafla fires were a major episode of rifting and volcanism in 1975–1984 [116]. The first years were dominated by rifting, dike injections, and minor eruptions. Five larger eruptions occurred in 1980–1984. The 1980 Krafla eruption started on July 11 on several discontinuous fissures on a 4 km long line. After the first day, it concentrated on the northernmost crater where the eruption continued until July 18. The tholeiitic basalt lava covered 5.3 km^2 and the volume is estimated to be 0.02 km^3. Encrustations were collected from lava surfaces on July 30, 1980. The observed mineral associations are as follows: (1) sulphur, salammoniac; (2) thenardite with aphthitalite and (3) ralstonite with other fluorides.

4.2. Italian volcanoes

Italy is one of the most volcanically active countries in Europe. Its recent and active volcanism is mainly connected with the convergence of the Eurasian and the African plate and the extension induced by the formation of the Tyrrhenian basin, which produced volcanism along the Tyrrhenian Sea, in Sicily and Sicily Channel, and on the Tyrrhenian Sea floor. This Plio-Quaternary magmatism exhibits an extremely variable composition and permits to distinguish various magmatic provinces, which differ in major and/or trace elements and/or isotopic compositions. The most important Italian localities showing a fumarolic activity belong to the Neapolitan area (Mt.Vesuvius and Campi Flegrei), to the Aeolian Arc (La Fossa Crater at Vulcano) and to the Sicily Province (Mt. Etna). In addition to these, a number of interesting areas with actual hydrothermal activity are present in Italy, but they have not been considered in this paper because nowadays they do not produce sublimates (**Figure 14**).

Figure 14. Map of the SE Tyrrhenian Sea with locations of volcanoes and fumarolic fields.

4.2.1. *The Campanian Plain*

The Neapolitan Province is located in the Campania Region, along the Tyrrhenian coastline, in one of the most densely populated active volcanic areas of the Earth. The geology of the area is prevalently represented by volcanics erupted from the Upper Pleistocene to present by Mt. Somma-Vesuvius on the east, by Ischia and Campi Flegrei volcanic fields on the west and by different ignimbrite eruptions (Campanian Ignimbrites) connected to fissural volcanism along fractures activated in the Campanian Plain. Campanian Plain is a waste graben bordered by Mesozoic carbonate platforms, which began to form in Late Pliocene but largely developed in Quaternary times. Its origin is related to the counter-clockwise rotation of the Italian Peninsula and the contemporaneous opening of the Tyrrhenian Sea, which generated a stretching and thinning of the continental crust accompanied by the consequent subsidence of the carbonate platform along most of the Tyrrhenian coast. The formation of the Campanian Plain was accompanied by the uplift of the central part of the Southern Appennines, in a regional stress regime generating NW-SE and NE-SW-trending faults and establishing the ideal conditions for magma to form and rise to the surface. With the exception of oldest volcanic rocks with calc-alkaline composition, erupted magmas younger than 400 ka are alkaline, with high potassium compositions erupted only at Vesuvius [117].

4.2.2. *The Phlegrean Fields*

Figure 15. As-sulphide fumarole at Pozzuoli, Campi Flegrei. Photograph by courtesy of Fabrizio Reale.

The Phlegrean Fields (Campi Flegrei: "The Burning Plain") are an area of active volcanism located about 25 km west of Vesuvius and 5 km west-southwest of Naples and consists of a large caldera formed about 35,000 years ago with the eruption of 80 km^3 of ash (the Campanian Tuff). The caldera is about 12 km in diameter and includes a series of mostly monogenetic pyroclastic vents. After the caldera formation, with a progressive decrease in volume of the emitted products, the eruptive centres migrated towards the centre of the caldera. There were two historic eruptions in the area. The 1198 phreatic eruption was at Solfatara, a pyroclastic cone formed about 4000 years ago and still in a fumarolic stage. The last one in 1538 produced the new pyroclastic cone of Monte Nuovo. It was explosive and generated pyroclastic flows, but also produced a lava lake and lava flows. After that, the Phlegrean area is characterized by fumarolic activity and periodic episodes of unrest involving seismic activity and slow ground motion (bradyseism) with uplifts. Spectacular fumaroles, with temperatures rising to 160°C, are mainly located at the Solfatara cone, but occur also along active fractures throughout the area both on land and in the Gulf of Pozzuoli. They belong to a widespread geothermal system, which is characterized by a geothermal gradient of up to 170°/km, numerous superimposed aquifers, and a zoning typical of hydrothermal mineral deposition [118]. The predominance of magmatic volatiles over the meteoric water in the volcanic fluids is suggested by isotopic compositions, as well as by the modelling of the fluxes of CO_2 and S. The mineralogy of Solfatara is generally characterized by sulphur and a sulphate paragenesis (alunite and alunogen with other sulphates) [119]. At La Bocca Grande vent, where the vent temperature reaches 160°C, arsenic sulfides (realgar and pararealgar) are deposited [120] (**Figure 15**).

4.2.3. *Mt. Somma-Vesuvius*

The present Somma-Vesuvius is a moderate size composite central volcano rising more than 1200 m asl. The activity of Monte Somma has been characterized by a series of sub-Plinian and Plinian eruptions alternating with long quiescent periods lasting from centuries to millennia, followed by periods of semipersistent activity characterized by lava effusions and low-energy explosive eruptions in the past 4 kyr. The Plinian AD 79 eruption [121] broke a long period of quiescence and represented one of better-known devastating events in the Mt. Vesuvius history. The recent Vesuvius cone has grown within the oldest Mt. Somma caldera possibly after this eruption. After it, the volcano featured two sub-plinian events in AD 472 and 1631 [121]. Throughout the following centuries, Vesuvius activity was characterized by periods of open-conduit activity with the alternating strombolian activity with violent eruptions. The eruption of 1906 is the largest explosive eruption of the twentieth century. The climax of the eruption was reached on the April 7th, when impressive lava fountains, accompanied by earthquakes, rose from the crater. At the end of the eruption, the top of the cone was truncated to form a vast crater. After this explosive event, the Vesuvius area was characterized by quiescent periods alternated by prolonged volcanic activity, effusions of lava and low energy explosions. The last eruption occurred in March 1944.

Mt. Vesuvius is presently affected by relatively low level volcanic-hydrothermal activity, which is mainly characterized by the following: (a) widespread fumarolic emissions that are accompanied by diffuse soil CO_2 degassing in the crater area [122]; (b) CO_2-rich groundwaters along the southern flank of Vesuvius and in the adjacent plain and (c) seismic activity with epicentres clustered inside the crater [123]. Fumarolic fluids discharging by the crater rim fumaroles are of relatively low temperatures (<75°C) and are mainly composed of atmospheric components. Fumaroles from the crater bottom have H_2O and CO_2 as the major components, followed by H_2, H_2S, N_2, CH_4, CO, and He (in order of decreasing content), and discharge temperature of about 95°C [123]. CH_4 and NH_3 contents suggest that the origin of these fluids could be from a high-temperature hydrothermal system located below the Vesuvius' crater [122]. Lacroix classified older Vesuvian fumaroles in four different types [9]: (1) HT (>300°C) with halite, sylvite, thenardite, Na-K carbonates, aphthitalite, sulphides, Cu-oxide and chlorides (alteration); (2) ("acid", T=300–100°C) with Mg-Fe-Al chlorides, sulphur, and realgar; (3) (around 300°C) containing salammoniac with ammonium sulphate and fluoride and (4) (sulphurous, LT around 100°C) with sulphur, gypsum, sassolite, K-Al-Fe sulphates, opal.

4.2.4. *Vulcano Island*

Vulcano is the southernmost of the seven islands that form the Aeolian archipelago. Its activity dates back about 120 Ka. The four main eruptive centres of the island (Vulcano Primordiale, Lentia, La Fossa cone, and Vulcanello) are formed by a progressive migration of the volcanic activity from SSE to NNW. In the middle of La Fossa caldera sits La Fossa cone, the active volcanic centre of the island, which formed during the last 5.5 ka through recurrent hydro-magmatic to volcanic explosive phases [124]. Since its last eruption in 1888–1890, Vulcano has remained in a fumarolic stage of varying intensity with shallow seismicity. Presently, active fumarole fields are concentrated in the northern section of the La Fossa crater and in the neighbourhood of the coast at Baia di Levante. The former show high variability in temperature and fluid compositions as a function of the volcanic activity, the latter are typical hydrothermal emissions.

Figure 16. Fumaroles on La Fossa Crater, Vulcano.

Over the last two decades, several models were proposed to explain the genesis of the Vulcano fumarole fluids, each of them involving the presence of a deep and a shallow component in the gas phase. Compositionally, the crater fumaroles are rich in CO_2 as the main component and have significant concentrations of HCl, SO_2, H_2S, HF, and CO [125]. The diffuse emissions at Baia di Levante are more typical of hydrothermal fluids, with higher CH_4 and H_2S contents than the crater fumaroles, lower CO concentrations, and no measurable amount of SO_2 [126]. Strong variations of the chemicophysical features of the gas output at the crater have been observed periodically. During these so-called "crisis", the crater fumaroles demonstrated a substantial increase in temperature from the usual range between 330 and 400°C and in the magmatic component of the total gas flux. The interest in the collection and analysis of the fumarolic minerals especially increased during the last thermal crisis, which started in 1988 and reached a maximum in 1993. A systematic research was started, supported by the Italian National Group of Volcanology. This led to the discovery of a large variety of rare phases and new minerals [8, 12, 13, 26–29, 31–34, 40, 49, 55, 66, 90, 91]. La Fossa crater became a mineralogical attraction and the large number of already described minerals from the locality [10] increases constantly making it one of the most prolific type locality in the world. The fumarolic mineral association observed on La Fossa Crater are (1) (LT + MT) sulphur, borates, borosilicates, sulphates (mainly hydrous, only in oxidizing conditions), halogenides and sulfohalogenides and (2) (HT >400–450°C) sulphides, sulphosalts, sulfochlorides (reducing conditions); anhydrous sulfates (oxidizing conditions) (**Figure 16**).

4.2.5. Mt. Etna

Mount Etna is one of the largest European volcanoes, world famous for its spectacular and frequent eruptions. Its main feature is the voluminous lava emission, occasionally associated with explosive activity from its four summit craters. Mt. Etna is a basaltic composite apparatus, with a basal diameter of 40 km and 3350 m of altitude, situated on the eastern border of Sicily. The Etnean volcanism is still not definitely understood in its geological context. Mt Etna is situated on the crossing of important regional fault systems trending NW-SE, NE-SW, and WSW-ESE and this probably facilitates the uprise of magma in this place. There is some evidence that Etna is but the most recent manifestation of volcanism fed from a very long-lived mantle source, having caused numerous earlier phases of mafic volcanism in the Monti Iblei, SE Sicily, from the late Triassic to the early Pleistocene. Mt. Etna has erupted many times in historical time and presently is constantly active with spectacular summit and flank eruptions, interspersed by periods of intermittent activity. The recent volcanic activity has been characterized by almost continuous summit eruptions of effusive and moderate explosive activity. Its most powerful historical eruption occurred in 122 BC [127]. This plinian summit eruption produced a large volume of pyroclastics (ash and lapilli), which fell in a sector on the southeast flank of the volcano, causing devastation in the city of Catania. Presently, the frequent eruptive activity heavily affects the morphology of the summit area that consists of a central crater (Voragine) surrounded by three active cones (Bocca Nuova, NE Crater, and SE Crater) and is cut by N-S-oriented fracture system, mainly related to the extensional stress produced by magma ascent [128]. From the four active craters in historical times, a large number of summit and flank eruptions have occurred [129] producing numerous composite lava fields and more than 300 scoria and spatter cones. The chemical composition of magmas produced in historical times is rather uniform, ranging from alkali basalt to basic mugearite.

The fumarolic activity at the crater area of Mt. Etna is variable and largely influenced by the activity of the volcano, which is continuous and changes frequently. Consequently, as the morphology of the summit area undergoes significant variations over time, also the localization and amplitude of gas discharging and temperature of fumaroles change. Generally, the fumaroles are aligned along dynamic fractures produced by extensional stress phenomena, but also around hornitos. Intense fumarolic activity is often present due to residual degassing during the cooling of erupted lava. The most of the gas discharging is from the surface of lava flows, but it also happens in the inner parts of volcanic lava tubes and caves in which the cooling of the gases often produces interesting deposition of sublimates and encrustations. After exsolving from magmas, the gases ascend along rock fractures sometimes reaching directly the surface (high-temperature fumarole). In other cases (low-temperature fumarole), they interact with peripheral hydrothermal systems and surficial acquifers, undergo contamination and change their pristine composition [130]. The observed main mineral associations are as follows: (1) (surface, HT) thenardite, halite and sylvine; (2) (surface, LT) sulphur, salammoniac, hydrous sulphates and (3) (in caves) halite, sylvite with thenardite, and hydrous sulphates.

4.3. Greek volcanoes

4.3.1. The Aegean (Hellenic) active volcanic arc

The Aegean volcanic arc is one of the tectonically most active regions of the Mediterranean area, extending from the mainland of Greece through the volcanic centers of Soussaki, Aegina, Methana, Poros, Milos, Santorini, Kos, Yali, and Nisyros, to the Bodrum peninsula in Turkey. In that area, the eastern Mediteranian lithosphere subducts under the Aegean and the Aegean microplate overrides the eastern Mediterranean [131]. The volcanism started 3.5 Ma ago and is still continued up today in the form of post-magmatic activity [132–135]. The Pliocene-Quaternary volcanic arc of the Aegean arose from the subduction of the African plate beneath the microplate of the Aegean-Anatolian [136] with simultaneous destruction of intermediate-Tethys oceanic crust. The Pliocene-Pleistocene volcanism in the arc of the Aegean Sea is

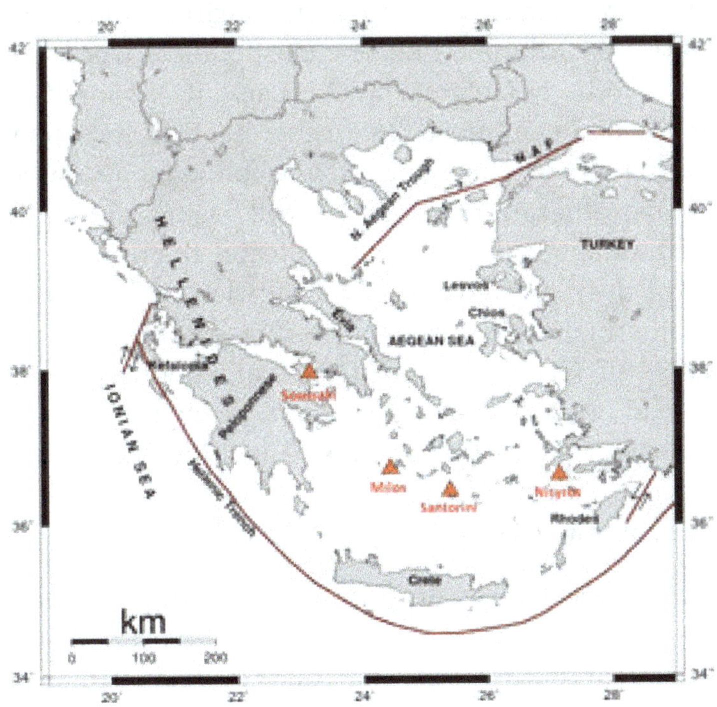

Figure 17. Map of Greece with the locations of the investigated volcanoes and fumarolic fields.

dominated by andesites and dacites, while in the central and eastern parts of the arc, the lower Quaternary volcanism is characterized by dacitic or rhyolitic composition (**Figure 17**).

4.3.2. *Soussaki volcano*

The Soussaki area represents the NW end of the active Aegean volcanic arc. The volcanic activity took place between 4.0 and 2.3 Ma ago. The lithological types occurring in the area are rocks of the dacitic and rhyolitic composition, remnants of the late Pliocene to Quaternary volcanic activity [132–135, 137]. Gas manifestations display typical geothermal compositions with CO_2 and water vapour as the main components and CH_4 and H_2S as minor species. The volcanic activity observed today manifests itself by emanation of warm fluids, while widespread fumarolic alteration, vapours, and warm (35–45°C) gas emissions are still observed. Reaction between the fluids and pre-existing rocks, consisting mainly of serpentinite, chert, marlstone, limestone, and subordinate rhyodacitic lava, has resulted in the formation of gypsum, sulphur, silica polymorphs, and Fe and Mg sulphates. Kaolinite, anhydrite, carbonates, and Ni, and K-Al sulfates are also present in some cases ([73], this work). The observed main mineral associations can be summarized as follows: (1) gypsum with quartz and carbonates and (2) Fe-Mg hydrous sulphates.

4.3.3. *Milos Island*

The Milos volcanic district is a wide volcanic archipelago comprising the islands of Milos, Kimolos, Antimilos, and Poliegos. The volcanic activity started about 5 million years ago and is now considered to be extinct. Milos Island is located at the western end of the Aegean Volcanic Arc. It is built mainly of calcalkaline volcanic rocks (tuffs, pumice flows, ignimbrites, pyroclastic flows, domes, and lava flows of andesitic-dacitic, and rhyolitic composition). The volcanic sequence is built on Miocene-Pliocene clastic and carbonate platform sediments, which unconformably overly a metamorphic basement. Following their emplacement, volcanic rocks were involved in an intense hydrothermal activity. The volcanism in the area developed with submarine activity character and was followed by a sub-aerial effusive phase. The distribution of the hydrothermal minerals derived basing on cores and cuttings from the two drill sites to a depth of about 1100 m [138] indicates the dilution of the K-, Na-, Cl-rich hydrothermal fluid of the deep reservoir by a Ca-, Mg-rich cold water at a shallower level. In some places, the hydrothermal activity is expressed by the occurrence of many hot springs (30–85°C), fumaroles (98–102°C), hot grounds (100°C at a depth of 30–40 cm) and submarine gas emissions, widespread on and around the island. Fyriplaka volcano is the expression of the most recent volcanic activity on the island and includes fumaroles, solfataras, and hot grounds. Where gas emanations are very high, the soils are altered and covered with thin layers of secondary minerals as alunite, magnesite, and sulphur.

We examined samples from the area of Amygdales and in the neighbourhood of Paleochori. The substrate rock is argillaceous, composed of SiO_2 polymorphs and kaolinite mixed with alunite, which is the dominating fumarolic mineral. The main fumarolic associations are as follows: (1) alunite with sulphur (plus quartz, opal, kaolinite) and (2) alunite and alunogen with Fe-hydrous sulphates and barite (plus sulphur, pyrite, quartz, opal, muscovite).

4.3.4. *Santorini volcano*

The volcanic field of Santorini consists of various islands (Thera, Therasia, Aspronisi, Palea Kameni, and Nea Kameni), the Christiania islands around 20 km to the southwest and the submerged Columbus volcano 7 km to the northeast. It is the most active part of the Aegean Volcanic Arc. Between Thera, Therasia, and Aspronisi, exists a sea-flooded caldera of about 13 km diameter. According to various geochronological data, the volcanic activity in the Santorini volcano started about 2 Ma ago. The field has had 12 major (1–10 km^3 or more of lava), and numerous minor, explosive eruptions over the last ~200 ka. A hypothesis the eruption mechanism [139] supposes a large water-filled crater, an extensional environment to facilitate downward penetration of water, and a hot silicic magma.

Figure 18. Fumarole on Nea Kameni, Santorini.

Nea Kameni fumaroles have a distinct hydrothermal nature with compositions dominated by H_2O, CO_2 and H_2 with minor H_2S, and SO_2 typically absent [140]. We investigated samples from the fumaroles at the Nea Kameni central crater. The altered rock consists of cristobalite, opal, plagioclase (around An50), gypsum, alunite or natroalunite and sporadic quartz, trydimite, hematite, and kaolinite. The sublimate is mostly sulphur that makes yellowish crusts, aggregates of isometric crystals or spear-like dendritic crystals. Hydrated sulphates are alunogen (mixed with the substrate or as thin platy crystals), rhomboclase, coquimbite, and tamarugite. Jarosite is also found in small quantities (**Figure 18**).

4.3.5. *Nisyros volcano*

Nisyros is the newest of the major active volcanoes of Greece, composed almost exclusively of volcanic rocks, with the oldest of them being little less than 160,000 years and the youngest reaching the limits of prehistory, about 20,000 years ago. In the wider area of Nisyros at the eastern margin of the volcanic arc, the first eruptions date 3.4 million years. Since then, small or large eruptions built up Nisyros and the islets Pirgousa, Pachia, Strogyli, and Giali. None of the eruptions of the volcano recorded in historical sources produced molten rock, as they are hydrothermal and originate from the overheated steam in the underground of the island. Seawater and rainwater penetrate the rocks of the island, are heated by magma and converted to superheated steam, which exerts tremendous pressure and causes a hydrothermal explosion when overcomes the weight and consistency of the caprock. Such explosions were recorded in Nisyros in historical times. In the southern part of the caldera floor, there are traces of 20 craters, 10 of them being well preserved. The largest and most striking crater is Stephanos. The latest hydrothermal craters are concentrated in the area of Lofos (=hill), a small post-dome that largely has been destroyed by hydrothermal explosions. Here are situated six well-preserved craters, the creation of three of them being recorded in historical sources. In October or November 1871, a powerful earthquake caused the beginning of a series of hydrothermal explosions, which until 1873 created two small craters: Polyvotis and Alexandos (or Flegethron). The last recorded hydrothermal explosion in Nisyros is the one of 1887, which created the crater of Small Polyvotis. Various amounts of volcanic gases come out from the intra-caldera soil area including also the main fumarolic craters of Kaminakia, Ramos, Stefanos, Lofos, and Flegethron. The gas species are H_2S, CO_2, CH_4, and CO [141–143](**Figure 19**).

Figure 19. Fumaroles in the Stephanos crater, Nisyros.

The only sublimate mineral observed at the vents at the edges of the Stefanos and Polyvotis craters is sulphur. It develops as granular aggregates or acicular spear-like crystals. On the floor of the Stefanos crater, numerous circular formations surround the "hot spots" of the fluid, vapour, and gas emanations. The central part of these formations consists of the reddish to yellowish sand formed by grains of quartz, labradoritic plagioclase, alunite, alunogen, and gypsum. They are surrounded by white rims containing dendritic crystals of alunogen and metaalunogen. On their outer part is a blackish zone with additional voltaite and tamarugite. The outermost rim consists of aggregates of microscopic yellowish globules of rhomboclase overgrown by silky needles of halotrichite.

5. Discussion

The list of the minerals found in European fumaroles impresses by the number of different species, especially taking in account that some prolific mineral groups (like silicates and phosphates) do not contribute to the list or contribute very little. Among the European fumarole localities, Vesuvius and Vulcano stand as the two volcanoes with the richest mineralogy. This status is due to the exceptional abundance of otherwise rare elements in their emanations: Cr, Mn, Ni, Cu, B, Tl, Pb, As, Se on Vesuvius, Ba, Au, Zn, B, Tl, Sn, Pb, As, Bi, Se, Te, Br, I on Vulcano. The presence of the rare elements that form rare minerals is, however, not the only reason for the diversity of fumarole mineralogy. Although they do not contain minerals of rare elements, the three Icelandic volcanoes Surtsey, Eldfell, and Hekla each have around 30 different minerals in their fumaroles. Moreover, their dominating mineral associations are different from the main associations of both Vesuvius and Vulcano, and are differing among themselves, although all five fumarolic systems are formed by the same combination of main gaseous species (H_2O, CO_2, HCl, HF, SO_2, and H_2S). This illustrates the importance of various factors that can influence the processes and through that the mineral world of fumaroles mentioned in the introduction. The intensive parameters, composition and temperature of volcanic gases and lava, can decisively be supplemented by other factors in forming or modifying the minerals in fumaroles. We will try to examine them in the following comparison of different fumarolic systems.

In general, the fumaroles can be classified according to their temperature of formation and mineral stability. Already in the nineteenth century, Lacroix made a classification of Vesuvius fumaroles [9] after their mineralogy, reflecting the temperature of formation. He distinguished the high-temperature "dry" fumaroles characterized by the K and Na salts and also some sulphides and oxides, the medium-temperature "acid" fumaroles characterized by the chlorides of Fe, Mg, Al, and Mn, fumaroles of even lower temperature characterized by salammoniac and other ammonium minerals, and the low-temperature fumarole producing sulphuric acid and steam with sulphur and gypsum as typical minerals. We shall try to classify the fumaroles investigated in this work in three categories: (1) HT, the high temperature (>400°C), (2) MT, medium temperature (200–400°C) and (3) LT, low temperature (<200°C). The given temperature ranges are approximate. HT fumaroles appear with the eruption of the volcano and are short living after its finish. Alternatively, they can be products of temperature increase in the volcanic system not leading to eruption, but again short-lived (Vulcano). MT and LT fumaroles might be active at the same time as the HT fumaroles, but at different places in the system where fumarolic gases travelled longer and got cooled and/or diluted by the atmosphere before they come to the surface. MT fumaroles prevail in the period of recession after the paroxysm and last roughly for decades, maybe centuries, while the shallow intruded magma cools down, or the degassing surface of the magma retreats to greater depths. LT fumaroles are the only ones present in the quiescence period with a deep thermal source and may transform eventually to a solfatara (mofeta), if the volcanic cycle is finished or very much prolonged.

Inspecting the mineralogy of European fumaroles, we can see that specific three (or four) minerals are typical in most cases for the HT fumaroles. These are halite, (meta) thenardite (with or without aphthitalite), and anhydrite. The mentioned four minerals are all stable to over 800°C. The fumarolic systems can be classified according to relative abundances of these minerals reflecting the Na(+K)/Ca and Cl/S proportions in the fumarolic paragenesis. In this classification, the alcalic fumaroles are those of Vesuvius (dominantly chloridic), Fimmvörðuháls and Askja (chloridic-sulphatic) and Krafla (sulphatic). The mixed alcalic-calcic are Surtsey and Etna (chloridic-sulphatic) and Hekla (sulphatic), the calcic is Eldfell (sulphatic). Outside the classification stands Vulcano, with a high-temperature association composed of sulphides/sulphosalts in combination with sulphochlorides. It can be locally expanded with an association of anhydrous sulphates where and when the atmospheric influence is large and the general reducing conditions change to oxidative.

The MT fumaroles reflect more the individual nature of each fumarolic system. First of all, the range of temperature, the relatively slow flux rates and the chemistry of fumarolic systems permit the deposition and the growth of a larger number of mineral phases, including products of reaction between gas, steam, or liquid and minerals previously formed. In addition, the often longer time of their existence and the longer and slower ascent of the fluid phase allows for a larger contribution of other influences on magmatic exhalations, that is, interaction with the wall rock and surficial fluids. Here, we can recognize the following associations: A, salammoniac in association with other ammonium minerals; B, metal chlorides (Fe-Al-Mg plus others); C, fluoride association with ralstonite as the dominating mineral, typically mixed with other aluminofluorides. Type A was registered on all recently active volcanoes except Surtsey and Fimmvörðuháls. On Eldfell and Hekla, it was forming only early following the eruption. On Vesuvius and especially Vulcano, it is a persistent feature, where the latter also demonstrates a rich variety of ammonium minerals not found on other places. The situation on Icelandic volcanoes suggests that nitrogen in the volcanic gases is one of the first species to be exhausted upon an eruptive episode. If it is true, it would mean that the shallow-degassing magmas of Vesuvius and Vulcano are constantly replenished in nitrogen or that it is there supplied from some other sources. Type B is found on Vesuvius and Vulcano. On the former, the chlorides are mixed with fluorides, on the latter with fluorides (boro- and silicofluorides), bromides and even an iodate. The type C is characteristic for a number of Icelandic volcanoes (Surtsey, Eldfell, Fimmvörðuháls, Hekla, Krafla). In Hekla fumaroles, silicofluorides appear together with aluminofluorides.

The dominating minerals in the LT fumaroles are sassolite, gypsum, alunogen, and sulphur (listed with decreasing thermal stability). The differences in thermal stability are practically of no importance in this group because all minerals form at temperatures that are around 100°C or even lower in this type of fumaroles. Here, also alunite can be named as an abundant component. It is a product of feldspar decomposition through action of sulphuric gases and solutions in solfataras and low-temperature fumaroles, together with opal, and in this respect does not represent a simple sublimate. It is stable to around 500°C and could be a component of high-temperature fumarole as well, but the process of formation makes it a typical representative of the low-temperature one. The LT fumaroles can be categorized by the contents of the above-mentioned minerals. The sulphur dominated fumaroles are found on Nisyros, Etna, Campi Flegrei, Vesuvius, Vulcano, and Krafla. Gypsum dominated are those of Soussaki, Surtsey, Eldfell, and Hekla. Vulcano and Vesuvius have low-temperature fumaroles that contain also abundant sassolite together with sulphur and gypsum. The second type of fumaroles on Nisyros and fumaroles on Milos are dominated by alunogen, and finally, the fumaroles on Santorini have comparable amounts of sulphur, gypsum and alunogen. In all of them, various kinds of hydrous sulphates appear, which can be used as fine-tuning indicators of temperature and humidity conditions.

The classifications given above can serve a general orientation purpose and understanding of the actual conditions in the fumarolic system under investigation. They are in no way clearly delimited categories and mixing of associations is a frequent phenomenon. It can be produced as a consequence of the gradual lowering of temperature and the flux rate, but often also a simultaneous formation of closely separated minerals at quite different temperatures occurs due to the high fluctuation of conditions on a small scale at the border between the volcanic gases and the atmosphere. On Vulcano, the high-temperature anhydrous sulphates that form under the locally oxidizing conditions are found in close association with sulphides and sulphosalts formed under the reducing conditions. Our detailed investigation of fumarolic profiles on Eldfell shows that within less than 20 cm the low-temperature mineral association of fluorides changed to the high-temperature association of anhydrous sulphates. The mixed associations can also be an artefact produced by the mineral instability at atmospheric conditions, and laboratory analyses taken alone can give misleading results. Our analyses of Eldfell material showed always hydrous Mg-Al sulphates accompanying the anhydrous ones taken from the levels featuring temperatures over their stability ranges. Their formation as atmospheric hydration products was confirmed by the field observations of the deliquescent behaviour of exposed excavated samples and through laboratory analyses of freshly crushed massive ones. The caveat about the above classification is that it presents a general picture from which numerous important details have been removed in order to give as clear as possible an overview. For a full characterization of individual fumarolic systems, the details are indispensable, and this emphasizes the importance of detailed studies.

The realm of fumarolic mineralogy presents us with interesting and important still unanswered questions. For example, the seeming contradiction of the experimental results that show sulphuric species to be the ones with the fastest and earliest devolatilization from silicic melts, and the observation that sulphates and elemental sulphur are the most persistent parts of fumaroles, leads to important enquiries into the sub-surface volcanogenic sulphuric cycle (see e.g. [144]). Another interesting largely unsolved problem is the understanding of reasons why sometimes large differences in mineral compositions of seemingly similar fumarolic systems exist. During the last century, the appearance of powerful methods of research on microscale has revealed a great number of new mineral species from fumarolic samples, typically rare and exotic, and contributed to the solid state science and the general mineralogical knowledge. We believe that the application of these methods to investigation of both minor and main components of fumaroles and integrated with research in other fields (like gas evolution and composition and gas-magma-rock interactions) can also bring answers to the most intriguing questions of the genesis and origins of particular fumaroles. Accomplish-ing this, they can become another important tool in the surveillance of active volcanic systems.

Acknowledgements

The authors are grateful to researchers and students with whom they collaborated in the investigations of European fumaroles: Filippo Vurro, Daniela Pinto, Donatella Mitolo, Eric Leonardsen, Sigurður S. Jónsson, Níels Óskarsson, Sigmundur Einarsson, Anna Katerinopoulou and Morten J. Jacobsen. We thank Helene Almind for the technical assistance during XRD measurements and Dora Balić Žunić for technical support during the field work and manuscript preparation. Parts of the research work were financially supported by the Danish Agency for Science Technology and Innovation and the National Group for Volcanology, Italy.

Author details

Tonči Balić-Žunić[1*], Anna Garavelli[2], Sveinn Peter Jakobsson[3], Kristjan Jonasson[3], Athanasios Katerinopoulos[4], Konstantinos Kyriakopoulos[4] and Pasquale Acquafredda[2]

*Address all correspondence to: toncib@snm.ku.dk

1 University of Copenhagen, Copenhagen, Denmark

2 University of Bari, Bari, Italy

3 Icelandic Institute of Natural History, Gardabaer, Iceland

4 University of Athens, Athens, Greece

References

[1] Jakobsson SP, Leonardsen ES, Balic-Zunic T, Jónsson SS. Encrustations from three recent volcanic eruptions in Iceland: the 1963–1967 Surtsey, the 1973 Eldfell and the 1991 Hekla eruptions. Fjölrit Náttúrufræðistofnunar (Special Publication of the Icelandic Institute of Natural History).2008;52:65 p.

[2] Ruste J. X-ray spectrometry. In: Maurice F, Meny L, Tixier R, editors. Microanalysis and Scanning Electron Microscopy. Orsay: Les Editions de Physique; 1979. pp. 215–267.

[3] Acquafredda P, Paglionico A. SEM-EDS microanalyses of microphenocrysts of Mediterranean obsidians: a preliminary approach to source discrimination. European Journal of Mineralogy. 2004;16:419–429.

[4] Pouchou JL, Pichoir F. Quantitative analysis of homogeneous or stratified microvolumes applying the model "PAP". In: Heinrich KFJ, Newbury DE, editors. Electron Probe Quantitation. New York: Plenum Press; 1991. pp. 31–75.

[5] Strunz H, Nickel EH. Strunz Mineralogical Tables. 9th ed. Stuttgart: E. Schweizerbart'sche Verlagsbuchhandlung (Nägele u. Obermiller); 2001. 870 p.

[6] Panichi U. Muthmann'scher Schwefel, auf der Insel Vulcano beobachtet. Zeitschrift für Krystallographie. 1915;54:393–394.

[7] Steudel R, editor. Elemental Sulphur and Sulphur-Rich Compounds I, Topics in Current Chemistry 230. 1st ed. Berlin, Heidelberg, New York: Springer; 2003. 166 p.

[8] Garavelli A, Mitolo D, Pinto D, Vurro F. Lucabindiite, $(K,NH_4)As_4O_6(Cl,Br)$, a new fumarole mineral from the "La Fossa" crater at Vulcano, Aeolian Islands, Italy. American Mineralogist. 2013; 98;470–477.

[9] Pelloux A. The minerals of Vesuvius. American Mineralogist. 1927;12:14–21.

[10] Campostrini I, Demartin F, Gramaccioli CM, Russo M. Vulcano: tre secoli di (Vulcano: three centuries of its mineralogy). Cremona: Ami Associazione Micromineralogica Italiana; 2011. 343 p.

[11] Fulignati P, Sbrana A. Presence of native gold and tellurium in the active high-sulfidation system of the La Fossa volcano (Vulcano, Italy). Journal of Volcanology and Geothermal Research. 1998;86:187–198.

[12] Garavelli A, Laviano R, Vurro F. Sublimate deposition from hydrothermal fluids at the Fossa crater – Vulcano, Italy. European Journal of Mineralogy. 1997;9:423–432.

[13] Pinto D, Garavelli A, Mitolo D. Balićžunićite, $Bi_2O(SO_4)_2$, a new fumarole mineral from La Fossa crater, Vulcano, Aeolian Islands, Italy. Mineralogical Magazine. 2014;78:1043–1055.

[14] Russo M. I minerali di formazione fumarolica della grande eruzione Vesuviana del 1906 (Minerals of fumarolic origin from the large 1906 eruption of Vesuvius). Istituto Nazionale di Geofisica e Vulcanologia, sezione Napoli – Osservatorio Vesuviano. Open File Report; 2006. 39 p.

[15] Parascandola A. Contributo alla mineralogia Flegrea (A contribution to the Phlegrean mineralogy). Bolletino della Società Geologica Italiana (Bulletin of the Italian Geological Society). 1951;70(3):527–532.

[16] Emelina AL, Alikhanian AS, Steblevskii AV, Kolosov EN. Phase diagram of the As-S system. Inorganic Materials. 2007;43(2):95–104.

[17] Gavezzotti A, Demartin F, Castellano C, Campostrini I. Polymorphism of As_4S_3 (tris-(μ_2-sulfido)-tetra-arsenic): accurate structure refinement on natural α- and β-dimorphites and inferred room temperature thermodynamic properties. Physics and Chemistry of Minerals. 2013;40:175–182.

[18] Roberts AC, Ansell HG, Bonardi M. Pararealgar, a new polymorph of AsS, from British Columbia. Canadian Mineralogist. 1980;18:525–527.

[19] Bonazzi P, Bindi L, Popova V, Pratesi G, Menchetti S. Alacranite, As_8S_9: structural study of the holotype and re-assignment of the original chemical formula. American Mineralogist. 2003;88:1796–1800.

[20] Campostrini I, Demartin F, Russo M. A new ammonium arsenite chloride from the Solfatara di Pozzuoli, Napoli, Italy. Rendiconit della Società Geologica (Statements of the Italian Geological Society). 2014;31(Suppl.1):309.

[21] Whitfield HJ. Crystal structure of the β-form of tetra-arsenic trisulphide. Journal of the Chemical Society Dalton Transactions. 1973;17:1737–1738.

[22] Demartin F, Gramaccioli CM, Campostrini I, Orlandi P. Demicheleite, BiSBr, a new mineral from La Fossa crater, Vulcano, Aeolian Islands, Italy. American Mineralogist. 2008;93:1603–1607.

[23] Demartin F, Gramaccioli CM, Campostrini I. Demicheleite-(Cl), BiSCl, a new mineral from La Fossa crater, Vulcano, Aeolian Islands, Italy. American Mineralogist. 2009;94:1045–1048.

[24] Demartin F, Gramaccioli CM, Campostrini I. Demicheleite-(I), BiS(I,Br,Cl), a new mineral from La Fossa crater, Vulcano, Aeolian Islands, Italy. Mineralogical Magazine. 2010;74(1):141–145.

[25] Henley RW, Mavrogenes J, Tanner D. Sulfosalt melts and heavy metal (As-Sb-Bi-Sn-Pb-Tl) fractionation during volcanic gas expansion: the El Indio (Chile) paleo-fumarole. Geofluids. 2012;12:199–215.

[26] Pinto D, Balić-Žunić T, Garavelli A, Makovicky E, Vurro F. Comparative crystal-structure study of Ag-free lillianite and galenobismutite from Vulcano, Aeolian Islands, Italy. The Canadian Mineralogist. 2006;44:159–175.

[27] Borodaev YS, Garavelli A, Garbarino C, Grillo SM, Mozgova NN, Organova NI et al. Rare sulfosalts from Vulcano, Aeolian Islands, Italy. III. Wittite and cannizzarite. The Canadian Mineralogist. 2000;38:23–34.

[28] Vurro F, Garavelli A, Garbardino C, Moëlo Y, Borodaev YS. Rare sulfosalts from Vulcano,Aeolian Islands, Italy. ll. Mozgovaite, $PbBi_4(S,Se)_7$, a new mineral species. The Canadian Mineralogist. 1999;37:1499–1506.

[29] Borodaev YS, Garavelli A, Garbardino C, Grillo SM, Mozgova NN, Uspenskaya TY, et al. Rare sulfosalts from Vulcano, Aeolian Islands, Italy. IV. Lillianite. The Canadian Mineralogist. 2001;39:1383–1396.

[30] Moëlo Y, Makovicky E, Mozgova NN, Jambor JL, Cook N, Pring A, et al. Sulfosalt systematics: a review. Report of the sulpfosalt sub-committee of the IMA Commission on Ore Mineralogy. European Journal of Mineralogy. 2008;20:7–46.

[31] Borodaev YS, Garavelli A, Garbarino C, Grillo SM, Mozgova NN, Paar WH, et al. Rare sulfosalts from Vulcano, Aeolian Islands, Italy. V. Selenian Heyrovskýite. The Canadian Mineralogist. 2003;41:429–440.

[32] Borodaev YS, Garavelli A, Kuzmina O, Mozgova NN, Organova NI, Trubkin NV, et al. Rare sulfosalts from Vulcano, Aeolian Islands, Italy. I. Se-bearing kirkiite, $Pb_{10}(Bi,As)_6(S,Se)_{19}$. The Canadian Mineralogist. 1998;36:1105–1114.

[33] Garavelli A, Mozgova NN, Orlandi P, Bonaccorsi E, Pinto D, Moëlo Y, et al. Rare sulfosalts from Vulcano, Aeolian Islands, Italy. VI. Vurroite, $Pb_{20}Sn_2(Bi,As)_{22}S_{54}Cl_6$, a new mineral species. The Canadian Mineralogist. 2005;43:703–711.

[34] Pinto D, Bonaccorsi E, Balić-Žunić T, Makovicky E. The crystal structure of vurroite, $Pb_{20}Sn_2(Bi,As)_{22}S_{54}Cl_6$: OD-character, polytypism, twinning, and modular description. American Mineralogist. 2008;93:713–727.

[35] Métrich N, Wallace PJ. Volatile abundances in basaltic magmas and their degassing paths tracked by melt inclusions. Reviews in Mineralogy and Geochemistry. 2008;69:363–402.

[36] Ustunisik G, Nekvasil H, Lindsley DH, McCubbin FM. Degassing pathways of Cl^-, F^-, H^-, and S-bearing magmas near the lunar surface: implications for the composition and Cl isotopic values of lunar apatite. American Mineralogist. 2015;100:1717–1727.

[37] Villemant B, Boudon G. H_2O and halogen (F, Cl, Br) behaviour during shallow magma degassing processes. Earth and Planetary Science Letters. 1999;168: 271–286.

[38] Barone G, Mazzoleni P, Priolo G. Catalogo delle mineralizzazioni secondarie riscontrate all'interno di alcune grotte vulcaniche etnee (A catalogue of the secondary mineralizations in some Etnean volcanic caves). In: Proceedings of XXII Congresso Nazionale di Speleologia "Condividere i dati", Sessione Documentatione – a1; 30 May – 2 June 2015; Pertosa-Auletta (SA). pp. 99–109.

[39] Oskarsson N. The chemistry of Icelandic lava incrustations and the latest stages of degassing. Journal of Volcanology and Geothermal Research. 1981;10:93–111.

[40] Coradossi N, Garavelli A, Salamida M, Vurro F. Evolution of Br/Cl ratios in fumarolic salammoniac from Vulcano (Aeolian Islands, Italy). Bulletin of Volcanology. 1996;58:310–316.

[41] Roberts AC, Venance KE, Seward TM, Grice JD, Paar WH. Lafossaite, a new mineral from the La Fossa Crater, Vulcano, Italy. Mineralogical Record. 2006;37:165–168.

[42] Campostrini I, Russo M. Lafossaite e dimorphite: due nuove specie per il Vesuvio (Lafossaite and dimorphite: two new mineral species from Vesuvius). Micro. 2012;3/2012:136–141.

[43] Palache C, Berman H, Frondel C. The System of Mineralogy of James Dwight Dana and Edward Salisbury Dana Yale University 1837–1892, Volume II. 7th ed. New York: John Wiley and Sons, Inc.; 1951. 1124 p.

[44] Cruciani G, Orlandi P, Pasero M, Russo M. First Italian occurrence of cumengeite from Vesuvius: crystal-structure refinement and revision of the chemical formula. Mineralogical Magazine. 2005;69(6):1037–1045.

[45] Jacobsen MJ, Balić-Žunić T, Mitolo D, Katerinopoulou A, Garavelli A, Jakobsson SP. Oskarssonite, AlF_3, a new fumarolic mineral from Eldfell volcano, Heimaey, Iceland. Mineralogical Magazine. 2014;78(1):215–222.

[46] Rosenberg PE. Stability relations of aluminium hydroxyl-fluoride hydrate, a ralstonite-like mineral, in the system AlF_3-Al_2O_3-H_2O-HF. The Canadian Mineralogist. 2006;44;125–134.

[47] Demartin F, Campostrini I, Castellano C, Russo M. Parascandolaite, $KMgF_3$, a new perovskite-type fluoride from Vesuvius. Physics and Chemistry of Minerals. 2014;41:403–407.

[48] Russo M, Campostrini I. Ammineite, matlockite and post 1944 eruption fumarolic minerals at Vesuvius. Plinius. 2011;37:312.

[49] Garavelli A, Vurro F. Barberiite, NH_4BF_4, a new mineral from Vulcano, Aeolian Islands, Italy. American Mineralogist. 1994;79:381–384.

[50] Bernauer F. Eine Gearksutit-Lagerstätte auf der Insel Vulcano (A gearksutite-deposit on the island Vulcano). Zeitschrift der Deutschen Gesellschaft für Geowissenschaften (Journal of the German Geoscientific Society). 1941;93:65–80.

[51] Balić-Žunić T, Garavelli A, Mitolo D, Acquafredda P, Leonardsen E. Jakobssonite, $CaAlF_5$, a new mineral from fumaroles at the Eldfell and Hekla volcanoes, Iceland. Mineralogical Magazine. 2012;76(3):751–760.

[52] Mitolo D, Garavelli A, Balić-Žunić T, Acquafredda P, Jakobsson SP. Leonardsenite, $MgAlF_5(H_2O)_2$, a new mineral species from Eldfell volcano, Heimaey Island, Iceland. The Canadian Mineralogist. 2013;51:377–386.

[53] Courbion G, Ferrey G. $Na_2Ca_3Al_2F_{14}$: a new example of a structure with "independent F" – a new method of comparison between fluorides and oxides of different formula. Journal of Solid State Chemistry. 1988;76:426–431.

[54] Demartin F, Gramaccioli CM, Campostrini I. Thermessaite, $K_2[AlF_3|SO_4]$, a new ino-aluminofluoride-sulfate from La Fossa crater, Vulcano, Aeolian Islands, Italy. The Canadian Mineralogist. 2008;46:693–700.

[55] Garavelli A, Mitolo D, Pinto D. Thermessaite-(NH_4), IMA 2011-077. CNMNC Newsletter No. 12, February 2012; page 152; Mineralogical Magazine. 2012;76:151–155.

[56] Effenberger H, Kluger F. Ralstonit: ein Beitrag zur Kenntnis von Zusammensetzung und Kristallstruktur (Ralstonite: a contribution to the knowledge of its composition and crystal structure). Neues Jahrbuch fuer Mineralogie. Monatshefte. 1984;1984:97–108.

[57] de Pape R, Ferey G. A new form of FeF_3 with the pyrochlore structure: soft chemistry synthesis, crystal structure, thermal transitions and structural correlations with the other forms of FeF_3. Materials Research Bulletin. 1986;21(8):971–978.

[58] Ferey G, Leblanc M, de Pape R. Crystal structure of the ordered pyrochlore $NH_4Fe^{(II)}Fe^{(III)}F_6$ structural correlations with $Fe_2F_5(H_2O)_2$ and its dehydration product $Fe_2F_5H_2O$. Journal of Solid State Chemistry. 1981;40:1–7.

[59] Vergasova LP, Semyonova TF, Epifanova VB, Filatov SK, Chubarov VM. Meniaylovite, $Ca_4AlSi(SO_4)F_{13}\cdot 12H_2O$, a new mineral of volcanic exhalations. Vulkanologiya i Seismologiya (Volcanology and Seismology). 2004;2:3–5.

[60] Demartin F, Gramaccioli CM, Campostrini I, Castellano C. Cossaite, $(Mg_{0.5},\square)Al_6(SO_4)_6(HSO_4)F_6\cdot 36H_2O$, a new mineral from La Fossa crater, Vulcano, Aeolian Islands, Italy. Mineralogical Magazine. 2011;75(6):2847–2855.

[61] Cossa A. Sulla hieratite, nuova specie mineralogica (About hieratite, new mineral species). Transunti dell'Accademia dei Lincei (Summaries of the Lincei Academy). 1881–1882;3(6):141–142.

[62] Gramaccioli CM, Campostrini I. Demartinite, a new polymorph of K_2SiF_6 from La Fossa crater, Vulcano, Aeolian Islands, Italy. The Canadian Mineralogist. 2007;45:1275–1280.

[63] Garavelli A, Balić-Žunić T, Mitolo D, Acquafredda P, Leonadsen E, Jakobsson SP. Heklaite, $KNaSiF_6$, a new fumarolic mineral from Hekla volcano, Iceland. Mineralogical Magazine. 2010;74:147–157.

[64] Demartin F, Gramaccioli CM, Campostrini I. Knasibfite, $K_3Na_4[SiF_6]_3[BF_4]$, a new hexafluorosilicate-tetrafluoroborate from La Fossa crater, Vulcano, Aeolian Islands, Italy. The Canadian Mineralogist. 2008;46:447–453.

[65] Russo M, Campostrini I, Demartin F. Fumarolic minerals after the 1944 Vesuvius eruption. In: Abstracts, Congresso SGI-SIMP; 10–12 September 2014; Milan. Rendiconti Societa Geologica Italiana. 2014;31(Suppl. 1):315.

[66] Mitolo D, Pinto D, Garavelli A, Bindi L, Vurro F. The role of the minor substitutions in the crystal structure of natural challacolloite, KPb_2Cl_5, and hephaistosite, $TlPb_2Cl_5$, from Vulcano (Aeolian Archipelago, Italy). Mineralogy and Petrology. 2009;96:121–128.

[67] Campostrini I, Demartin F, Gramaccioli CM. Hephaistosite, $TlPb_2Cl_5$, a new mineral species from La Fossa crater, Vulcano, Aeolian Islands, Italy. The Canadian Mineralogist. 2008;46:701–708.

[68] Demartin F, Gramaccioli CM, Campostrini I. Brontesite, $(NH_4)_3PbCl_5$, a new product of fumarolic activity from La Fossa Crater, Vulcano, Aeolian Islands, Italy. The Canadian Mineralogist. 2009;47:1237–1243.

[69] Demartin F, Campostrini I, Gramaccioli CM. Panichiite, natural ammonium hexachlorostannate(IV), $(NH_4)_2SnCl_6$, from La Fossa Crater, Vulcano, Aeolian Islands, Italy. The Canadian Mineralogist. 2009;47:367–372.

[70] Demartin F, Gramaccioli CM, Campostrini I. Steropesite, Tl_3BiCl_6, a new thallium bismuth chloride from La Fossa Crater, Vulcano, Aeolian Islands, Italy. The Canadian Mineralogist. 2009;47:373–380.

[71] Demartin F, Campostrini I, Castellano C, Gramaccioli CM. Argesite, $(NH_4)_7Bi_3Cl_{16}$, a new mineral from La Fossa Crater, Vulcano, Aeolian Islands, Italy: A first example of the $[Bi_2Cl_{10}]^{4-}$ anion. American Mineralogist. 2012;97:1446–1451.

[72] Parascandola A. I minerali di Vesuvio nella eruzione del marzo 1944 e quelli formati durante l'attuale periodo di riposo (Vesuviam minerals from the March 1944 eruption, and those formed during the present quiescent period). Bollettino della Società Geologica Italiana (Bulletin of the Italian Geological Society). 1952;70:523–526.

[73] Kyriakopoulos K, Kanakis Sotiriou R, Stamatakis M. The authigenic minerals formed from volcanic emanations at Soussaki, West Attica peninsula, Greece. The Canadian Mineralogist. 1990;28:363–368.

[74] Demartin F, Gramaccioli CM, Campostrini I. Clinometaborite, natural beta-metaboric acid from La Fossa Crater, Vulcano, Aeolian Islands, Italy. The Canadian Mineralogist. 2011;49:1273–1279.

[75] Murashko MN, Pekov IV, Krivovichev SV, Chernyateva AP, Yapaskurt VO, Zadov AE, et al. Steklite, $KAl(SO_4)_2$: a finding at the Tolbachik Volcano, Kamchatka, Russia, validating its status as a mineral species and crystal structure. Geology of Ore Deposits. 2013;55(7):594–600.

[76] Demartin F, Gramaccioli CM, Campostrini I. Pyracmonite, $(NH_4)_3Fe(SO_4)_3$, a new ammonium iron sulfate from La Fossa Crater, Vulcano, Aeolian Islands, Italy. The Canadian Mineralogist. 2010;48:307–313.

[77] Demartin F, Campostrini I, Castellano C. Aluminopyracmonite, IMA 2012-075. CNMNC Newsletter No. 15, February 2013, page 9. Mineralogical Magazine, 2013;77:1–12.

[78] Balić-Žunić T, Garavelli A, Acquafredda P, Leonardsen E, Jakobsson SP. Eldfellite, $NaFe(SO_4)_2$, a new fumarolic mineral from Eldfell volcano, Iceland. Mineralogical Magazine. 2009;73:51–57.

[79] Campostrini I, Demartin F, Gramaccioli CM. Vulcano: ein aussergewohnlicher Fundpunkt von neuen und seltenen Mineralien. MineralienWelt. 2010;21(3):40–57.

[80] Burke EAJ. A mass discreditation of GQN minerals. The Canadian Mineralogist. 2006;44:1557–1560.

[81] Mofaddel N, Bouaziz R, Mayer M. Le polymorphisme du sulfate de sodium anhydre et les phases intermediaires, glaserite et aphtitalite, dans le binaire Na_2SO_4-K_2SO_4. Thermochimica Acta. 1991;185:141–153.

[82] Okada K, Ossaka J. Structures of potassium sodium sulphate and tripotassium sodium disulphate. Acta Crystallographica. 1980;B36:919–921.

[83] Garavelli A, Grasso MF, Vurro F. Sublimates and fumarolic incrustations at Mt. Etna from 1993 to 1996. Acta Vulcanologica. 1997;9:87–89.

[84] Demartin F, Castellano C, Campostrini I. Therasiaite, $(NH_4)_3KNa_2Fe^{2+}Fe^{3+}(SO_4)_3Cl_5$, a new sulphate chloride from La Fossa Crater, Vulcano, Aeolian Islands, Italy. Mineralogical Magazine. 2014;78(1):203–213.

[85] Mees F, Hatertand F, Rowe R. Omongwaite, $Na_2Ca_5(SO_4)_6 \cdot 3H_2O$, a new mineral from recent salt lake deposits, Namibia. Mineralogical Magazine. 2008;72(6):1307–1318.

[86] Autenrieth H, Braune G. Ein neues Salzmineral, seine Eigenschaften, sein Auftreten und seine Existenzbedingungen im System der Salze ozeanischer Salzablagerungen (A new salt mineral, its properties, its occurence and the conditions for its existence in the system of salts of oceanic salt deposits). Naturwissenschaften. 1958;45(15):362–363.

[87] Demartin F, Campostrini I, Castellano C, Gramaccioli CM, Russo M. D'ansite-(Mn), $Na_{21}Mn^{2+}(SO_4)_{10}Cl_3$ and d'ansite-(Fe), $Na_{21}Fe^{2+}(SO_4)_{10}Cl_3$, two new minerals from volcanic fumaroles. Mineralogical Magazine. 2012;76(7):2773–2783.

[88] Demartin F, Gramaccioli CM, Campostrini I, Pilati T. Aiolosite, $Na_2(Na_2Bi)(SO_4)_3Cl$, a new sulfate isotypic to apatite from La Fossa crater, Vulcano, Aeolian Islands, Italy. American Mineralogist. 2010;95:382–385.

[89] Demartin F, Gramaccioli CM, Campostrini I. Adranosite, $(NH_4)_4NaAl_2(SO_4)_4Cl(OH)_2$, a new ammonium sulfate chloride from La Fossa crater, Vulcano, Aeolian Islands, Italy. The Canadian Mineralogist. 2010;48:315–321.

[90] Mitolo D, Demartin F, Garavelli A, Campostrini I, Pinto D, Gramaccioli, et al. Adranosite-(Fe), $(NH_4)_4NaFe_2(SO_4)_4Cl(OH)_2$, a new ammonium sulfate chloride from La Fossa crater, Vulcano, Aeolian Islands, Italy. The Canadian Mineralogist. 2013;51:57–66.

[91] Garavelli A, Pinto D, Mitolo D, Bindi L. Leguernite, $Bi_{12.67}O_{14}(SO_4)_5$, a new Bi oxysulfate from the fumarole deposit of La Fossa crater, Vulcano, Aeolian Islands, Italy. Mineralogical Magazine. 2014;78:1629–1645.

[92] Demartin F, Castellano C, Gramaccioli CM. Campostriniite, $(Bi^{3+},Na)_3(NH_4,K)_2Na_2(SO_4)_6 \cdot H_2O$, a new sulfate isostructural with görgeyite, from La Fossa Crater, Vulcano, Aeolian Islands, Italy. Mineralogical Magazine. 2015;79(4):1007–1018.

[93] Vergouwen L. Eugsterite, a new salt mineral. American Mineralogist. 1981;66:632–636.

[94] Slyusareva MN. Hydroglauberite, a new mineral of the hydrous sulphate group. American Mineralogist. 1970;55:321.

[95] Alpers CN, Jambor JL, Nordstrom DK. Sulfate minerals – crystallography, geochemistry, and environmental significance. Reviews in Mineralogy and Geochemistry. 2000;40:22.

[96] Sarp H, Deferne J. Chessexite, a new mineral. American Mineralogist. 1984;69:406.

[97] Thordarson Th, Höskuldsson Á. Postglacial volcanism in Iceland. Jökull. 2008;58:197–228.

[98] Harðarson BS, Fitton JG, Hjartarson Á. Tertiary volcanism in Iceland. Jökull. 2008;58:161–178.

[99] Jakobsson S, Jónasson K, Sigurðsson IA. The three igneous rock series in Iceland. Jökull. 2008;58:117–138.

[100] Þórarinsson S, Einarsson Þ, Sigvaldason GE, Elísson G. The submarine eruption off the Vestman Islands 1963–1964. A preliminary report. Bulletin of Volcanology. 1964;27:435–445.

[101] Þórarinsson S. Sitt af hverju um Surtseyjargosið. (Some facts about the Surtsey eruption.) Náttúrufræðingurinn (Naturalist). 1966;35:153–181.

[102] Þórarinsson S. Síðustu þættir Eyjaelda. (The last phases of the Surtsey eruption.) Náttúrufræðingurinn (Naturalist). 1969;38:113–135.

[103] Jakobsson SP, Jónsson SS, Leonardsen ES. Encrustations from lava caves in Surtsey, Iceland. A preliminary report. Surtsey Research Progress Report. 1992;10:73–78.

[104] Þórarinsson S, Steinþórsson S, Einarsson Þ, Kristmannsdóttir H Óskarsson N. The eruption on Heimaey, Iceland. Nature. 1973;241:372–375.

[105] Jakobsson SP, Pedersen AK, Rønsbo JG, Larsen LM. Petrology of mugearite-hawaiite: early extrusives in the 1973 Heimaey eruption, Iceland. Lithos. 1973;6:203–214.

[106] Sigurðsson O. Jarðeldar á Heimaey. (Eruption in Heimaey.) Týli. 1974;4:5–26.

[107] Sigmundsson G, Hreinsdóttir S, Hooper A, Árnadóttir T, Pedersen R, Roberts MJ, et al. Intrusion triggering of the 2010 Eyjafjallajökull eruption. Nature. 2010;468:426–430.

[108] Sigmarsson O, Vlastelic I, Andreasen R, Bindeman I, Devidal J-L, Moune S, et al. Remobilization of silicic intrusion by mafic magmas during the 2010 Eyjafjallajökull eruption. Solid Earth. 2011;2:271–281.

[109] Jónasson K, Jakobsson SP. Encrustations from the 2010 Fimmvörðuháls eruption. In: 30th Nordic Geological Winter Meeting: Programme and Abstracts, 9–12. January 2012. Reykjavík: Geoscience Society of Iceland; 2012.

[110] Þórarinsson S. Course of events. In: The Eruption of Hekla 1947–1948. Vísindafélag Ísl. 1976;4(1): 1–33.

[111] Schythe JC. Hekla og dens sidste udbrud, den 2den September 1845 (Hekla and its latest eruption, the 2nd September 1845). Kjöbenhavn: Bianco Lunos Bogtrykkeri; 1847. 154 p.

[112] Guðmundsson Á., Óskarsson N, Grönvold K, Sæmundsson K, Sigurðsson O, Stefánsson R, et al. The 1991 eruption of Hekla, Iceland. Bulletin of Volcanology. 1992;54:238–246.

[113] Sigvaldason GE. Some geochemical and hydrothermal aspects of the 1961 Askja eruption. Beiträge zur Mineralogie und Petrographie (Contributions to Mineralogy and Petrography). 1964;10:263–274.

[114] Þórarinsson S. Sigvaldason GE. The eruption in Askja, 1961. A preliminary report. American Journal of Science. 1962;260:641–651.

[115] Þórarinsson S. Askja on fire. Reykjavík: Almenna bókafélagið; 1963. 54 p.

[116] Björnsson A. Dynamics of crustal rifting in NE Iceland. Journal of Geophysical Research. 1985;90:10151–10162.

[117] Piochi M, Pappalardo L, De Astis G. Geochemical and isotopical variations within the Campanian Comagmatic Province: implications on magma source composition. Annals of Geophysics. 2004;47(4):1485–1499.

[118] Piochi M, Mormone A, Balassone G, Strauss H, Troise C, De Natale G. Native sulfur, sulfates and sulfides from the active Campi Flegrei volcano (southern Italy): genetic environments and degassing dynamics revealed by mineralogy and isotope geochemistry. Journal of Volcanology and Geothermal Research. 2015;304:180–193.

[119] Sgavetti M, Pompilio L, Roveri M, Manzi V, Valentino GM, Lugli S, et al. Two geologic systems providing terrestrial analogues for the exploration of sulfate deposits on Mars: initial spectral characterization. Planetary and Space Science. 2009;57:614–627.

[120] Russo M. Realgar e pararealgar della solfatara di Pozzuoli (Napoli) (Realgar and pararealgar from the Pozzuoli solfatara (Neaples). Micro. 2004;2004:25–28.

[121] Rolandi G, Paone A, Di Lascio M, Stefani G. The 79 AD eruption of Somma: the relationship between the date of the eruption and the southeast tephra dispersion. Journal of Volcanology and Geothermal Research. 2007;169:87–98.

[122] Chiodini G, Marini L, Russo M. Geochemical evidence for the existence of high-temperature hydrothermal brines at Vesuvio volcano, Italy. Geochimica et Cosmochimica Acta. 2001;65:2129–2147.

[123] Caliro S, Chiodini G, Avino R, Minopoli C, Bocchino B. Long time-series of chemical and isotopic compositions of Vesuvius fumaroles: evidence for deep and shallow processes. Annals of Geophysics. 2011;54:137–149.

[124] De Astis G, Lucchi F, Dellino P, La Volpe L, Tranne C A, Frezzotti ML, et al. Geology, volcanic history and petrology of Vulcano (central Aeolian archipelago). Geological Society, London, Memoirs. 2013;37:281–349.

[125] Paonita A, Favara R, Nuccio PM, Sortino F. Genesis of fumarolic emissions as inferred by isotope mass balances: CO_2 and water at Vulcano Island, Italy. Geochimica et Cosmochimica Acta. 2002;66:759–772.

[126] Capasso G, D'Alessandro W, Favara R, Inguaggiato S, Parello F. Interaction between the deep fluids and the shallow groundwaters on Vulcano Island (Italy). Journal of Volcanology and Geothermal Research. 2001;108:187–198.

[127] Coltelli M, Del Carlo P, Vezzoli L. The discovery of a Plinian basaltic eruption of Roman age at Etna volcano, Italy. Geology. 1998;26:1095–1098.

[128] Neri M, Acocella V, Behncke B. The role of the Pernicana fault system in the spreading of Mt. Etna (Italy) during the 2002–2003 eruption. Bulletin of Volcanology. 2004;66:417–430.

[129] Branca S, Del Carlo P. Eruptions of Mt. Etna during the past 3200 years: a revised compilation integrating the historical and stratigraphical records. In: Bonaccorso A, Calvari S, Coltelli M, Negro CD, Falsaperla S, editors. Mt. Etna: Volcano Laboratory. Washington: American Geophysical Union; 2004. pp. 1–22.

[130] Liotta M, Paonita A, Caracausi A, Martelli M, Rizzo A, Favara R. Hydrothermal processes governing the geochemistry of the crater fumaroles at Mount Etna volcano (Italy). Chemical Geology. 2010;278:92–104.

[131] Papazachos BC, Comninakis PE. Geophysical and tectonic features of the Aegean Arc. Journal of Geophysical Research. 1971;76:8517–8533.

[132] Nicholls JA. Petrology of Santorini volcano, cyclades, Greece. Journal of Petrology. 1971;12(1):67–119.

[133] Keller J. Mediterranean island arcs. In: Thorpe RS, editor. Andesites. New York: Wiley; 1982. pp. 307–325.

[134] Fytikas M, Innocenti F, Manetti P, Mazzuoli R, Peccerillo A, Villari L. Tertiary to quaternary evolution of volcanism in the Aegean region. Geological Society London Special Publications. 1984;17(1):687–699.

[135] Fytikas M, Innocenti F, Kolios N, Manetti P, Mazzuoli R, Poli G, et al. Volcanology and petrology of volcanic products from the island of Milos and neighbouring islets. Journal of Volcanology and Geothermal Research. 1986;28:297–317.

[136] Pe-Piper G, Piper DJW. Volcanism at subduction zones; Aegean area. Bulletin of the Geological Society of Greece. 1972; 9:113–144.

[137] Pe-Piper G, Hatzipanagiotou K. The Pliocene volcanic rocks of Crommyonia, western Greece and their implications for the early evolution of the South Aegean arc. Geological Magazine. 1997;134:55–66.

[138] Liakopoulos A, Katerinopoulos A, Markopoulos T, Bouleègue, J. A mineralogical petrographic and geochemical study of samples from wells in the geothermal field of Milos Island (Greece). Geothermics. 1991;20(4):237–256.

[139] Tarney J, Barr SR, Mitropoulos P, Sideris K, Katerinopoulos A, Stouraiti H. Santorini: geochemical constraints on magmas sources and eruption mechanisms. The European Laboratory Volcanoes. 1996;EUR18161EN:89–112.

[140] Bagnato E, Tamburello G, Aiuppa A, Sprovieri M, Vougioukalakis GE, Parks M. Mercury emissions from soils and fumaroles of Nea Kameni volcanic centre, Santorini (Greece). Geochemical Journal. 2013;47:437–450.

[141] Chiodini G, Cioni R, Leonis C, Marini L, Raco B. Fluid geochemistry of Nisyros Island, Dodecanese, Greece. Journal of Volcanology and Geothermal Research. 1993;56:95–112.

[142] D'Alessandro W, Brusca L, Kyriakopoulos K, Michas G, Papadakis G. Hydrogen sulphide as a natural air contaminant in volcanic/geothermal areas: the case of Soussaki, Corinthia (Greece). Environmental Geology. 2009;57(8):1723–1728.

[143] D'Alessandro W, Brusca L, Martelli M, Rizzo A, Kyriakopoulos K. Geochemical characterization of natural gas manifestations in Greece. Bulletin of the Geological Society of Greece. 2010;43(5):2327–2337.

[144] Kiyosu Y, Kurahashi M. Origin of sulfur species in acid sulfate-chloride thermal waters, northeastern Japan. Geochimica et Cosmochimica Acta. 1983;47:1237–1245.

3

Radiative Impacts of Volcanic Aerosol in the Arctic

Cindy L. Young and Jennifer W. Telling

Abstract

High latitude volcanic eruptions are high-frequency and intensity events capable of releasing large amounts of aerosols into the environment. Studies have shown that the Arctic is particularly sensitive to radiative perturbations due to aerosols, and a high sensitivity to volcanic aerosols would be expected. Despite the potential for volcanic aerosols to significantly perturb the Arctic radiation balance, the radiative impacts of volcanic aerosols in the Arctic are poorly understood and have received less attention than the effects of other aerosol types that are often present in the region, both natural and anthropogenic. A novel review of this topic is presented in detail in this chapter, focusing on the current state of the knowledge and the natural complexities involved with the problem, the important research tools, and the improvements that can be made over the status quo. The Arctic environment is both unique and complicated, and the perturbations caused by volcanic aerosol need to be examined in a regional context. An introduction to remote sensing and data collection in the Arctic is provided because there are often specific challenges, including high surface reflectivities, persistent meteorological clouds, the lack of winter daylight, and harsh conditions that hamper both in situ and remote data collection. Methods for tracking both aerosol and gas plumes in the Arctic that can help mitigate these issues are introduced. In addition to the physical constraints of data collection presented by the Arctic environment, volcanic aerosol is a complex mixture of varying aerosol compositions and sizes. Dealing with the nature of volcanic aerosol for optical calculations is further described, leading into a detailed discussion of the radiative impacts of volcanic plumes in the atmosphere. Radiative forcing comparisons of other aerosol types with comparable plume characteristics (e.g., thicknesses and optical depths) suggest that aerosol layers composed of significant proportions of volcanic ash can dominate the aerosol forcing in the region. Similar comparisons for ash deposits with other types of deposits that can be present in the region emphasize the ability of volcanic ash to produce large, and in some cases extreme, loadings that reduce albedo, which can have profound impacts on the Arctic radiation balance and hydrological cycle. The strengths and shortcomings of volcanic ash transport and dispersion models are reviewed and recommendations are made for future research that would strengthen the use of these models in Arctic environments. In particular, ash aggregation (or the sticking together of ash particles) is often not considered fully in transport modeling, and the consequences of this are discussed. Finally, we present a review of secondary volcanic impacts to oceans and ecosystems that have not been constrained in an Arctic context but are potentially important to the Arctic environment and the global CO_2 cycle.

Keywords: Arctic, volcanic eruptions, ash, sulfate, remote sensing, transport modeling

1. Introduction

There is a strong need to understand the role natural aerosols play in modulating Arctic climate. The radiative impacts of smoke [1] and dust [2] have been considered. However, volcanic aerosols continue to be largely overlooked in the Arctic environment. The Arctic environment has a high sensitivity to radiative perturbations, as demonstrated by many previous studies of other types of aerosols (i.e., smoke, dust, and haze), and may be strongly sensitive to volcanic aerosols. This chapter defines the "Arctic region" to include the true Arctic (north of 66.5°N) and the sub-Arctic (50–66.5°N), because both areas are more sensitive to radiative perturbations than lower latitudes and aerosol can be easily transported between regions.

Volcanic eruptions are capable of producing a huge, sporadic aerosol signal, lasting from minutes to years [3], and high northern latitude eruptions can distribute aerosols over large areas, as evidenced by the 2010 eruption of Eyjafjallajökull in Iceland, which halted air transportation in much of Europe. The Alaska Volcano Observatory (AVO) reports that Alaskan volcanoes alone have had an average eruption frequency of two per year over the past 40 years. Depending on the time of year, volcanic aerosol may be present along with other aerosol types. Despite the high frequencies and intensities of volcanic eruptions, volcanic aerosols in the Arctic are relatively neglected in assessments of Arctic aerosol.

Volcanoes in the Arctic have been the source of many climatically important eruptions in the last several centuries. The June 1783 eruption of Laki, a volcanic fissure in Iceland, caused a drop in global temperatures [4], drought, and famine [5]. The June 1912 Novaerupta-Katmai eruption near Kodiak, Alaska, was the most powerful eruption of the 20th century, causing surface cooling in the Northern Hemisphere throughout the summers of 1912 and 1913 and a maximum surface cooling of −0.9°C for September 1912 [6]. Recent volcanic eruptions in the region include Redoubt (Alaska, USA) in March to April 2009, Sarychev (Russia) in June 2009, Shiveluch (Russia) in September 2009, Eyjafjallajökull (Iceland) in April to May 2010, and Plosky Tolbachik (Russia) in 2012 to 2013. Recent and ongoing eruptions in the Arctic have fortunately been small to medium-sized events. Although small to mid-sized volcanic eruptions are less extreme events, they occur more frequently and provide a more regular stream of ash and gases to the environment than larger eruptions.

Volcanic aerosols can be external mixtures of ash, sulfates, and hydrometeors and/or internal mixtures of ash coated with sulfates, water, and/or ice [7]. Ash and sulfates are considered the dominant aerosol components of most eruptive plumes (e.g., [8]). In volcanic eruptions, sulfates are formed from a reaction between emitted volcanic sulfur dioxide (SO_2) and water and can remain in the stratosphere for up to 3 years [9], causing cooling at the surface and warming in the stratosphere [10]. The lifetime of fine ash (aerodynamic diameter <2.5 μm) in the stratosphere is on the order of a few weeks [11]. Consequently, sulfates are usually the only aerosol included in global climate perturbation estimates from a particular eruption [12]. However, volcanic ash and sulfate aerosols in the troposphere are important on a regional level and have comparable lifetimes (approximately days to weeks). Therefore, both ash and sulfate must be included in assessing the regional radiative impact of volcanic aerosol.

Volcanic aerosols can reflect and absorb shortwave (SW) radiation, and they scatter, absorb, and emit radiation in the longwave (LW) part of the spectrum. The interaction of volcanic aerosols with electromagnetic radiation can cause warming or cooling of the surface and atmosphere depending on the reflectivity of the underlying surface, the solar zenith angle (SZA), the optical properties of the aerosol layer, the vertical structure of temperature and humidity, and cloud characteristics. Ash deposits can lower the albedo of highly reflective snow and ice surfaces and may perturb the Arctic radiation budget and cause early snowmelt, analogous to dust deposits [13] and soot deposits [14]. Little consideration has been given to the radiative impacts due to volcanic ash deposits in snow (e.g., [15, 16]), and even fewer studies focus on the surface radiative impacts of volcanic ash deposits from Arctic eruptions or map the entire extent of the deposit area (e.g., [17, 18]). Additionally, ash [19, 20] and sulfates [21] can serve as cloud condensation nuclei and ice nuclei. **Figure 1** illustrates the broad impacts volcanic aerosols can have on the Arctic environment.

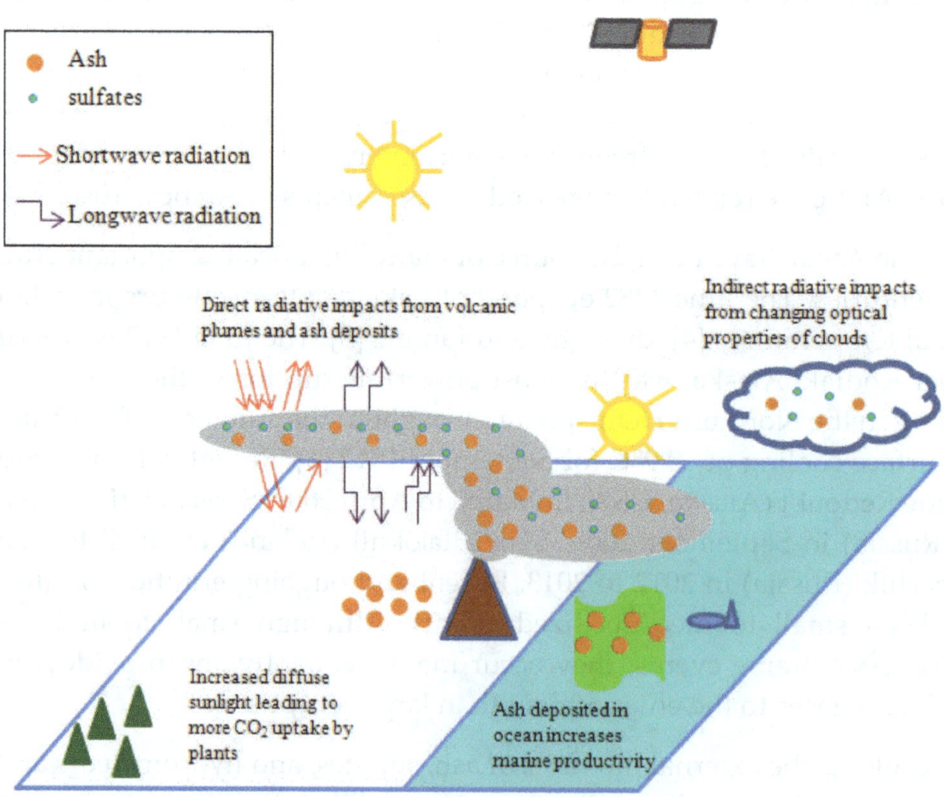

Figure 1. Broad impacts of volcanic aerosols on the Arctic environment.

This chapter will review the current state of the science behind the radiative impacts of volcanic aerosol in the Arctic and what remains to be explored. The work will touch on techniques involving satellite remote sensing, field measurements, and different types of modeling (radiative transfer, ash dispersion and transport, and atmospheric chemistry and circulation) that are used to help constrain the complexities of this problem. This chapter will focus mainly on the direct radiative impacts and then will briefly touch on other effects volcanic aerosols can have on the Arctic environment.

2. Radiative impacts of volcanic aerosol plumes

Determining the radiative influence volcanic aerosols have on the Arctic environment is a challenging problem because of the complexities of volcanic aerosols and the few measurements of their physical, optical, and chemical properties, which must be known to calculate radiative transfer and spectral refractive index. Particle counters on balloons and aircrafts have measured size distributions of stratospheric sulfate aerosol several weeks to months after an eruption (e.g., [22]), but these measurements are not helpful in determining the size distributions of less aged plumes of sulfate and/or ash. Models have been developed that include the sulfate formation and aging process [23] and have been added into general circulation models (GCMs) [11]. Size distributions for volcanic ash are dangerous to measure in situ and are usually measured on ash fall samples, which are not representative of atmospheric size distributions because of sorting that takes place during transport [24]. The role of plume aging on altering the composition and size distribution of volcanic aerosol is important to consider and is expected to substantially influence the radiative impacts.

In addition to SZA, surface albedo, and optical properties of aerosol, there are more aerosol layer-specific characteristics that must be known for radiative calculations. For an aerosol plume, these include aerosol optical depth (AOD), physical thickness, and vertical placement of the layer in the atmosphere. Sun photometers and ground-based lidars can help determine AOD, and ground-based lidars can also be used to obtain thickness and vertical placement. However, the coverage of ground-based sensors is limited. Due to the limitations of ground-based sensors and field measurements as well as the remote locations of Arctic volcanoes, satellite remote sensing is essential for monitoring aerosol from volcanic eruptions. The NASA afternoon satellite constellation A-Train provides a unique opportunity to examine eruptions and the evolution of volcanic plumes. The A-Train consists of six polar orbiting satellites [Aqua, Aura, Cloud-Aerosol Lidar and Infrared Pathfinder Satellite Observations (CALIPSO), CloudSat, GCOM-W1, and OCO-2] flying in close configuration, each equipped with different sensors measuring in wavelength ranges from the ultraviolet (UV) to radio. Combining the data retrieved from several sensors allows for an unprecedented view of volcanic eruptive plumes.

The Moderate Resolution Imaging Spectroradiometer (MODIS) instrument flying aboard Aqua and Terra satellites, the Ozone Monitoring Instrument (OMI) on the Aura spacecraft, and the Cloud-Aerosol with Orthogonal Polarization (CALIOP) aboard the CALIPSO platform are sensors capable of detecting volcanic aerosols. The MODIS instrument provides true color images of ash plumes and deposits and AOD of volcanic plumes. The OMI provides a UV aerosol index (AI) that can detect the presence of UV-absorbing aerosols, such as ash, dust, smoke, and SO_2 emissions. CALIPSO provides the vertical plume structure, which is useful in determining the placement of the plume in the atmosphere, the plume top height, and the plume thickness. In addition to these platforms, the Atmospheric Infrared Sounder (AIRS) aboard the NASA Aqua satellite views the Earth in the infrared (IR) and is sensitive to the SO_2 absorption band at ~7.3 μm [25]. AIRS and OMI observations can often complement each other, as AIRS is not dependent on solar UV and OMI is less sensitive to water vapor, which hinders AIRS observations near the tropopause. However, in the Arctic, environmental conditions, such as meteorological clouds, little to no winter daylight, and high surface reflectivities, often hamper retrievals for passive instruments measuring in the visible and UV, such as MODIS and OMI, as well as space-based lidars, such as CALIPSO. It is because of these challenging environmental conditions that modeling of the eruptive plume transport, dynamics, and dispersion is essential to accompany, and sometimes supplement, satellite data retrievals in the Arctic region.

2.1. Tracking volcanic plumes

Tracking the movement, areal extent, and evolution of volcanic plumes is not only relevant for radiative purposes but also from an aviation safety and public health standpoint. It is possible to track both gaseous and particulate species in the volcanic plume using satellite remote sensing and transport modeling. Two of the most common volcanic gases, H_2O and CO_2, are also major atmospheric constituents, making them hard to distinguish from the background. Volcanic SO_2, however, has a relatively low background concentration, making it an ideal trace gas for many volcanic eruptions. SO_2 is observable in both the UV and thermal IR (TIR). For use as a tracer in Arctic environments, the absorption bands in both the UV and TIR are critical. During the Arctic winter, the lack of daylight hinders UV observations, as solar radiation is the dominant source of UV, whereas TIR measurements can still be collected. Conversely, low thermal contrast between the volcanic plume and the ambient environment makes observations in the TIR difficult, diminishing the potential for TIR instruments to observer older, cooler plumes. Due to the strong absorption of water in the TIR, high water vapor content in the lower troposphere can also limit TIR observations of SO_2 [25]. In most of the world, this means that TIR instruments are best used to observe plumes above 3 km [25]; however, the generally low relative humidity in the Arctic can mitigate this issue.

A combination of plume observations from both UV and TIR observations can allow plumes to be tracked for days or even weeks depending on the size of the eruption. As a volcanic plume matures and moves away from its source, dilution from atmospheric mixing and conversion of SO_2 to sulfate aerosol make plumes more difficult to separate from the surrounding atmosphere. Atmospheric models, such as NOAA HYSPLIT [26, 27], which predict the direction and altitude of the plume movement, can complement the use of AIRS, CALIPSO, MODIS, and OMI data to improve and prolong plume tracking metrics.

Tracking both Arctic and sub-Arctic plumes is crucial, as even plumes in the sub-Arctic region can be easily transported into the Arctic, possibly amplifying their environmental impacts in this sensitive region. SO_2 from the initial activity from the 2012 to 2013 eruption of Plosky Tolbachik Volcano, in the Kamchatka Peninsula, could be observed for a week as it moved northwest from the source and then across the Siberian Sea before heading south again [28]. Despite its longevity, the Plosky Tolbachik plume had remarkably little influence on the Arctic environment because it was predominately transported in the troposphere, over water, during the winter, and had significantly more SO_2 than ash [28, 29].

Due to the lower neutral buoyancy height of particles versus that of gases and the possibility of wind shears, particles may be concentrated at lower levels in the atmosphere and can move in different directions than the gases. Tracking ash in volcanic plumes can be accomplished with the use of UV and visible satellite remote sensing (e.g., OMI AI and MODIS AOD) as well as using IR brightness temperature difference (BTD) techniques (e.g., [30, 31]), employing instruments such as AIRS, Geostationary Operational Environment Satellite (GOES), Advanced Very High Resolution Radiometer (AVHRR), and MODIS. Often called the "split window technique," this method takes advantage of the distinctive negative BTDs that result from volcanic ash/sulfates in bands centered at 11 and 12 μm and can be used to estimate optical depth, particle sizes, and masses of ash/sulfate that match the observed BTD from varying these parameters in a radiative transfer model [30]. This method assumes spherical particle shapes, a thin plume parallel to a homogenous surface, and a predetermined range of particle sizes. Fixed refractive indices are often used (selection of these are described further in Section 2.2), but the sensitivity as demonstrated by Wen and Rose [30] is higher for the assumed size distribution than the refractive indices. Meteorological clouds can give similar BTD values to ash and sulfate, and the BTD must be used along with other information (e.g., true color images) to support a definitive particle detection. Dense eruptive plumes can cause issues for the retrieval algorithm, since it was developed for a semitransparent plume, and are better suited for single-band studies. Positive BTDs have been used to infer the presence of ice in the eruptive plume, and similar radiative transfer calculations were done to estimate the size distributions and masses of ice [31]. However, ice is often not the dominant particle in eruptions that occur above 40°latitude, in contrast to those that occur closer to the tropics, because the entrained tropospheric air has a lower water vapor content at higher latitudes [7]. Particle depolarization ratios from CALIPSO may also be used to confirm the presence of nonspherical particles within the plume, but lidars are more suited to studying plume profiles than areal extents. Ash transport models are another tool for understanding plume transport and ash deposition and are discussed later in this chapter (Section 3.1).

2.2. Constraining the optical properties of volcanic plumes

To develop a microphysical model, the compositional types of volcanic aerosols present in the eruptive plume and their refractive indices, size distributions, and relative abundances with respect to other aerosol types present must be ascertained. Refractive indices can be measured in a laboratory for a variety of aerosol types and compositions. The Optical Properties of Aerosols and Clouds (OPAC) data set [32] compiles refractive indices for several typical aerosols under different atmospheric humidity conditions. To represent volcanic sulfate, sulfuric acid solutions of ~70% are often used (e.g., [12]), but at Arctic ambient air temperature and relative humidity these concentrations tend to be lower (~40–50%) [33]. Because the temperature and relative humidity in a fresh volcanic plume are higher than that of the surrounding air, sulfuric acid solutions closer to 70% are a more realistic estimate for a fresh volcanic plume than the ambient air mixtures. The refractive indices of ashes with varying silica content have also been measured (e.g., [34]) and are commonly used to represent an ash component in microphysical models.

Other quantities that must be included in the microphysical model (i.e., size distributions and relative abundances) cannot be measured directly, especially in the case of "fresh" plumes. Most microphysical models treat volcanic aerosols with a lognormal size distribution, similar to the form from Kearney and Watson [35]:

$$n_c(r) = \frac{1}{\sqrt{2\pi\sigma}} e^{-\frac{(\ln(r)-\mu)^2}{2\sigma^2}} \tag{1}$$

where n_c is the particle number concentration, r is the particle radius, and σ is the variance of the size distribution or the log of the standard deviation:

$$\mu = (\ln \text{Reff}) - 2.5\sigma^2 \tag{2}$$

where Reff is the effective radius [35].

Sulfates are typically nanometer-scale particles and occur only in the fine-mode fraction (aerodynamic diameter <2.5 μm), but ash sizes can vary greatly from coarse to fine. As the plume ages, sulfate aerosols grow larger in size and more numerous, as the effective radius for ash becomes smaller and larger grain sizes are scavenged. A more sophisticated model, such as a GCM, may be able to estimate and track the evolution of size distributions and the relative abundances of the aerosol types given a set of initial conditions and may even treat sulfate formation from SO_2 [11]. However, these estimates are difficult to validate and initial conditions may add uncertainty, especially in the Arctic. When available, satellite data may be used to deduce size distributions (e.g., [30]), and ratios of fine- and coarse-mode aerosols can be constrained to determine the proper proportions to externally mix aerosol types (e.g., using MODIS fine-mode fraction retrievals). Literature investigations of size distributions for representative eruptions can be used to study end-member cases of fresh and aged plumes under Arctic conditions [29, 35].

An additional consideration when determining the optical properties of volcanic aerosol is whether to consider particle nonsphericity of ash and ice in the calculations. This can be done employing the T-matrix method. However, because the relative errors in radiative flux calculations for using the single scattering properties of spherical particles to approximate those of nonspherical dust particles are low [36], optical properties calculated with the Mie theory are often still used.

2.3. Comparing the radiative impacts of Arctic aerosol plumes

The direct aerosol radiative forcing (DARF) of two moderately thin volcanic layers from a mid-sized volcanic eruption have been compared to those for other aerosol types [29]. Two volcanic layers were chosen to represent a young, ash-rich/sulfate-poor plume and an older, sulfate-rich/ash-poor plume. Radiative modeling was used to obtain upward and downward fluxes. Net fluxes (Fnet) for both the SW and LW components at the top of the atmosphere (TOA) and at the surface were computed by subtracting the upward flux from the downward flux. The total net flux at TOA and at the surface were then calculated by adding the respective SW and LW net fluxes. DARF efficiency (DARFE), defined as the change in the net flux with respect to the change in AOD (550 nm), was computed as

$$DARFE = \frac{\Delta F_{net}}{\Delta AOD_{550\ nm}} \tag{3}$$

To calculate DARFE for an aerosol layer, the surface DARFE was subtracted from the DARFE at TOA. The units of DARFE are W m^{-2} AOD^{-1}, and a similar definition of DARFE to that of Stone et al. [1] was selected to consistently compare radiative impacts for all aerosol types. Calculations of DARF were done by multiplying DARFE by the layer AOD, resulting in units of W m^{-2}.

Volcanic eruptions are special events because of their ability to create aerosol layers many times thicker and more optically opaque than sources of other aerosol types frequently found in the region. In considering the nature of volcanic eruptions, it is expected that they could produce larger vertical loadings and dominate over the radiative impacts of other aerosols. Although many factors influence the regional radiative impacts, similar surface albedo, AOD, SZA, and plume thickness for each aerosol type were chosen to study the effects of different aerosol compositions and size distributions alone (for more information, see Table 4 in Young et al. [29]). A comparison of DARFs for two volcanic aerosol compositions [29], mineral dust

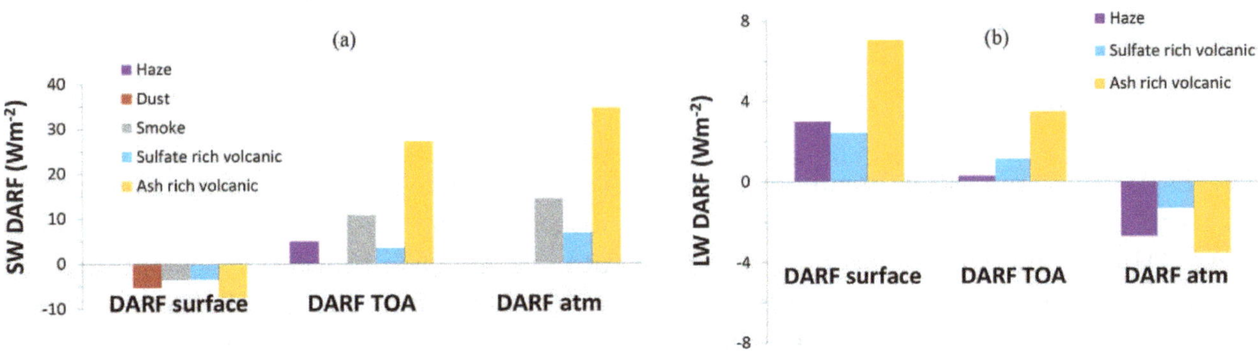

Figure 2. (a) SW DARFs [37], dust [2], smoke [1], and ash- and sulfate-rich volcanic aerosols [29]. Values of forcing for dust were only available at the surface. (b) LW DARFs for haze [38] and ash- and sulfate-rich volcanic aerosols (taken from Young et al. [29]).

[2], wildfire smoke [1], and haze [37, 38] is presented in **Figure 2** (taken from Young et al. [29]).

The ash-rich volcanic mixture, followed by mineral dust, attenuates the most incoming solar radiation at the surface and also absorbs more solar radiation in the layer than the other aerosol types (**Figure 2a**). The ash-rich mixture is the most important aerosol type with regards to the LW radiative impacts (**Figure 2b**). It should be noted that the LW forcing for smoke is negligible due to small particle sizes [39] and is not shown in **Figure 2b**. LW forcings of dust and volcanic aerosols are important because those particles are of sufficient size to interact with LW radiation. In the Arctic, LW radiation can dominate the forcing during seasons when the sun is low and can change the sign of the total forcing, but the LW component is rarely considered in radiative assessments of Arctic aerosols. Comparisons of both SW and LW components suggest that a thin ash-rich volcanic layer can dominate the aerosol radiative impacts in the Arctic.

3. Radiative impacts of volcanic deposits

Similar information to plume radiative modeling is needed to calculate the radiative impacts of ash deposited onto ice and snow, but knowledge on ash deposit loadings and areal extents also must be ascertained. Ground measurements are helpful but usually do not map out the entire extent of deposits or provide an adequate spatial resolution of samples. Volcanic ash transport models can assist in characterizing ash plumes and deposits. Past research has employed GCMs to simulate the dispersion and deposition of volcanic ash by assigning an initial flux amount of ash into the atmosphere (e.g., [11]) without treating the source conditions of the eruption (e.g., the initial distribution of ash in the eruptive column). Failing to consider the dynamics of the eruption and initial eruptive column will impact the regional transport and deposition of ash. In efforts to account for previously neglected eruption dynamics, a preprocessing tool was developed to determine the initial ash fields from source conditions for input into mesoscale atmospheric chemistry models [40]. As an alternative to GCMs, volcanic ash transport and dispersion models (VATDMs) compute initial ash fields directly from eruption source condition input parameters, consider a full-sized spectrum of ash, and can include nonspherical ash particles.

3.1. Constraining VATDMs

VATDMs can be presented in either Eulerian (e.g., Fall3D and TEPHRA) or Lagrangian (e.g., Puff and HYSPLIT for volcanic ash) reference frames. Volcanic advection-diffusion models (e.g., [41]) are Eulerian models that can predict atmospheric ash concentrations and ground deposit loadings by considering advection, turbulent diffusion, and gravitational settling, making them useful tools in assessments of the radiative impacts of ash deposits. In general, advection-diffusion models require input of source conditions [mass flow rate (MFR), source type, and plume temperature], ash properties (initial size distribution, particle density, and shape), and transport conditions (meteorological data and diffusion coefficients).

Commonly, MFR is estimated by comparing eruption plume heights to those predicted with simple plume theory for a particular eruption rate (e.g., [42]), assuming there are no variations in plume height during the eruptive activity. Error can be introduced when determining the height of the plume, which is often deduced from seismic data or reported by pilots in the area. Another technique that is often used involves erupted masses determined from mapped deposits and the seismic durations of eruptive events [43, 44], but this technique relies on the assumption that MFR is constant over the entire duration of the eruptive episode. The success of this method hinges upon a thorough and timely sampling of the eruptive deposits, including only ash deposited from the relevant eruptive event. Because of the large uncertainties that go into any MFR calculation, independent estimates of MFR for a given eruptive event can vary over orders of magnitude [43–45]. There are not many good estimates of MFR for Arctic eruptions due to the remote locations of many volcanoes, making MFR a poorly constrained parameter to which advection-diffusion models can exhibit a high degree of sensitivity [46].

The volcanic column source type determines how ash concentrations will be initially allocated within the atmosphere. For example, the Fall3D model [41] handles three different source types: plume, point, and Suzuki. The plume source describes the distribution of ash within a rising, hot, buoyant plume in which ash concentrations increase with height, up to a height of neutral buoyancy with the atmosphere. The point source dispenses all ash from a single point and height within the atmosphere. The Suzuki source allows for the height of the largest ash concentration and how the total mass is distributed around it to be chosen based on the strength of the eruption (i.e., more violent eruptions distribute the highest concentration of ash more closely to the maximum column height).

Due to the dangers involved with measuring the size distributions of ash in plumes in situ, the initial size distributions used in VATDMs are measured from deposits in the field. These measurements are often depleted in fine ash because fine particles can be swept to great distances from the volcano, where they cannot be collected and measured [47]. Particle densities and shapes are also typically measured from field deposits and, along with the size, also affect the settling rate of ash. Particle density varies with vesicularity, which can be a function of particle size, and particle shapes can range widely from spherical to nonspherical [48]. Advection-diffusion models, such as Fall3D, can exhibit high sensitivity to particle shape [46], which is a parameter that eruption-specific and extensive field deposit sampling could better constrain relatively easily.

Transport-related parameters (i.e., wind fields and diffusion coefficients) govern the advection and diffusion portions of the model. The diffusion coefficients represent the dispersion of ash due to small-scale turbulent motion without directly modeling the scale of this process and impact the width of ash plumes and deposits [41, 49]. The values of diffusion coefficients can be challenging to determine and are often tuned to match model results with observations, thereby accounting for some turbulent processes that are not directly modeled. Most of the transport processes for large eruptive columns occur outside the boundary layer, causing the vertical diffusion coefficient to be negligible compared to the horizontal diffusion coefficient [49]. Despite the uncertainty and wide range of possible horizontal diffusion coefficients, the model is typically not as sensitive to this parameter as it is MFR, some source types, and particle shape [46].

To constrain VATDMs in the Arctic, where data are sparse, Young et al. [46] developed a methodology to use the most data-rich event and obtain the best-fit model parameters by solving the inverse problem [46]. These best-fit parameters were then used to extrapolate those for other less studied eruptive episodes. This methodology was applied by Young et al. [46] to the 2009 eruption of Redoubt Volcano in Alaska, USA, and would be applicable to other eruptions for which a small period of the explosive activity was better characterized than most of the explosive events. Although this technique resulted in good to moderate model agreement with the available satellite and field data, the model had particular difficulty reproducing ash deposit loadings measured by Schaefer and Wallace (2012) in the far field (200–300 km from the vent) for even the most studied eruptive episode [50]. Other works employing Eulerian models to simulate ash deposition from the same eruptive event had similar issues for many model runs in the far field using a similar model set-up [44, 51]. Potential sources of error for all studies include distinguishing between ash layers from different eruptive events in the field [43], issues with wind field resolution through high elevation regions, and ash aggregation effects (discussed below in detail) that are not currently treated by any volcanic transport model.

Aggregation models, such as the one developed by Telling et al. [52], should be incorporated in VATDMs. Additionally, the process of comminution, which is the break-up of particles due to collisions, is not considered. Comminution is likely important in the initial conduit, where particle densities and energies are high and more bounce events between particles occur than aggregation events [53] and also, potentially, in the lowest region of the volcanic column, where particle densities are highest. Aggregation is more critical to consider when attempting to reconcile deposition and long-range atmospheric transport models with field and satellite data.

3.2. Ash aggregation studies for future VATDM improvements

Aggregation, or sticking, between volcanic particulate is a key source of error in any volcanic transport model. Volcanic ash is defined as any solid particulate with a diameter less than 2 mm in size [24]. Depending on the eruption type, size, and duration, fine ash (1–10 μm) typically has a lifetime on the order of days [11, 54], whereas fine ash (0.1–1 μm) can remain in the atmosphere for months to years in the case of stratospheric eruptions [11]. This can be a particular concern for Arctic eruptions because the boundary between the troposphere and stratosphere is lower in the Arctic environment. Typically, the Arctic tropopause occurs between 8 and 10 km, depending on season, whereas the tropopause occurs between 12 and 16 km over the tropics.

Marzano et al. [55] cited the lack of information regarding particle aggregation as a key source of error in their efforts to model the 2004 eruption of Grimsvötn. More recently, the 2010 eruption of Eyjafjallajökull, Iceland, produced an abundance of both fine ash (raising concerns for aircraft travel and respiratory health impacts) and coarse ash [54]. However, aggregation processes ultimately removed much of the fine ash close to Eyjafjallajökull, dampening the regional hazards [56] and demonstrating the importance of these processes on both regional and global scales [19, 56].

Gilbert and Lane [57] provided a number of classifications for aggregates, but the two broadest categorizations are simply dry and wet. Dry aggregates are bound by particle charging [57], which can be a result of either triboelectric charging, resulting from particle-particle interactions in the flow, or fractoemission, which occurs in the conduit during fragmentation [58]. Wet aggregation arises from the presence of water in a volcanic flow, forming when liquid-coated particles collide and stick together or when dry aggregates are scavenged by a water droplet [57]. In general, the Arctic environment is drier than that in the mid-latitudes and tropics, and this is particularly true in the winter over land. Consequently, electrostatic aggregation is a particularly important consideration in the Arctic, although wet aggregation is still likely dominant in a fresh eruptive plume because water is abundant and atmospheric mixing has not yet cooled the plume.

Numerous models have been proposed to treat aggregation in volcanic and atmospheric simulations. The most simplistic of these typically assume that a certain percentage of ash will aggregate, that only wet or dry aggregation processes are important but not both [19], or that particle aggregation will mimic water droplet coalescence. More complex formulations treat aggregation as a function of Stokes number, with aggregation efficiency decreasing as the Stokes number increases [59]; however, this accounts only for wet aggregation. This is a serious shortcoming for any model used in an Arctic setting because the environment over land is typically dry, especially in winter, and aggregation processes would be poorly modeled as soon as the volcanic plume began to dilute. Research has also been done to experimentally determine how much water is needed to promote wet, over dry, aggregation [52, 57] and to determine the degree of charging required to form electrostatic aggregates [52, 58]. The delineation between wet and dry aggregation zones occurs at ~71% relative humidity [20, 52] and the wet and dry aggregation zones, which each have different behavioral characteristics, can be defined accordingly. Telling et al. [52] developed these experimental relationships into a series of equations that define aggregation efficiency as a function of the collisional energy of two particles for the wet and dry cases separately.

Considering aggregation in radiative calculations would call for an updated microphysical model of ash, which is discussed by Textor et al. [19]. In both dry and wet cases, aggregation would influence how the size distribution evolves with time. The formation of heavier aggregates from smaller particles would lead to aggregates with different shapes than the individual particles and change the gravitational settling rates of ash. Wet aggregation involves the addition of a water layer, which could also be frozen into an ice layer and would modify the refractive index of the ash.

3.3. Constraining a radiative model for volcanic ash deposits

Radiative modeling of volcanic ash deposits on snow and ice requires some similar knowledge to that of plume modeling (i.e., SZA, underlying surface reflectance, ash effective radius, and refractive index) but also must include additional snow layer properties (i.e., snow layer thickness, snow density for each layer, and snow effective radius) and particle mass mixing ratios of ash in snow. A spectral or monochromatic underlying surface reflectivity can be used, but the model would not be sensitive to this parameter if the snow depth is large.

Even the most detailed depositional maps are often not resolved enough to study the radiative impacts due to the entire extent of the deposit, especially in the aerosol size range (radius ≤50 μm), because the measured area typically includes only deposits proximal to the vent. Modeled depositional loading fields can be converted into mixing ratios using the density of snow and the mixing depth of ash in the snowpack. As reported by Young et al. [46], the shape of the ash particles is important in advection-diffusion modeling, with spherical particles having better agreement with field measurements in most areas, whereas nonspherical particles produce loadings that agree better with field data for locations ≥210 km from the vent [46]. Advection-diffusion models can also do fairly well at reproducing deposited ash median radius compared to field data (except in areas ≤12 km from the vent, where aerosol-sized particles are less abundant), which is useful for optics calculations [18].

Assuming only aerosol-sized ash particles are being considered, the optical properties of the ash may be calculated with the Mie theory for a lognormal size distribution and deposited ash mean effective radii and standard deviation fields determined from transport modeling. This assumption can be made because the larger particles are relatively low in abundance and are concentrated in regions close to the vent [43]. Ash refractive indices for a given composition can be taken from previous laboratory studies, where available (e.g., [34]). Although the inclusion of nonspherical particles is important for transport modeling [46], spherical particles and the Mie theory may be used in calculating the optical properties of ash because relatively little error arises from approximating particles as equal volume to projected area spheres in the radiative flux calculations [17, 36].

The snow properties required by the radiative model can be measured in the field, but in the absence of field data and adequate spatial/temporal coverage data sets, such as the Natural Resources Conservation Service (NRCS) Snow Telemetry (SNOTEL), may be used to estimate snow layer thickness and density. The refractive indices of ice are often taken from the literature (e.g., [60]). Snow effective grain size radius can vary between 50 and 1100 μm (e.g., [61]), with new snow having smaller grains. If snow effective radius profiles are unknown for the time and region of interest, smaller grain sizes may be used to represent fresher snow and larger sizes for older snow.

At the time of eruption, ash would be deposited on top of the snowpack, which could be blanketed by snowfall or another ash layer [62]. The covering of ash deposits with new snow may substantially increase the surface albedo, and the layering structure of snow and ash should be considered. It is likely that measurements of this kind have not been made throughout the entire extent of the ash deposits, and a representative layering scheme may need to be constructed from what is known. However, new snowfall onto ash may not have a long-lived effect on albedo because, as the snow melts, water will flow downward through pore spaces in the snow, and ash will become concentrated at the top of the snowpack. A concern might also be blowing winds spreading freshly fallen ash, but after melting starts wet ash particles will stick to the snow.

3.4. Radiative impacts of aerosol deposits in the Arctic

The radiative impacts of several types of aerosol deposits (i.e., black carbon, volcanic ash, and dust) in snow measured in the field are compared to those for volcanic ash deposits from the mid-sized 2009 eruption of Redoubt Volcano, modeled by Young et al. [18]. High surface reflectivity and cloudy conditions prevented the retrieval of satellite albedos for this eruption, and there were also no known measurements of albedo in the field, both of which are typical issues surrounding Arctic eruptions. In keeping with isolated observations of complete ash cover in many regions, significant albedo reductions were modeled for these areas [18]. **Table 1** breaks down the particulate species and the quantities compared by Young et al. [18].

Aerosol species	Reference	Loadings (ppb)	Percent surface albedo reduction	SW surface forcing (W m^{-2})	Percent increase in melt rate compared to pure snow
Ash[a]	Young et al. [18]	7×10^4–1×10^8	0–85	0–96	220–330
Ash[b]	Young et al. [18]	$\leq1.6\times10^7$	X	X	140–160
Black carbon	Clarke and Noone [62]	5–50	1–3	X	X
Dust	Skiles et al. [64]	2×10^5–4×10^6	X	35–70	X
Ash	Dadic et al. [16]	1×10^0–1×10^6	0–37	X	X
Ash	Driedger [15]	X	X	X	≥90

Please note that due to the nature of the measured deposits not all of the quantities are directly comparable (as will be discussed in the text) but are presented here to illustrate the extents of impacts possible from various particles. Calculations were done for a layer deposited on both new and old snow, and the ranges reported for albedo reduction and forcing include both cases from the edges of the deposits to the vent [18]. Melt rate increases were calculated at the maximum loadings only to report an upper bound. Information that was not reported is denoted by an X.

a Ash layer covering loadings from the vent to the edges of the deposits.

b Ash layer calculations only considering low to intermediate loadings.

Table 1. Measured radiative impacts of black carbon, volcanic ash, and dust deposits on snow compared to those modeled by Young et al. [18] for volcanic ash.

For a similar particle size and concentration, black carbon is far more absorbing than ash [63], but at the large loadings that may be produced by a mid-sized eruption ash would cause the most significant albedo reductions. To compare volcanic eruptions, it is important to look at albedo changes from similar loadings. The greatest deposit loading used in Young et al. [18] is two orders of magnitude larger than that measured by Dadic et al. [16] in Antarctica. It is possible that the deposits measured by Dadic et al. [16] were from a smaller eruption (or a mid-sized to large eruption under windier conditions), the study had only measured ash in more distal locations, or greater mixing of ash and snow had taken place. Young et al. [18] reported an albedo reduction of ~37% for ash deposited on new snow when considering only loadings at more distal locations (~200 km from the vent) and concentrations of ~2.8×10^7 ppb [18]. This albedo reduction is comparable to that of Dadic et al. [16] for the area of the largest ash concentration measured. These studies may not be directly comparable, however, due to potentially different size distributions in deposits at the locations of interest.

The SW surface net fluxes calculated by Young et al. [18] were made at a daily mean SZA for late March and can be viewed as 12-hour daily averages [18]. Skiles et al. [64] also computed springtime daily mean forcings for dust deposits on snow in Colorado. The SW imaginary part of the refractive index for ash [34] is similar to that of dust [65], but the mean daily forcings from dust would be much smaller in the Arctic than in Colorado. This is because the average Arctic dust concentrations are about two to three orders of magnitude lower (i.e., [66]), and the incident solar radiation is lessened at higher latitudes. Using the method of Skiles et al. [64] to calculate minimum forcings due to the direct effects of ash deposition onto snow alone, Young et al. [18] computed a range from the edges of the deposit to the vent of ~0 to 96 W m^{-2}.

The restricted degree-day radiation balance approach described in Melloh [67] was used by Young et al. [18] to estimate a range of melt rates from SW surface net fluxes. Although no significant increases in melt rate were found for the edges of the deposits, regions with larger ash loadings had daily melt rates that were substantially greater than pure snow (~220–320%). However, these areas included ash layer thicknesses exceeding 0.3 cm, which will start to insulate the snow at this thickness [15], making these estimates of snowmelt an upper bound. Considering only intermediate loadings $\leq1.6\times10^7$ ppb with layer thicknesses <0.3 cm achieved more conservative maximum melt rates ~140% to 160% larger than ash-free snow. These estimates are generally in agreement with conservative maximum snowmelt increases of at least 90% measured for snow plots manually covered with ash by Driedger [15]. Melting in regions where snow is present year-round can lead to an increase in albedo that results in further snowmelt. In areas that typically melt in the summer, early or accelerated snowmelts have the capacity to reduce runoff in later months and deplete water resources [68].

4. Other impacts

4.1. Indirect radiative effects

The indirect aerosol effect refers to the ability of aerosols to cause radiative perturbations by altering the microphysical properties (e.g., changing the refractive index, particle number, and/or size distribution), lifetime, and coverage of meteorological clouds, thereby also influencing the hydrological cycle [69]. The Intergovernmental Panel on Climate Change (IPCC) reports that the current understanding of the indirect effects of aerosols on clouds is low [70]. Studies have shown that sulfate from the eruption of Mount Pinatubo in the Philippines produced a global impact on cirrus formation and evolution [21] through increases in aerosol number, homogeneous ice nucleation rate and ice crystal number, and ice water path [21, 71]. Additionally, it was found that ash can uptake water efficiently and can serve as cloud condensation nuclei [20] and ice nuclei [19]. The indirect effects in the Arctic may be different than in other regions because of the special properties of the surface and atmosphere, and the Arctic is more sensitive to aerosol effects [69]. The aerosol indirect effects due to volcanic eruptions may be particularly significant, as eruptions supply a large number of particles (i.e., ash, ice, and sulfates) compared to the low particle levels in the ambient Arctic atmosphere. More research is needed on the subject of the indirect radiative effects of volcanic aerosols in the Arctic.

4.2. Effects on ecosystems

The effect that volcanic eruptions have on ecosystems has been well documented. Forests [72], oceans [73–75], and even large global cycles, such as CO_2 variability typically controlled by the El Niño-Southern Oscillation (ENSO) [76, 77], can be influenced by volcanic eruptions. This section examines a few of these effects in further detail.

4.2.1. Increased diffuse radiation to plants

Beyond the direct radiative impacts of volcanic eruptions, volcanic ash and aerosol can have long-lasting secondary effects. The ash-rich eruption of Tambora Volcano, Indonesia, in 1815, resulted in the well-known "year without summer" in 1816 [78]. Stratospheric ash loading reduced direct solar radiation globally, leading to widespread cooling, crop failure, and slowed forest growth [72, 78]. However, Gu et al. [79] found that large aerosol-rich eruptions, such as Mount Pinatubo, can have a quite different effect. Although the 1991 eruption of Mount Pinatubo did decrease the direct solar radiation as expected, it also increased the diffuse solar radiation [79]. Gu et al. [79] showed that this increase in diffuse solar radiation led to an increase in forest photosynthesis, which, in turn, led to a temporary decline in the amount of atmospheric CO_2 [76]. Although neither of these eruptions occurred in the Arctic, both the Tambora and Mount Pinatubo events were felt globally for years afterward [69, 79] and potentially had a disproportionate effect on the Arctic environment either by promoting additional growth of vegetation in the tundra or by reducing the already short Arctic growing season. Further study is needed to better understand the effects of large volcanic eruptions on the Arctic environment, which is home to thousands of plant species.

4.2.2. Increased ocean productivity

Plate tectonics dictate that many of Earth's explosive volcanism sources are located near oceans and the effects of volcanic eruptions on ocean ecosystems have only been examined in a few studies. Olgun et al. [80] found that eruptions from Mount Etna, Italy, have contributed appreciable amounts of nitrogen, phosphorus, silicon, iron, and zinc to the Mediterranean Ocean between 2001 and 2007. Durant et al. [75] found enhanced amounts of calcium, sodium, and iron in lakes following the 2008 eruption of Chaitén Volcano, Chile. The iron contribution, in particular, made by volcanic deposits, has been particularly interesting to studies of ocean chemistry. Much of the ocean away from continental boundary regions is iron poor [73]. Deposits of volcanic ash, which becomes coated in minerals, deliver iron, a necessary nutrient for phytoplankton [73, 74], to otherwise iron-poor regions of the ocean. Phytoplankton blooms were noted after the 2008 eruption of Kasatochi Volcano (Alaska) [74], Mount Spurr (Alaska), Arenal Volcano (Costa Rica), and Sakurajima Volcano (Japan) [73]. The phytoplankton blooms were large enough to be seen by MODIS from space [73, 74].

The effect on Earth's ecosystem does not end with increased phytoplankton growth, though. Similar to the effect seen in forests, the large phytoplankton blooms cause a temporary but significant decrease in atmospheric CO_2 [73, 81]. The decrease in CO_2 caused by moderate to large volcanic eruptions is on par with that caused by ENSO and can mitigate or exacerbate the effects of ENSO depending on when during the ENSO cycle an eruption occurs [76, 77]. Modeling studies by Rothenberg et al. [81] and Frölicher et al. [77] both showed that the additional CO_2 uptake results in further global cooling that is particularly strong in the Arctic region.

5. Conclusions

Volcanic aerosol, although sporadically present, can have a profound influence on the Arctic environment. There is a potential for volcanic aerosol to provide a sizeable contribution to the radiative effects and even exceed the impacts of other types of aerosol when significant amounts of volcanic ash are present, as in young volcanic plumes and volcanic ash deposits. The composition, thickness, and AOD of volcanic plumes vary greatly and are, in many cases, difficult to constrain. Meteorological clouds also often hamper satellite retrievals, making transport modeling necessary for plume tracking.

The development of multiphase models (e.g., [82]) to study eruption dynamics may assist in creating better microphysical models for volcanic ash and have the capacity to be modified to treat the formation and transport of other volcanic aerosols, such as sulfates and ice. The deployment of balloons with particle counters attached would also further our understanding of the evolution of aerosol size distributions in plumes, which are hazardous to fly through and difficult to fully characterize with deposit measurements. There is a need for better constrained eruption-specific model input parameters (i.e., MFR, source type, and nonsphericity of ash particles). Discrepancies in modeled and measured loadings tend to be larger farther from the vent, which could indicate issues with wind field resolution or aggregation and comminution processes. Existing aggregation models, such as the one developed by Telling et al. [52], should be incorporated in volcanic ash dispersion and transport models. Additionally, the process of comminution, which typically only occurs in the volcanic conduit, is not considered. These improvements to eruption source and dispersion parameters will provide more accurate ash deposition estimates, which are critical for understanding how volcanic plumes and deposits interact with the Arctic environment. Other plume-related simplifications, which are often made in radiative transfer modeling and which need to be addressed, include the partitioning of volcanic aerosol types at different altitudes and sulfate, water, and ice coatings on ash.

Loadings of ash and the optical properties of ash and snow are important to radiative calculations. Future studies should consider improving ash loading estimates and the acquisition of eruption-specific snow and ash properties for radiative modeling. The calculation of global radiative effects could be made possible by a more intimate coupling of volcanic eruption source conditions, transport, and deposition with radiative transfer and global circulation models. Improvements that could be made through fieldwork include measurements of albedo change and snow ablation rates, which could be done through coordination with local volcano observatories. Alternatively, laboratory simulations on a smaller scale could be conducted when fieldwork is not possible.

Additionally, volcanic aerosols may have strong impacts on ecosystems and the carbon cycle. Because of the important radiative effects of volcanic aerosols in the Arctic, it is recommended that volcanic aerosols be included in future assessments of the Arctic regional radiation budget to facilitate efforts in understanding the radiative impacts of natural aerosols on the Arctic environment. Because of the persistence of cloud cover in this region, the radiative impacts of meteorological clouds in the presence of volcanic plumes and deposits should be considered. The effects of volcanic aerosol on other permanently and seasonally snow-covered environments might also be considered; these include glaciers at lower latitudes (e.g., in the Andes) and in Antarctica.

Author details

Cindy L. Young[1*] and Jennifer W. Telling[2]

*Address all correspondence to: cindy.young@eas.gatech.edu

1 School of Earth and Atmospheric Sciences, Georgia Institute of Technology, Atlanta, GA, USA

2 National Center for Airborne Laser Mapping (NCALM), University of Houston, Houston, TX, USA

References

[1] Stone, R.S., Anderson, G.P., Shettle, E.P., Andrews, E., Loukachine, K., Dutton, E.G., Schaaf, C., Roman, M.O. Radiative impact of boreal smoke in the Arctic: Observed and modelled. J. Geophys. Res. Atmos. 2008;113(D14) pp. 1–17 . DOI: 10.1029/2007JD009657

[2] Stone, R.S., Anderson, G.P., Andrews, E., Dutton, E.G., Shettle, E.P. Incursions and radiative impact of Asian dust in northern Alaska. Geophys. Res. Lett. 2007;34:L14815. DOI: 10.1029/2007gl029878

[3] Simkin, T., Siebert, L. Volcanoes of the World. 2nd ed. Tucson: Geoscience Press; 1994. 369 pp.

[4] Thordarson, T., Self, S. Atmospheric and environmental effects of the 1783–1784 Laki eruption: A review and reassessment. J. Geophys. Res. 2003;108(D1):4011.

[5] Frogner-Kockum, P.C., Herbert, R.B., Gislason, S.R. A diverse ecosystem response to volcanic aerosols. Chem. Geol. 2006;231(1):57–66.

[6] Hildreth, W., Fierstein, J. The Novarupta-Katmai eruption of 1912: Largest eruption of the twentieth century. In: Centennial Perspectives. U.S. Department of the Interior, U.S. Geological Survey; 2012. Reston, Virginia USA, 259 pp.

[7] Rose, W.I., Bluth, G.J.S., Watson, I.M. Ice in volcanic clouds: When and where? In: Proceedings of the 2nd International Conference on Volcanic Ash and Aviation Safety. Washington, DC: OFCM, 2004.

[8] Andersson, S.M., Martinsson, B.G., Friberg, J., Brenninkmeijer, C.A.M., Rauthe-Schöch, A., Hermann, M., van Velthoven, P.F.J., Zahn, A. Composition and evolution of volcanic aerosol from eruptions of Kasatochi, Sarychev and Eyjafallajokull in 2008–2010 based on CARIBIC observations. Atmos. Chem. Phys. 2013;13(4):1781–1796.

[9] Carslaw, K.S., Boucher, O., Spracklen, D.V., Mann, G.W., Rae, J.G.L., Woodward, S., Kulmala, M. Atmospheric aerosols in the earth system: A review of interactions and feedbacks. Atmos. Chem. Phys. 2009;9:10087–11183.

[10] Robock, A. Volcanic eruptions and climate. Rev. Geophys. 2000;38:191–219.

[11] Niemeier, U., Timmreck, C., Graf, H.F., Kinne, S., Rast, S., Self, S. Initial fate of fine ash and sulphur from large volcanic eruptions. Atmos. Chem. Phys. 2009;9(22):9043–9057. DOI: 10.5194/acp-9-9043-2009

[12] Stenchikov, G.L., Kirchner, I., Robock, A., Graf, H.F., Antuna, J.C., Grainger, R.G., Lambert, A., Thomason, L. Radiative forcing from the 1991 Mount Pinatubo volcanic eruption. J. Geophys. Res. Atmos. 1998;103:13837–13857.

[13] Painter, T.H., Deems, J.S., Belnap, J., Hamlet, A.F., Landry, C.C., Udall, B. Response of Colorado River runoff to dust radiative forcing in snow. Proc. Natl. Acad. Sci. U. S. A. 2010;107(40):17125–17130. DOI: 10.1073/pnas.0913139107

[14] Flanner, M.G., Zender, C.S., Randerson, J.T., Rasch, P.J. Present-day climate forcing and response from black carbon in snow. J. Geophys. Res. Atmos. 2007; pp. 1–17 112(D11). DOI: 10.1029/2006JD008003

[15] Driedger, C. Effect of ash thickness on snow ablation. In: Lipman, P.W., Christiansen, R.L., editors. The 1980 Eruptions of Mount St. Helens. Washington, DC, 1981. pp. 757–760.

[16] Dadic, R., Mullen, P.C., Schneebeli, M., Brandt, R.E., Warren, S.G. Effects of bubbles, cracks, and volcanic tephra on the spectral albedo of bare ice near the Transantarctic

Mountains: Implications for sea glaciers on Snowball Earth. J. Geophys. Res.: Earth Surface. 2013;118(3):1658–1676.

[17] Flanner, M.G., Gardner, A.S., Eckhardt, S., Stohl, A., Perket, J. Aerosol radiative forcing from the 2010 Eyjafjallajökull volcanic eruptions. J. Geophys. Res. Atmos. 2014. pp. 9481–9491, Vol. 119 DOI: 10.1002/2014JD021977

[18] Young, C.L., Sokolik, I.N., Flanner, M.G., Dufek, J. Surface radiative impacts of ash deposits from the 2009 eruption of Redoubt Volcano. J. Geophys. Res. Atmos. 2014;119(19) pp. 11387–11397 .

[19] Textor, C., Graf, H.F., Herzog, M., Oberhuber, J.M., Rose, W.I., Ernst, G.G.J. Volcanic particle aggregation in explosive eruption columns. Part I: Parameterization of the microphysics of hydrometeors and ash. J. Volcanol. Geotherm. Res. 2006;150:359–377. DOI: 10.1016/j.jvolgeores.2005.09.007

[20] Lathem, T.L., Kumar, P., Nenes, A., Dufek, J., Sokolik, I.N., Trail, M., Russell, A. Hygroscopic properties of volcanic ash. Geophys. Res. Lett. 2011;38(L11802) pp. 1–4 . DOI: 10.1029/2011GL047298

[21] Liu, X., Penner, J.E. Effect of Mt. Pinatubo H_2SO_4/H_2O aerosol on ice nucleation in the upper troposphere using a global chemistry and transport model (IMPACT). J. Geophys. Res. 2002;107(4141) pp. 1–18 . DOI: 10.1029/2001JD000455

[22] Jager, H., Deshler, T. Lidar backscatter to extinction, mass and area conversions for stratospheric aerosols based on midlatitude balloonborne size distribution measurements. Geophys. Res. Lett. 2002;29 pp. 1–4 . DOI: 10.1029/2002gl015609

[23] Stier, P., Feichter, J., Kinne, S., Kloster, S., Vignati, E., Wilson, J., Ganzeveld, L., Tegen, I., Werner, M., Balkanski, Y., Schulz, M., Boucher, O., Minikin, A., Petzold, A. The aerosol-climate model echam5-ham. Atmos. Chem. Phys. 205;5:1125–1156.

[24] Rose, W.I., Durant, A.J. Fine ash content of explosive eruptions. J. Volcanol. Geotherm. Res. 2009;186(1–2):32–39. DOI: 10.1016/j.jvolgeores.2009.01.010

[25] Prata, A.J., Bernardo, C. Retrieval of volcanic SO_2 column abundance from atmospheric infrared sounder data. J. Geophys. Res. Atmos. 2007;112 pp. 1–17 . DOI: 10.1029/2006JD007955

[26] Draxler, R.R., Rolph, G.D. NOAA Air Resources Laboratory, College Park, MD. HYSPLIT (Hybrid Single-Particle Lagrangian Integrated Trajectory) Model access via NOAA ARL READY Website [Internet]. Available at: http://www.arl.noaa.gov/HYSPLIT.php

[27] Rolph, G.D. Real-time Environmental Applications and Display System (READY) Website. NOAA Air Resources Laboratory, College Park, MD. [Internet]. Available at: http://www.ready.noaa.gov

[28] Telling, J., Flower, V.J.B., Carn, S.A. A multi-sensor satellite assessment of SO_2 emissions from the 2012–13 eruption of Plosky Tolbachik Volcano, Kamchatka. J. Volcanol. Geotherm. Res. 2015;307:98–106. DOI: 10.1016/j.volgeores.2015.07.010

[29] Young, C.L., Sokolik, I.N., Dufek, J. Regional radiative impact of volcanic aerosol from the 2009 eruption of Mt. Redoubt. Atmos. Chem. Phys. 2012;12(8):3699–3715. DOI: 10.5194/acp-12-3699-20.

[30] Wen, S., Rose, W.I. Retrieval of sizes and total masses of particles in volcanic clouds using AVHRR bands 4 and 5. J. Geophys. Res. Atmos. 1994;99(D3):5421–5431.

[31] Rose, W.I., Delene, D.J., Schneider, D.J., Bluth, G.J.S., Krueger, A.J., Sprod, I., McKee, C., Davies, H.L., Ernst, G.G.J. Ice in the 1994 Rabaul eruption cloud: Implications for volcano hazard and atmospheric effects. Nature. 1995;375(477–479).

[32] Hess, M., Koepke, P., Schult, I. Optical properties of aerosols and clouds: The software package OPAC. B. Am. Meteorol. Soc. 1998;79:831–844.

[33] Yue, G.K., Poole, L.R., Wang, P.H., Chiou, E.W. Stratospheric aerosol acidity, density, and refractive-index deduced from SAGE-II and NMC temperature data. J. Geophys. Res. Atmos. 1994;99:3727–3738.

[34] Pollack, J.B., Toon, O.B., Khare, B.N. Optical properties of terrestrial rocks and glasses. Icarus. 1973;19:372–389.

[35] Kearney, C.S., Watson, I.M. Correcting satellite-based infrared sulphur dioxide retrievals for the presence of silicate ash. J. Geophys. Res. Atmos. 2009;114(D22208) pp. 1–12 . DOI: 10.1029/2008jd011407

[36] Fu, Q., Thorsen, T.J., Su, J., Ge, J.M., Huang, J.P. Test of Mie-based single-scattering properties of non-spherical dust aerosols in radiative flux calculations. J. Quant. Spectrosc. Radiat. Transfer. 2009;110(14):1640–1653. DOI: 10.1016/j.jqsrt.2009.03.010

[37] Quinn, P.K., Shaw, G., Andrews, E., Dutton, E.G., Ruoho-Airola, T., Gong, S.L. Arctic haze: Current trends and knowledge gaps. Tellus B. 2007;59:99–114. DOI: 10.1111/j. 1600-0889.2006.00238.x

[38] Ritter, C., Notholt, J., Fischer, J., Rathke, C. Direct thermal radiative forcing of tropospheric aerosol in the arctic measured by ground based infrared spectrometry. Geophys. Res. Lett. 2005;32:L23816. DOI: 10.1029/2005gl024331

[39] Myhre, C.L., Toledano, C., Myhre, G., Stebel, K., Yttri, K.E., Aaltonen, V., Johnsrud, M., Frioud, M., Cachorro, V., De Frutos, A., Lihavainen, H., Campbell, J.R., Chaikovsky, A.P., Shiobara, M., Welton, E.J., Torseth, K. Regional aerosol optical properties and radiative impact of the extreme smoke event in the European Arctic in spring 2006. Atmos. Chem. Phys. 2007;(7):5899–5915.

[40] Stuefer, M., Freitas, S.R., Grell, G., Webley, P., Peckham, S., McKeen, S.A., Egan, S.D. Inclusion of ash and SO_2 emissions from volcanic eruptions in WRF-Chem: Development and some applications. Geosci. Model Dev. 2013;6(2):457–468. DOI: 10.5194/ gmd-6-457-2013

[41] Costa, A., Macedonio, G., Folch, A. A three-dimensional Eulerian model for transport and deposition of volcanic ashes. Earth Planetary Sci. Lett. 2006;241(3–4):634–647. DOI: 10.1016/j.epsl.2005.11.019

[42] Wilson, L., Sparks, R.S.J., Huang, T.C., Watkins, N.D. Control of volcanic column heights by eruption energetic and dynamics. J. Geophys. Res. 1978;83(NB4):1829–1836.

[43] Wallace, K.L, Schaefer, J.R., Coombs, M.L. Character, mass, distribution, and origin of tephra-fall deposits from the 2009 eruption of Redoubt Volcano, Alaska—Highlighting the significance of particle aggregation. J. Volcanol. Geotherm. Res. 2013;259:145–169. DOI: 10.1016/j.jvolgeores.2012.09.015

[44] Mastin, L.G., Schwaiger, H., Schneider, D.J., Wallace, K.L., Schaefer, J., Denlinger, R.P. Injection, transport, and deposition of tephra during event 5 at Redoubt Volcano, 23 March, 2009. J. Volcanol. Geotherm. Res. 2013;259:201–213. DOI: 10.1016/j.jvolgeores.2012.04.025

[45] Schmehl, K.J., Haupt, S.E., Pavolonis, M.J. A genetic algorithm variational approach to data assimilation and application to volcanic emissions. Pure Appl. Geophys. 2012;169(3):519–537. DOI: 10.1007/s00024-011-0385-0

[46] Young, C.L., Dufek, J., Sokolik, I.N. Assessment of depositional ash loading from the 2009 eruption of Mt. Redoubt. J. Volcanol. Geotherm. 2014;274:122–138.

[47] Rose, W.I., Chesner, C.A. Dispersal of ash in the great Toba eruption, 75 ka. Geology. 1987;15(10):913–917. DOI: 10.1130/0091-7613(1987)15<913:DOAITG>2.0.CO;2

[48] Bonadonna, C., Genco, R., Gouhier, M., Pistolesi, M., Cioni, R., Alfano, F., Hoskuldsson, A., Ripepe, M. Tephra sedimentation during the 2010 Eyjafallajokull eruption (Iceland) from deposit, radar, and satellite observations. J. Geophys. Res. Sol. Ea. 2011;116 pp. 1–20. DOI: 10.1029/2011JB008462

[49] Bonadonna, C., Connor, C.B., Houghton, B.F., Connor, L., Byrne, M., Laing, A., Hincks, T.K. Probabilistic modelling of tephra dispersal: Hazard assessment of a multiphase rhyolitic eruption at Tarawera, New Zealand. J. Geophys. Res. Sol. Ea. 2005;110(B3) pp. 1–21 . DOI: 10.1029/2003/JB002896

[50] Schaefer, J.R. The 2009 eruption of Redoubt Volcano, Alaska, with contributions by Bull, K.P., Cameron, C.E., Coombs, M.L., Diefenbach, A.K., Lopez, T., McNutt, S.R., Neal, C.A., Payne, A.L., Power, J.A., Schneider, D.J., Scott, W.E., Snedigar, S.F., Thompson, G., Wallace, K.L., Waythomas, C.F., Webley, P.W., Werner, C.A. Report of Investigations RI 2011-5, State of Alaska, Department of Natural Resources, Division of Geological and Geophysical Surveys, Fairbanks, Alaska. 2011;5:45.

[51] Steensen, T., Stuefer, M., Webley, P., Grell, G., Freitas, S. Qualitative comparison of Mount Redoubt 2009 volcanic clouds using the PUFF and WRF-Chem dispersion models and satellite remote sensing data. J. Volcanol. Geotherm. Res. 2013;259:235–247. DOI: 10.1016/j.jvolgeores.2012.02.018

[52] Telling, J.W., Dufek, J., Shaikh, A. Ash aggregation in explosive volcanic eruptions. Geophys. Res. Lett. 2013;40(10):2355–2360. DOI: 10.1002/grl.50376

[53] Dufek, J., Manga, M., Patel, A. Granular disruption during explosive volcanic eruptions. Nat. Geosci. 2012;5:561–564.

[54] Gislason, S.R., Hassenkam, T., Nedel, S., Bovet, N., Eiriksdottir, E.S., Alfredsson, H.A., Hem, C.P., Balogh, Z.I., Dideriksen, K., Oskarsson, N., Sigfusson, B., Larsen, G., Stipp, S.L.S. Characterization of Eyjafjallajokull volcanic ash particles and a protocol for rapid risk assessment. Proc. Natl. Acad. Sci. U. S. A. 2011;108:7307–7312. DOI: 10.1073/pnas.1015053108

[55] Marzano, F., Barbieri, S., Picciotti, E., Karlsdóttir, S. Monitoring subglacial volcanic eruption using ground-based C-band radar imagery. IEEE Trans. Geosci. Remote Sensing. 2010;48:403–414.

[56] Taddeucci, J., Scarlato, P., Montanaro, C., Cimarelli, C., Del Bello, E., Freda, C., Andronico, D., Gudmundsson, M.T., Dingwell, D.B. Aggregation-dominated ash settling from the Eyjafjallajkull volcanic cloud illuminated by field and laboratory high-speed imaging. Geology. 2011;39(9):891–894. DOI: 10.1130/G32016.1

[57] Gilbert, J.S., Lane, S.J. The origin of accretionary lapilli. Bull. Volcanol. 1994;56:398–411. DOI: 10.1007/BF00326465

[58] Gilbert, J.S., Lane, S.J., Sparks, R.S.J., Koyaguchi, T. Charge measurements on particle fallout from volcanic plume. Nature. 1991;349:598–600. DOI: 10.1038/349598a0

[59] Costa, A., Folch, A., Macedonio, G. A model for wet aggregation of ash particles in volcanic plumes and clouds: 1. Theoretical formulation. J. Geophys. Res. 2010;115 pp. 1–14. DOI: 10.1029/2009JB007175

[60] Warren, S.G., Brandt, R.E., Grenfell, T.C. Visible and near-ultraviolet absorption spectrum of ice from transmission of solar radiation into snow. Appl. Opt. 2006;45(21): 5320–5334.

[61] Painter, T.H., Dozier, J., Roberts, D.A., Davis, R.E., Green, R.O. Retrieval of subpixel snow-covered area and grain size from imaging spectrometer data. Remote Sensing Environ. 2003;85(1):64–77.

[62] Clarke, A.D., Noone, K.J. Soot in the Arctic snowpack: A cause for perturbations in radiative transfer. Atmos. Environ. 1985;19(12):2045–2053.

[63] Wang, L., Li, Z., Tian, Q., Ma, Y., Zhang, F., Zhang, Y., Li, D., Li, K., Li, L. Estimate of aerosol absorbing components of black carbon, brown carbon, and dust from ground-based remote sensing data of sun-sky radiometers. J. Geophys. Res. Atmos. 2013;118:6534–6543. DOI: 10.1002/jgrd.50356

[64] Skiles, S.M., Painter, T.H., Deems, J.S., Bryant, A.C., Landry, C.C. Dust radiative forcing in snow of the Upper Colorado River Basin: 2. Interannual variability in radiative forcing and snowmelt rates. Water Resources Res. 2012;48(7) pp. 1–11 .

[65] Smith, A.J.A., Grainger, R.G. Does variation in mineral composition alter the short-wave light scattering properties of desert dust aerosol? J. Quant. Spectrosc. Radiat. Transfer. 2014;133:235–243. DOI: 10.1016/j.jqsrt.2013.08.005

[66] Zdanowicz, C.M., Zielinski, G.A., Wake, C.P. Characteristics of modern atmospheric dust deposition in snow on the Penny Ice Cap, Baffin Island, Arctic Canada. Tellus B. 1998;50(5):506–520. DOI: 10.1034/j.1600-0889.1998.t01-1-00008.x

[67] Melloh, R.A. A synopsis and comparison of selected snowmelt algorithms. Hanover, NH: Cold Regions Research and Engineering Lab; 1999. No. CRREL-99-8.

[68] Qian, Y., Gustafson, W.I., Leung, L.R., Ghan, S.J. Effects of soot-induced snow albedo change on snowpack and hydrological cycle in western United States based on Weather Research and Forecasting chemistry and regional climate simulations. J. Geophys. Res. Atmos. 2009;114(D3) pp. 1–19 . DOI: 1029/2008JD011039

[69] Sokolik, I.N., Curry, J.A., Radionov, V. Interactions of Arctic aerosols with land-cover and land-use changes in northern Eurasia and their role in the Arctic climate system. In: Gutman, G., Reissell, A., editors. Arctic Land-Cover and Land-Use in a Changing Climate: Focus on Eurasia. Springer; New York, USA, 2011.

[70] Anisimov, O.A., Vaughan, D.G., Callaghan, T.V., Furgal, C., Marchant, H., Prowse, T.D., Vilhjálmsson, H., Walsh, J.E. Polar regions (Arctic and Antarctic). In: Parry, M.L., Canziani, O.F., Palutikof, J.P., Van der Linden, P.J., Hanson, C.E., editors. Climate Change 2007: Impacts, Adaptation and Vulnerability. Contribution of Working Group II to the Fourth Assessment Report of the Intergovernmental Panel on Climate Change. Cambridge University Press; Cambridge, England, 2007. pp. 653–685.

[71] Lohmann, U., Karcher, B., Timmreck, C. Impact of the Mount Pinatubo eruption on cirrus clouds formed by homogeneous freezing in the ECHAM4 GCM. J. Geophys. Res. Atmos. 2003;108(4568) pp. 1–13 . DOI: 10.1029/2002jd003185

[72] Yadav, R. Basin specificity of climate change in western Himalaya, India: Tree-ring evidences. Curr. Sci. 2007;92(10):1424–1429.

[73] Duggen, S., Croot, P., Schacht, U., Hoffmann, L. Subduction zone volcanic ash can fertilize the surface ocean and stimulate phytoplankton growth: Evidence from biogeochemical experiments and satellite data. Geophys. Res. Lett. 2007;34(L01612) pp. 1–5 . DOI: 10.1029/2006GL027522

[74] Langmann, B., Zaksek, K., Hort, M., Duggen, S. Volcanic ash as fertilizer for the surface ocean. Atmos. Chem. Phys. 2010;10:3891–3899.

[75] Durant, A.J., Villarosa, G., Rose, W.I., Delmelle, P., Prata, A.J., Viramonte, J.G. Long-range volcanic ash transport and fallout during the 2008 eruption of Chaitén Volcano, Chile. Phys. Chem. Earth. 2012;45–46:50–64. DOI: 10.1016/j.pce.2011.09.004

[76] Jones, C.D., Cox, P.M. Modeling of the volcanic signal in the atmospheric CO_2 record. Global Biogeochem. Cycles 2001;15(2):453–465. DOI: 10.1029/2000GB001281

[77] Frölicher, T.L., Joos, F., Raible, C.C., Sarmiento, J.L. Atmospheric CO_2 response to volcanic eruptions: The role of ENSO, season, and variability. Global Biogeochem. Cycles 2013;27:239–251. DOI: 10.1002/gbc.20028

[78] Oppenheimer, C. Climatic, environmental and human consequences of the largest known historic eruption: Tambora Volcano (Indonesia) 1815. Prog. Phys. Geog. 2003;27(2):230–259. DOI: 10.1191/0309133303pp379ra

[79] Gu, L., Baldocchi, D.D., Wofsy, S.C., Munger, J.W., Michalsky, J.J., Urbanski, S.P., Boden, T.A. Response of a deciduous forest to the Mount Pinatubo eruption: Enhanced photosynthesis. Science. 2003;299(5615):2035–2038.

[80] Olgun, N., Duggen, S., Andronico, D., Kutterolf, S., Croot, P.L., Giammanco, S., Censi, P., Randazzo, L. Possible impacts of volcanic ash emissions of Mount Etna on the primary productivity in the oligotrophic Mediterranean Sea: Results from nutrient-release experiments in seawater. Mar. Chem. 2013;152:32–42. DOI: 10.106/j.marchem. 2013.04.004

[81] Rothenberg, D., Mahowald, N., Lindsay, K., Doney, S.C., Moore, J.K., Thornton, P. Volcano impacts on climate and biogeochemistry in a coupled carbon-climate model. Earth Syst. Dyn. 2012;3:121–136. DOI: 10.5194/esd-3-121-2012

[82] Dufek, J., Bergantz, G.W. Suspended load and bed-load transport of particle-laden gravity currents: The role of particle-bed interaction. Theor. Comp. Fluid Dyn. 2007;21:119–145. DOI: 10.1007/s00162-007-0041-6

4

Submarine Volcanism of the Cabo de Gata Magmatic Arc in the Betic-rif Orogen, SE Spain: Processes and Products

Carles Soriano, Ray A.F. Cas, Nancy R. Riggs and Guido Giordano

Abstract

Volcanic eruptions in subaqueous settings have been traditionally characterized by the study of ancient deposits and, more recently, by indirect observation of the sea floor with different geophysical means. Subaqueous volcanism is largely governed by the physical properties of water and the way water interacts with magma. Among the characteristic products of subaqueous volcanism are hyaloclastite breccias of dense clasts and of pumiceous clasts produced by the quench fragmentation of hot magma in effusive eruptions. Pumice breccias driven by fragmentation of magma in explosive eruptions are not infrequent. The Miocene volcanic zone of Cabo de Gata in southeastern Spain provides excellent exposures where to test the current understanding on subaqueous volcanism. In particular, submarine lavas with a coherent core and an outer carapace of vesicular hyaloclastite together with pumice breccias and crystal tuffs of the El Barronal Formation provide clues to understand transient conditions during explosive and effusive eruptions. Debris avalanches deposits are rather common in Cabo de Gata, such as those of the Los Frailes Formation and the Cerro Estorvillas Formation, and help to understand the instability processes of submarine volcanic edifices and the resultant mass flows. Interbedding of volcanic rocks with shallow water sedimentary rocks allows inferring water depth conditions for volcanism and the subsidence history of the volcano-sedimentary basin.

Keywords: hyaloclastite, submarine explosions, debris-avalanche, subsidence, infralittoral

1. Introduction

The Cabo de Gata volcanic zone is located in southeastern Spain and is a portion of the Neogene volcanic arc of the Betic-rif Orogen, an arcuate orocline formed by westward retreat of an east dipping slab Orogen, (**Figure 1**). The zone includes a wide variety of volcanic facies, ranging from explosive to effusive and recording transient conditions during eruptions. Volcanic rocks of Cabo de Gata are calc-alkaline in composition and were erupted from submarine vents and deposited in submarine settings, although rare facies record the transition to subaerial settings. Therefore, volcanism in Cabo de Gata is essentially submarine, although some volcanic edifices and products may have risen from above sea level. In terms of rock chemistry, eruptive style, age and depositional setting, volcanic successions in Cabo de Gata can be compared with similar examples from the volcanic arc of Japan. The Cabo de Gata volcanic zone constitutes the best exposed, best preserved and most voluminous record of the volcanic arc of the Betic-Rif Orogen and is certainly the most important record of submarine volcanism in the western Mediterranean area.

Despite its significance, it has not been until very recently that systematic studies with a modern facies to processes approach were undertaken in the Cabo de Gata volcanic zone and for this reason this area has remain virtually unknown for most volcanologists. Previous studies were mainly focused on petrological aspects of the volcanic rocks and on the Rodalquilar gold epithermal deposits associated with volcanic rocks, without providing a comprehensive stratigraphic framework aimed to understand the processes and products of Cabo de Gata volcanism [1–6]. Comparison of Cabo de Gata volcanic rocks with other Miocene volcanic zones such as the Iblean Mountains in Sicily, Sulcis in Sardegna and volcanic islands in the Aegean Sea may provide a more comprehensive view on volcanism of similar age occurring in different geotectonic scenarios of the narrow Mediterranean area.

Here, we first present an introductory section on the principles of subaqueous volcanism and then a review of our research undertaken during the last five years in the Cabo de Gata volcanic zone. In particular, we focus on those facies and successions of facies that reveal transient conditions of eruptive styles during eruptions and on submarine volcanic debris-avalanche deposits, which constitute a unique opportunity for a direct study of this kind of deposits in the geological record. We also focus on the facies and the stratigraphic successions of sedimentary rocks interbedded with volcanic rocks, aiming to characterize the bathymetric conditions for sediment deposition and, indirectly, for volcanism. Using the evidence provided by facies analyses we typify the characteristic eruptive models in Cabo de Gata and their link to the tectonics of the Neogene volcano-sedimentary basin in which volcanic rocks are exposed.

2. Basic principles of subaqueous volcanism

Understanding volcanic processes in subaqueous environments has always been challenging because first, in most cases, they cannot be directly observed, and second the physical properties of water and the way that water interacts with erupting or intruding magma are still not fully understood. Despite many contributions to the subject, the major debate still seems to revolve around the factors that relate water depths and explosive eruptions.

In modern environments, one of the greatest limitations has been the inability to observe processes and seafloor volcanic topography as well as the distribution of different facies or deposit types at scales that provide understanding of the characteristics, extent, scale and relations of processes and deposits. Local submersible and camera observations have been helpful, but are very limiting. However, recent advances in imaging technology, such as high-resolution side-scan sonar and bathymetry technology mounted on autonomous underwater vehicles (AUVs) such as Sentry, allow very high-resolution Digital Elevation Model imagery to be produced over large areas and whole submerged volcanoes that will open up a new world of understanding of subaqueous volcanic processes. In addition, new generation photographic videoing and sampling facilities on remotely operated vehicles (ROVs) such as Jason and Medea provided hitherto unparalleled opportunities to learn about modern sea floor processes, including volcanism.

In ancient volcanic successions, the principal approach for understanding volcanological processes and settings remains detailed mapping of the relations between facies and analysis of their characteristics to assess original water depths and the process origins of the facies. This has given rise to an ongoing need to quantify the physical processes associated with subaqueous eruptions, including the constraints on processes imposed by subaqueous environments, in order to understand the origins of different deposit types.

The most recent review of submarine volcanic processes has been by [7] (see for a listing of other reviews and major research contributions) in which they provided a summary of the essential physical properties of water and magma, as well as a conceptual overview of the processes and deposit characteristics. To illustrate the concepts, they cited examples from the Cabo de Gata volcanic succession that has been well described by [8–12] and is summarized below in this paper. Recently, [13] quantified many of the physical processes involved in the formation of volcanic glass and the quench fragmentation of volcanic glass in subaqueous settings based on the materials-science literature. These authors then related the characteristics of hyaloclastite deposits to those physical processes and rates of processes. In this section, we briefly review the main processes, principles and deposit types in subaqueous environments.

The major difference between subaerial volcanism and subaqueous volcanism is the aqueous medium in the latter and the different ways that magma and water can interact to influence the style of eruption and the types of deposits. Water has a significant impact on the cooling rates of magma that is erupted into it, as well as creating ambient confining-pressure constraints on the way that exsolving magmatic volatiles and superheated water behave. Water is a high-enthalpy substance. It has a high heat capacity (4.187 kJ/kg K) and high thermal conductivity (0.58 W/m Kelvin at 25°C for water and 0.61 W/mK at 25°C, and 0.68 W/mK at 120°C for seawater [14]). Water is therefore very thermodynamically responsive to changes in temperature, which allow it to readily absorb and release heat transmitted to it by magma, via conduction and radiative-heat transfer. This is particularly important when magma at temperatures from 700 to 1200°C comes into contact with liquid water (<20°C). Water acts as a heat sink and can cause the magma to instantaneously cool through the glass transition temperature, causing glass to form. If the rate of heat loss or cooling rate is high, the magma/glass is thermally shocked, causing contractional tensile stresses to form in the glass. If these exceed the tensile strength of the glass, it shatters in situ, leading to a formation of a network of contractional fractures that propagate inwards from the cooling margin of the magma body.

This process is called quench fragmentation and the breccia of glass debris that forms is called hyaloclastite [7, 13]. This process can occur at any water depth.

However, water can very quickly change state over a limited temperature range (0–100°C) when superheated by contact with magma (700–1200°C) and over a range of external pressures that are determined by water depth. The hydrostatic pressure gradient in a body of water is 1 bar (0.1 MPa)/10 m water depth. In particular, liquid water can readily transform into steam, which has a much lower thermal conductivity (0.016 W/m K at 125°C) and specific heat capacity (1.996 kJ/kg K [15]). If this occurs at the interface between magma and water in a process called film boiling (also known as the Leidenfrost Effect), a vapour film forms at the interface, which can insulate the magma from heat loss. Sustained and efficient film boiling can thus reduce the rate of heat loss or cooling of the magma, moderate the rate of contraction, and so minimize thermal tensile stresses in the cooling glass. As a result, the magma may remain coherent and be preserved as a coherent lava or a coherent intrusion in the case where magma intrudes water-saturated sediments.

The super-heating of liquid water at low confining pressures causes a phase transformation to steam and a consequent major volume expansion. At high, instantaneous, rates of expansion such transformations are explosive, driving phreatic and phreatomagmatic volcanic explosions. This is most likely to occur at shallow water depths up to several hundred metres [16].

However, at increasing water depths and hydrostatic pressures the hydrostatic pressure approaches the critical pressure of water. The critical point of water is the pressure (and temperature) at which there is no distinction between the liquid and the gaseous phases; the fluid is called a supercritical fluid and has the properties of both the gaseous and the liquid forms. For pure water, the critical pressure is ~218–221 bars or 21.8–22.1 MPa and for seawater, it is about 300 bars or 30 MPa [17]. Since the hydrostatic pressure gradient in water is 1 bar or 0.1 MPa/10 m water depths, at water depths of 2 or more kms steam that forms at the interface between magma and the water mass is highly compressed and cannot expand explosively unless it can be instantaneously decompressed, which is difficult to do at those water depths. The modern seafloor record and the geological record tell us that voluminous explosive eruptions (cf. subaerial plinian eruptions) that occurred at water depths of more than 1 km are rare, although there are many examples of deposits of variably vesiculated pumiceous debris that originated in shallower water but were then resedimented into deeper water by various processes.

On this point, however, it is important to stress that not all pumice is explosive in origin. Coherent pumice domains can be well developed in subaerial lavas, formed through slow volatile exsolution and vesicle growth at nonexplosive rates in lavas with a subcritical volatile content to drive explosive eruption. In relatively deep subaqueous environments, the hydrostatic confining pressure will hinder both the initial exsolution of volatiles and the vesicle-growth rates. Growth rates may, however, be high enough to form well-vesiculated lavas with coherent pumice domains, but not high enough to drive explosions [7, 13]. Quench fragmentation of such coherent pumice domains can produce large volumes of pumice hyaloclastite.

Another factor that suppresses the intensity of submarine explosive eruptions is the bulk modulus of water, which for pure water is 2.15 × 109 Pa and for seawater is 2.34 × 109 Pa. The bulk modulus of a fluid is a measure of its resistance to uniform compression, i.e., its compressibility or the degree to which it deforms. In comparison, the adiabatic bulk modulus of air is orders of magnitude less, at 1.42 × 105 Pa. Because water is much less compressible and deformable than air, explosions into water are much more suppressed, restricted and much less intense than in air [7].

We now focus on the principal findings of [13] about the principles involved in the fracture behavior of volcanic glass, based on assessment of the ceramics and glass literature in materials science.

- Cooling contraction cracks propagate perpendicular to the cooling surfaces, which initially are the margins of the magma body at its interface with water or water-saturated sediments.
- The distance between cooling cracks depends on the cooling rate, the temperature differential and gradient between magma and water and the level of thermal stress. This determines the first-order grain size of the resultant hyaloclastite breccia and explains why grain size is smaller at the margins of a magma body compared with the interior, where cooling rates and temperature gradients are lower.
- Once a cooling crack forms, it also becomes a cooling surface and other cracks will form perpendicular to it and then others perpendicular to the second-order fractures, etc.
- The spacing between propagating cracks is usually equal to the length of the crack.
- When a crack approaches another crack, it will try to change direction and approach at right angles.
- When a crack intersects another crack, it will usually stop because it cannot propagate across the space.
- Crack formation results in release of both thermal and mechanical energy and cracks may stop if the tensile stress falls below the strength of the glass, which increases as glass cools.
- Crack shape, and thus fragment shape, depends on cooling rate, the crack propagation velocities and the orientation of the cooling surfaces. At high crack velocities, which occur when the temperature differential is high, cracks tend to be straight and form under what is called the critical crack regime. Crack velocities can be >600 ms^{-1}. At crack velocities greater than 0.25 times the speed of sound in glass, cracks may bifurcate and form wedge to splinter-shaped fragments. Low-temperature differentials lead to low crack velocities (subcritical regime (<cm s^{-1}) and cracks may become curviplanar.
- Water not only plays a physical role in facilitating crack propagation by causing cooling contraction tensile stresses at the crack tip but also plays a chemical role in breaking chemical bonds in the glass in a propagating crack tip.
- Heterogeneous glass, with dispersed crystals, lithic fragments and vesicles, experiences complex fracture formation because of the variable properties of the various components of the glass.

- Glass with abundant dispersed crystals is stronger than aphyric glass, is more resistant to quench fragmentation and, should it fragment, it is likely to form coarse hyaloclastite breccia.
- The presence of abundant vesicles in glass weakens it and makes it more prone to quench fragmentation.
- Pumice hyaloclastite formed through nonexplosive quench fragmentation should be quite common in submarine settings. If the deposit is still in situ, it may still preserve jigsaw fit textures. However, if it is resedimented, it should be bedded and preserve characteristics of a variety of subaqueous mass flow processes (e.g., grain-flow, turbidity currents, debris flow, debris avalanches and rafted suspension transport).

These principles and processes help to explain many of the characteristics of quench-fragmented hyaloclastite deposits, as described above, but also including the following:

- Equant, blocky clast aggregates are likely to have formed under steady, slow, subcritical crack-propagation conditions. Equant clasts with curviplanar margins are likely to have formed at low temperatures and low temperature differentials.
- Elongate, splinter- and wedge-shaped clasts are likely to have formed under high crack velocity in critical crack-growth regimes at high temperature differentials.
- Perlite cracking may be a product of thermal stressing and quenching under slow-cooling, subcritical conditions.
- Fine hyaloclastite breccia is the product of a fast cooling rate and high temperature differential, which is why hyaloclastite grainsize may vary from fine at the margins of a body to coarser in the interior and even grade into a coherent interior.
- Jig-saw fit textures are indicative of in situ, nonexplosive, cooling contraction quench fragmentation.
- Clast-rotated textures indicate postcrack formation movement as a result of continued lava or intrusion movement or downslope resedimentation.
- The grainsize of quench-fragmented crystalline glass hyaloclastite is likely to be coarser than for aphyric glass hyaloclastite.
- Pumice hyaloclastite should not be unusual in subaqueous settings and not all pumice is pyroclastic in origin.

In the following sections, we illustrate how some of the processes of subaqueous volcanism, in particular those regarding magma fragmentation and transport and deposition of volcanic debris, are recorded in the submarine volcanic rocks of Cabo de Gata.

3. Geological setting of Cabo de Gata

The Betic-Rif is a complex orogenic system whose geodynamic evolution has been widely debated (e.g., for recent reviews [18–20]). This complexity partly arises from the need to reconcile the structures and kinematics of geodynamic processes operating at different scales and times in the narrow area of the western Mediterranean with the large-scale and long-standing geodynamic scenario of African and Eurasian plate convergence.

The Betic-Rif Orogen is an arcuate mountain belt with a northern branch formed by the Betic cordillera in the Iberian Peninsula and a southern branch formed by the Rif chain in northern Africa. The orogen displays a tight arc with a westward concavity in the Gibraltar strait (**Figure 1**). Igneous rocks of tholeitic, calc-alkaline, shoshonitic and ultrapotassic compositions show enrichment in K with time and are exposed in the internal areas of the belt [19, 21, 22]. Igneous rocks are subduction-related in a broad sense and are partly coeval with the extension tectonics developed in the internal areas of the Betic-Rif during Neogene. Calc-alkaline rocks in particular, display a clear arc-related geochemical signature and are synchronous with the opening of the back-arc Alborán basin (**Figure 1**). Arc curvature, arc magmatism and back-arc extension were formed during Miocene time by the westward rollback of a narrow eastward-subducting slab fragmented from the African plate [20, 22, 23]. Westward retreat of the subducted slab was accompanied by thinning of the continental crust, formation of oceanic crust and extension, leading to the formation of Neogene intramontane basins in the internal areas of the Betic-Rif [20, 22, 24]. Extension ceased in Late Miocene and deformation in the internal areas of the Betic-Rif was dominated by the kinematics of African and Eurasian plate convergence. NNW convergence of Africa with Iberia compressed the Alborán basin and reactivated suitably oriented faults in the intramontane basins. The internal areas of the Betic-Rif became a diffuse plate boundary between Africa and Eurasia dominated by wrench tectonics processes that are still active, as shown by recent seismicity of the Carboneras fault and other similar strike-slip faults [20, 22, 25].

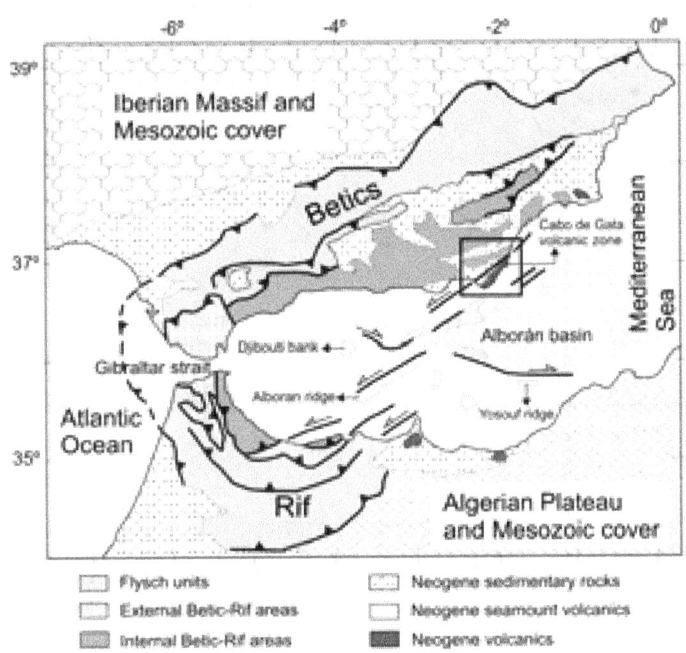

Figure 1. Geologic map of the Betic-Rif Orogen in the Western Mediterranean with location of the Cabo de Gata volcanic zone (modified from [24, 26]). Box shows location of **Figure 2**.

Due to the complex geodynamic evolution of the Betic-Rif Orogen and to the wrench tectonics established in the internal areas since Late Miocene time, the volcanic arc of the Betic-Rif appears distributed along fault-bounded seamounts in the Alborán Sea and onshore in volcano-sedimentary basins on both branches of the mountain belt (**Figure 1**). The Cabo de Gata volcanic zone corresponds to a portion of this volcanic arc exposed in the eastern part of the Almería-Níjar basin (**Figure 2**). This basin is structurally controlled by the Carboneras fault, which divides the basin into a western through dominated by Late Miocene to Pleistocene sedimentary rocks and an eastern through dominated by Miocene volcanic rocks (**Figure 2**). The Carboneras fault is an active structure with a complex kinematic history that includes contemporaneous strike-slip and dip-slip movements since Early Miocene time [27–29]. Basement to the Neogene succession of the Almería-Níjar basin are Paleozoic to Jurassic metamorphic rocks of the Maláguide, Alpujárride and Nevado-Filábride complexes that crop out in the mountain ranges bounding the basin (Sierra Cabrera, Sierra Alhamilla) and along the Carboneras fault strands (**Figure 2**). Gravimetric data and magnetic modeling suggest that volcanic rocks underlie Late Miocene to Pleistocene sedimentary rocks in the western part of the Almería-Níjar basin. The residual gravimetric anomaly together with well log data suggest that Neogene infill of the basin increases to the southwest (SW) and that the Neogene-basement boundary dips to the SW [30, 31]. The structure of the Almería-Níjar basin, of other intramontane basins of the Betic-Rif and of submarine promontories in the Alborán Sea consists of low-amplitude open folds with kilometer-scale wavelengths related to stepped normal faults, which are in turn associated with regional strike-slip faults [30, 32, 33].

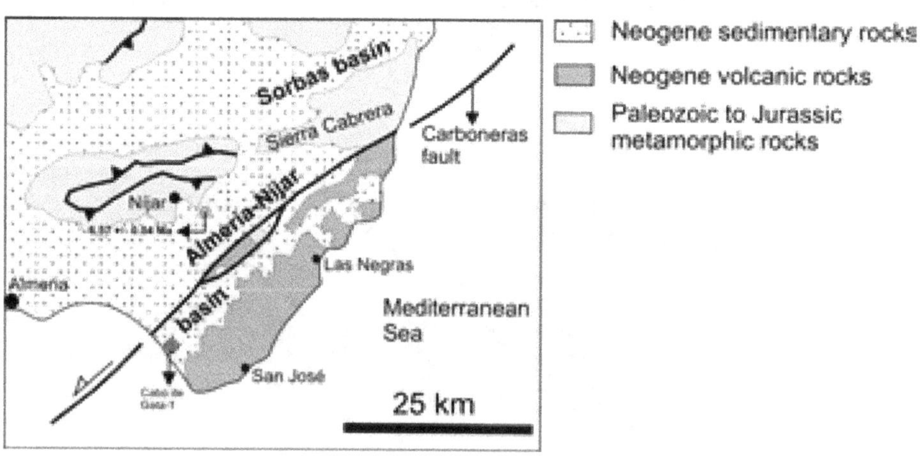

Figure 2. Simplified geologic map of the Almeria-Níjar basin and the Cabo de Gata volcanic zone (modified from [34]). Ar^{40}/Ar^{39} age corresponds to the Cerro del Hoyazo dacite [35].

4. Stratigraphy, geochronology, and structure of the Cabo de Gata volcanic zone

The Cabo de Gata volcanic zone is an area of semiarid climate in which rock exposure is virtually continuous both along the shoreline and the inland. Maximum elevation is nearly

500 m above sea level and the original morphologies of volcanic edifices are poorly preserved due to erosion. Nevertheless, excellent exposures of stratigraphic sections along marine cliffs and inland allow a precise reconstruction of the volcanic stratigraphy. Based on preserved exposures of deposits, Cabo de Gata can be understood as a volcanic field formed by dispersed small-scale lava domes and larger dome complexes and by larger volcanic seamounts. The stratigraphic succession of Cabo de Gata consists of volcanic rocks interbedded with sedimentary rocks, mainly carbonate and siliciclastic deposits. Bedding of volcanic and sedimentary rocks, dip of tabular lavas and overall disposition of volcanic bodies are subhorizontal or shallowly inclined to the northeast (NE), providing a general upward stratigraphic polarity toward the NE.

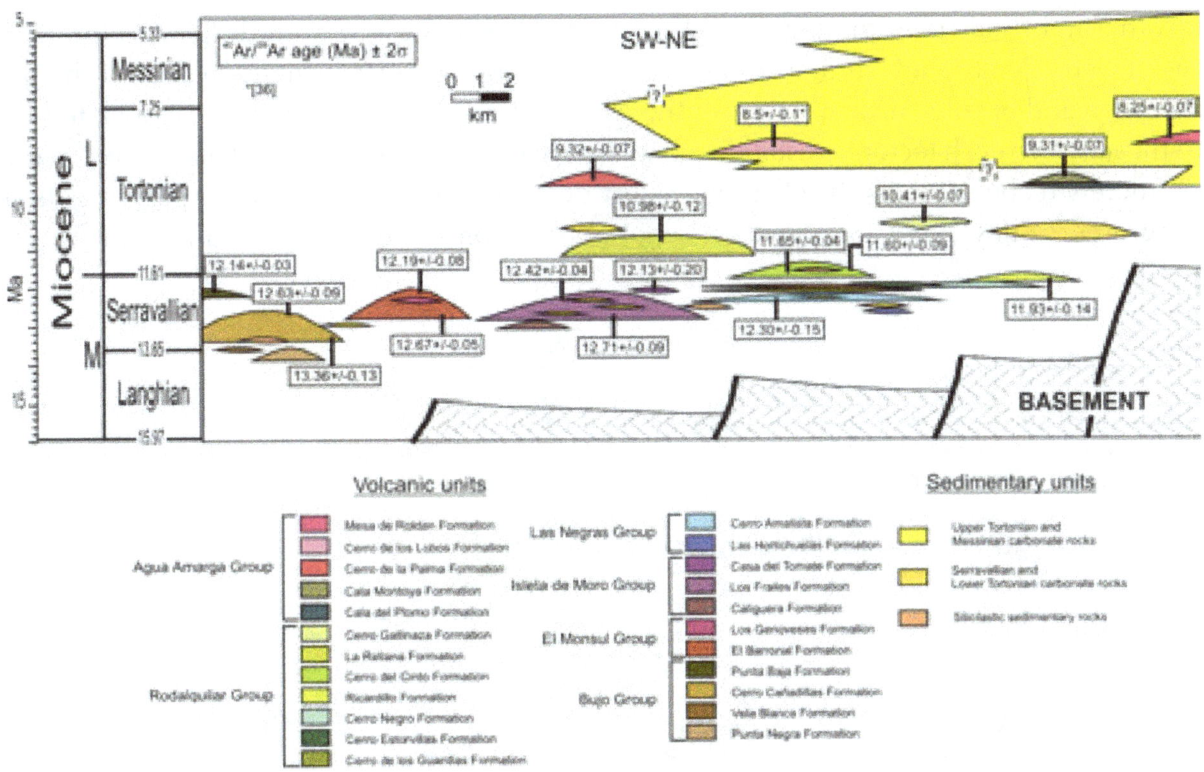

Figure 3. Chronostratigraphic scheme of the Cabo de Gata volcanic zone with distribution of the different lithostratigraphic units (modified from [10]). All Ar^{40}/Ar^{39} ages from [22] except [36].

The volcanic stratigraphy shown in **Figure 3** is a revision of that in [10] with some minor changes regarding the aggregation of formations into groups, in particular the Agua Amarga and Rodalquilar groups. Description, discussion and extended data on the Ar^{40}/Ar^{39} ages of formations are available in [22]. The lithostratigraphic units distinguished are formations formally defined that were subsequently divided into informal subunits when required [10]. The criteria used to divide the stratigraphic succession of Cabo de Gata into formations and to aggregate them into groups is based on lithology, stratigraphic position, age and geochemical affinity. Formations, groups and the informal units within formations are bounded by unconformities of different hierarchy. Volcanic units are interbedded with sedimentary units throughout the Cabo de Gata region with an overall trend of thicker and older volcanic units toward the SW and thicker and younger sedimentary units toward the NE (**Figure 3**). Upper Tortonian to Messinian sedimentary rocks cap the volcano-sedimentary succession toward the NE. Sedimentary units are laterally discontinuous, in particular toward the SW of Cabo de Gata, which may result in the local amalgamation of thick piles of volcanic rocks separated by unconformities.

The main structural features in Cabo de Gata are subvertical normal faults that affect volcanic and sedimentary units and can be grouped into three sets based on their orientation: dominant NW-SE- to N-S-trending faults and subordinate NE-SW-trending faults (**Figure 4**). Fanning stratal dips and differences in bed thickness across fault blocks are thought to indicate that displacement along some faults is syndepositional to the emplacement of volcanic and sedimentary rocks [32]. Epithermal ore deposits of the Rodalquilar mine area result from hydrothermal fluids precipitated at high temperatures along N-S-trending faults [6]. Structures in Cabo de Gata are similar to those of other Neogene intramontane basins of the Betic Rif Orogen and of volcanic seamounts of the Alborán Sea. These structures are associated with the still active wrench tectonics established in the area since Late Miocene. This tectonic regime is ultimately responsible for the observed uplift-downlift displacements along some faults and for the present-day exposure above sea level of the submarine volcano-sedimentary successions of Cabo de Gata.

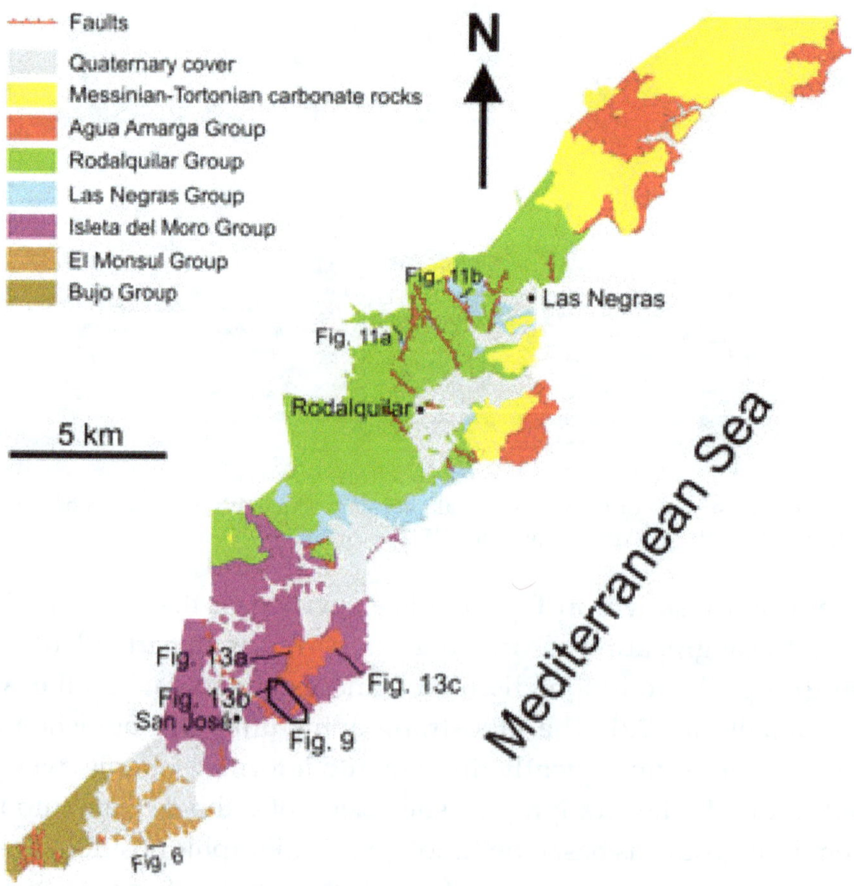

Figure 4. Simplified geologic map of the Cabo de Gata volcanic zone with the distribution of groups and location of figures (modified from [22]).

The chronostratigraphic position of volcanic units based on radio-isotopic ages indicates that volcanism in Cabo de Gata started in Serravalian and ended in Tortonian times (**Figure 3**). During this time, volcanism migrated toward the NE while hiatuses in the volcanic activity became increasingly significant (**Figure 5**). Geophysical and well-log data reveal the occurrence of volcanic rocks beneath the oldest dated unit of Cabo de Gata [30, 31]. Hence, given the Messinian age of the Cerro del Hoyazo dacite [35], the volcanic activity in the whole Almeria-Níjar basin may have encompassed Middle to Late Miocene time (**Figure 5**). Volcanic formations and informal subunits within them correspond to volcanic cycles in a wide sense, meaning that they include all the deposits associated with the volcanic activity during an eruptive period. The volcanic units are separated from each other either by distinct unconformities or by sedimentary rocks, indicating hiatuses in the volcanic activity and erosion of volcanic edifices and deposits. In the following sections, the relevant features of Cabo de Gata volcanism are illustrated with selected examples of deposits and stratigraphic successions.

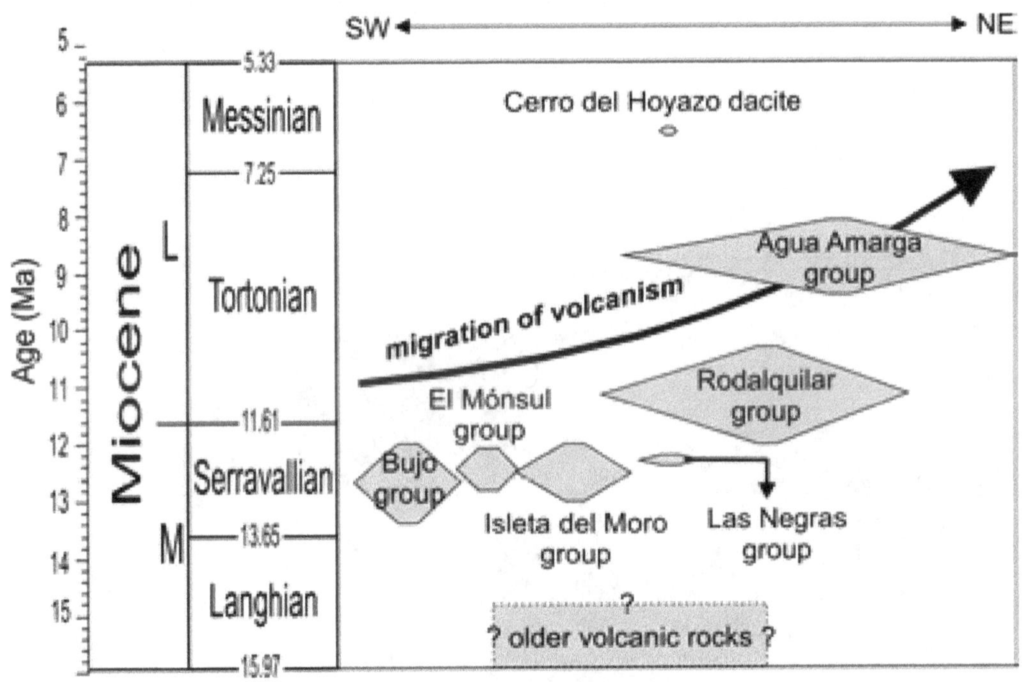

Figure 5. Chronologic timetable of the volcanic activity in the Almería-Níjar basin based on Ar^{40}/Ar^{39} ages of dated rocks approximately distributed from SW-NE (modified from [22]).

5. Transitions between explosive and effusive conditions during submarine eruptions

Transitions from explosive to effusive activity or vice versa are common during individual eruptions in most subaerial volcanoes (Teide, Soufrière Hills, Mount St. Helens, Monte Pilato, etc.). Transient conditions during eruptions have been either witnessed in active volcanoes or are well documented through the study of volcanic deposits. In subaqueous settings, however, their study is certainly more problematic and the usually incomplete and fragmental nature of many submarine successions hampers a proper understanding of how these transitions are recorded. Here, transitions from explosive to effusive eruptive style during individual eruptions are illustrated with examples from the volcanic succession of the El Barronal Formation in southwestern Cabo de Gata (**Figure 4**). An extended description of the lihofacies succession of this unit is available in [9].

The El Barronal Formation consists of a succession of lavas interbedded with volcaniclastic rocks that is intruded by subvertical dikes (**Figure 6**). The lower boundary of this unit is an unconformity with the Cerro Cañadillas Formation while the upper limit is not exposed. The whole El Barronal Formation is composed of up to five lava units. The El Barronal Formation has been Ar^{40}/Ar^{39}-dated at 12.19 Ma for lava unit 4 and 12.67 Ma for lava unit 3 [9]. Interbedded between these two lava units is the rhyolitic dome of the Los Genoveses Formation and also carbonate rocks with shell fragments and polymictic siliciclastic rocks with rounded pebbles from the metamorphic basement of the Neogene succession of Cabo de Gata (**Figure 3**).

Figure 6. Panoramic view of the stratigraphic succession of the El Barronal Formation at Playa del Barronal. Note lava unit 3 pinching out into bedded volcaniclastic succession to the west (see **Figure 4** for location).

Lavas, volcaniclastic rocks and dikes of the El Barronal Formation have identical andesitic composition and mineralogy. Lavas are formed by a coherent core grading into an outer carapace of in situ to clast-rotated hyaloclastite breccia. The coherent core of lavas has colonnade columnar joints and entablature columnar joints usually forming rosette structures (**Figure 7A**). The outer carapace consists of massive hyaloclastite with dense clasts grading outward into massive and flow-banded hyaloclastite with vesicular clasts (~40% vesicles). Volcaniclastic rocks are bedded and consist of massive breccia with dense and vesicular clasts, diffusely bedded pumice-rich breccia, cross-bedded crystal-rich sandstone and thinly bedded fine tuffaceous sandstone. These facies types contain outsized clasts and have rapid vertical and lateral transitions between them (**Figure 8**).

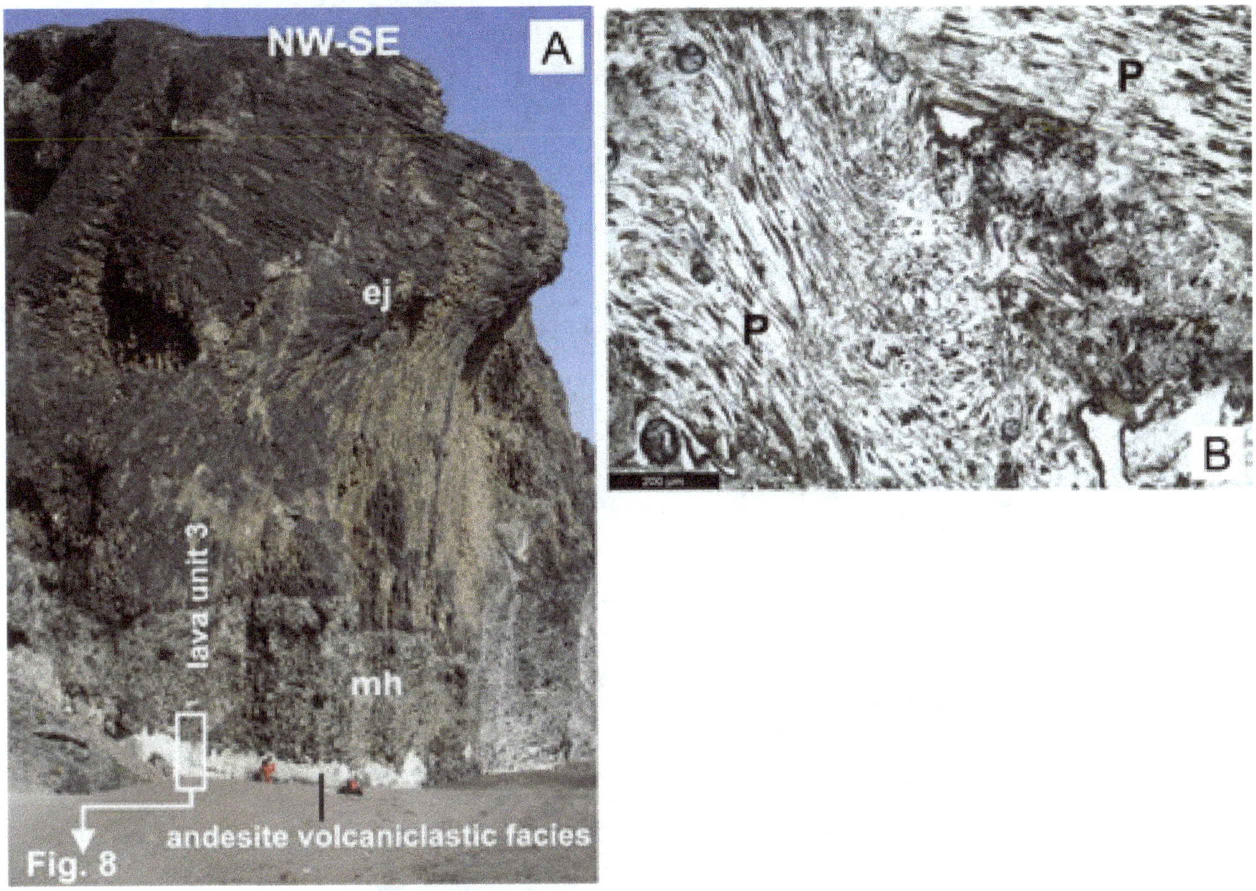

Figure 7. A: lava unit 3 of the El Barronal Formation with entablature jointed facies (ej) showing fan-like joints and grading downward into massive hyaloclastite with dense clasts (mh) and **B**: photomicrograph of finely bedded tuffaceous sandstone with tube-pumice clasts (P).

Volcaniclastic facies contain a variety of clasts indicating provenance from different sources: juvenile components (pumice clasts, glass shards and crystals), clasts derived from underlying lavas (dense rock clasts and vesicular clasts) and exotic well-rounded andesite cobbles likely derived from sources above wave base. Juvenile components are rare in massive breccia, whereas they are present in different amounts in the rest of volcaniclastic facies. Textural evidence of juvenile clasts (i.e., vesicularity >60%, tube-pumice clasts and bubble-wall shards) suggests fragmentation of magma in the conduit by magmatic explosions, whereas the blocky shape of crystals and lithics can be attributed to some degree of magma-water interaction in the fragmentation processes (**Figure 7B**). Finer-grained volcaniclastic facies show distorted bedding, small-scale faults, folds and dish structures, suggesting that they were water-saturated at emplacement. Soft-sediment deformation structures are invariably located at the contact between volcaniclastic facies and upper lavas indicating that volcaniclastic facies were wet and poorly consolidated at the emplacement of upper lavas (**Figure 8**). These contact relations suggest that the time encompassed between the deposition of volcaniclastic facies and the emplacement of upper lavas was short and that lavas and volcaniclastic rocks were likely syn-eruptive.

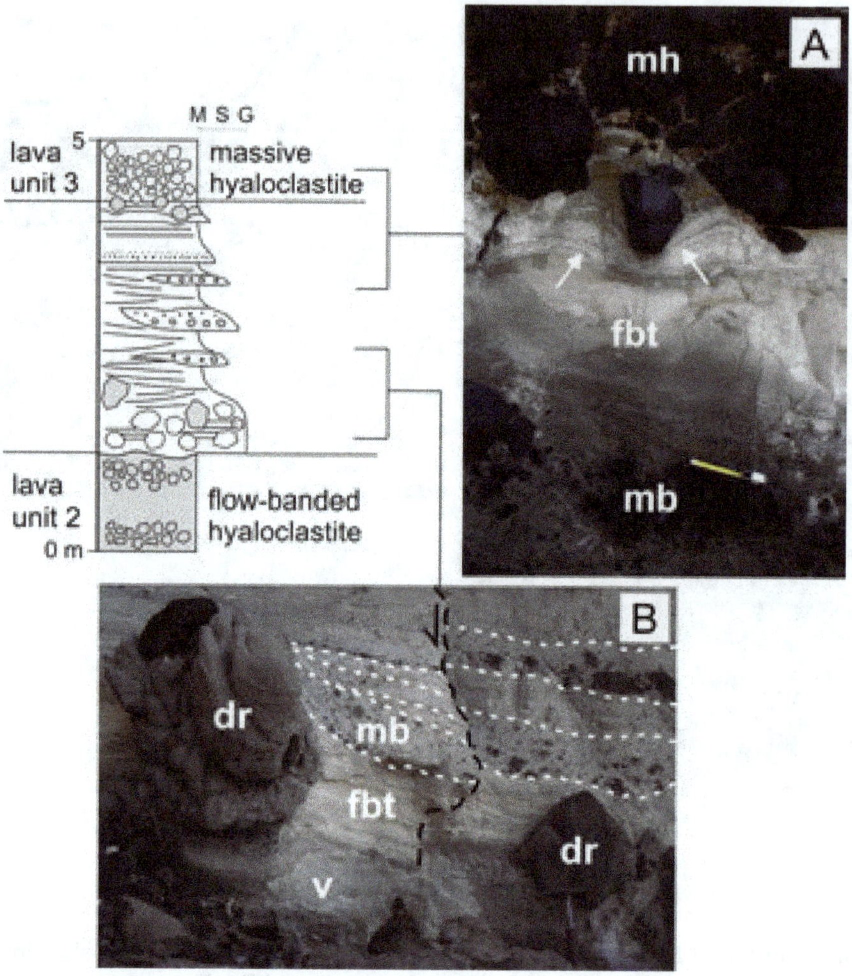

Figure 8. Stratigraphic log and contact relations of the volcaniclastic deposits located between lava unit 2 and 3 of the El Barronal Formation. mh, massive hyaloclastite; mb, massive breccia; tbt, thinly bedded fine tuffaceous sandstone; dr, dense rock clast; v, vesicular clast (see **Figure 7A** for log location).

The El Barronal Formation can be characterized as a succession of eruptive cycles, each of which started with explosive eruptions that yielded deposition of volcaniclastic facies and ended with the effusive emplacement of lavas. A hiatus in the andesite volcanic activity of El Barronal, with eventual deposition of carbonate sediments and emplacement of rhyolitic magma, separates each volcanic cycle. Explosive activity was complex as suggested by the rapid lateral and vertical transitions among volcaniclastic facies and by coexistence of textural evidences indicating fragmentation of magma by magmatic explosions and by magma-water interaction processes. The latter would have enhanced fragmentation efficiency yielding the grain size of finer volcaniclastic facies. Explosive eruptions may have mixed together the different types of juveniles, lithics and clasts from above-wave-base settings too. Steam-driven explosions would have been responsible for the fragmentation of blocks and smaller clasts and the deposition of massive breccia.

6. Submarine volcanic debris-avalanche deposits

Collapse of volcanic edifices and deposition of debris-avalanche deposits in submarine settings are common processes. They often partially affect ocean-island volcanoes such as Hawaii, Tenerife, Reunion and Augustine and also submarine volcanic edifices such as lava domes and stratovolcanoes. Most of the debris avalanches under the sea are inaccessible for obvious reasons and their study has to be undertaken indirectly by means of different geophysical techniques. To our knowledge, there is only one volcanic submarine debris avalanche deposit described in the literature that has been studied in detail by direct access to onshore exposures [37].

The volcanic debris avalanche deposits of Cabo de Gata are well exposed and fairly well constrained in terms of their submarine depositional setting. Therefore, they constitute a unique opportunity for a direct study of this kind of deposits. Here we summarize the main characteristics of two debris avalanche deposits in Cabo de Gata, one of which has been already described in a former work [8] and is the subject of ongoing research.

6.1. The debris avalanche deposits of the Los Frailes Formation

Submarine volcanic debris avalanche deposits crop out to the northeast of San José village in southern Cabo de Gata (**Figure 4**). These deposits are stratigraphically located in informal subunits 2 and 4 of the Los Frailes Formation. Ar^{40}/Ar^{39} ages of sampled blocks from the debris avalanche deposits in subunits 2 and 4 are 12.71 Ma and 12.42 Ma, respectively.

The debris avalanche deposit of subunit 2 is particularly well exposed and is bounded by upper and lower carbonate sedimentary rocks that are laterally discontinuous and contain oysters, corals, echinoderms, algae, and undifferentiated shell fragments (**Figure 9**). This debris avalanche deposit is laterally associated to the northwest with dacite lavas showing a flow-banded and columnar-jointed coherent core that grades outward into in situ to clast-rotated hyaloclastite breccia (**Figure 9**). Lavas and debris-avalanche deposit have identical dacite composition and mineralogy. The debris-avalanche deposit of subunit 2 is massive and monomictic and has an internal organization consisting of block facies "floating" in matrix facies or mixed facies (**Figure 10A**), which is consistent with the classical descriptions of debris avalanche deposits in the literature [38–40]. Some blocks are up to 10^4 cubic meters, although most of them are usually about an order of magnitude smaller and most blocks have a flow-banded and polyhedral-jointed interior (**Figure 10B**). The basal contact of megablocks is sharp, planar, and often fractured with dislocations showing stepped geometry. Shear planes with sigmoid shape form anastomosed lenses and are common at the base of megablocks (**Figure 10C**). Matrix consists of a clast-supported framework of centimeter- to decimeter-size angular clasts with randomly oriented internal flow banding (**Figure 10B**). The upper surface of the debris avalanche deposit in subunit 2 is flat, whereas the lower surface is more irregular, gently dipping to the SE in the northwestern part (**Figure 9**). The maximum thickness of this deposit is nearly 100 m. If a map view of the upper surface of the deposit was exposed rather than an oblique cross-section of the whole deposit, the positive topography built up by megablocks would have likely resulted in the classical hummocky morphology of most debris avalanche deposits [38, 41, 42].

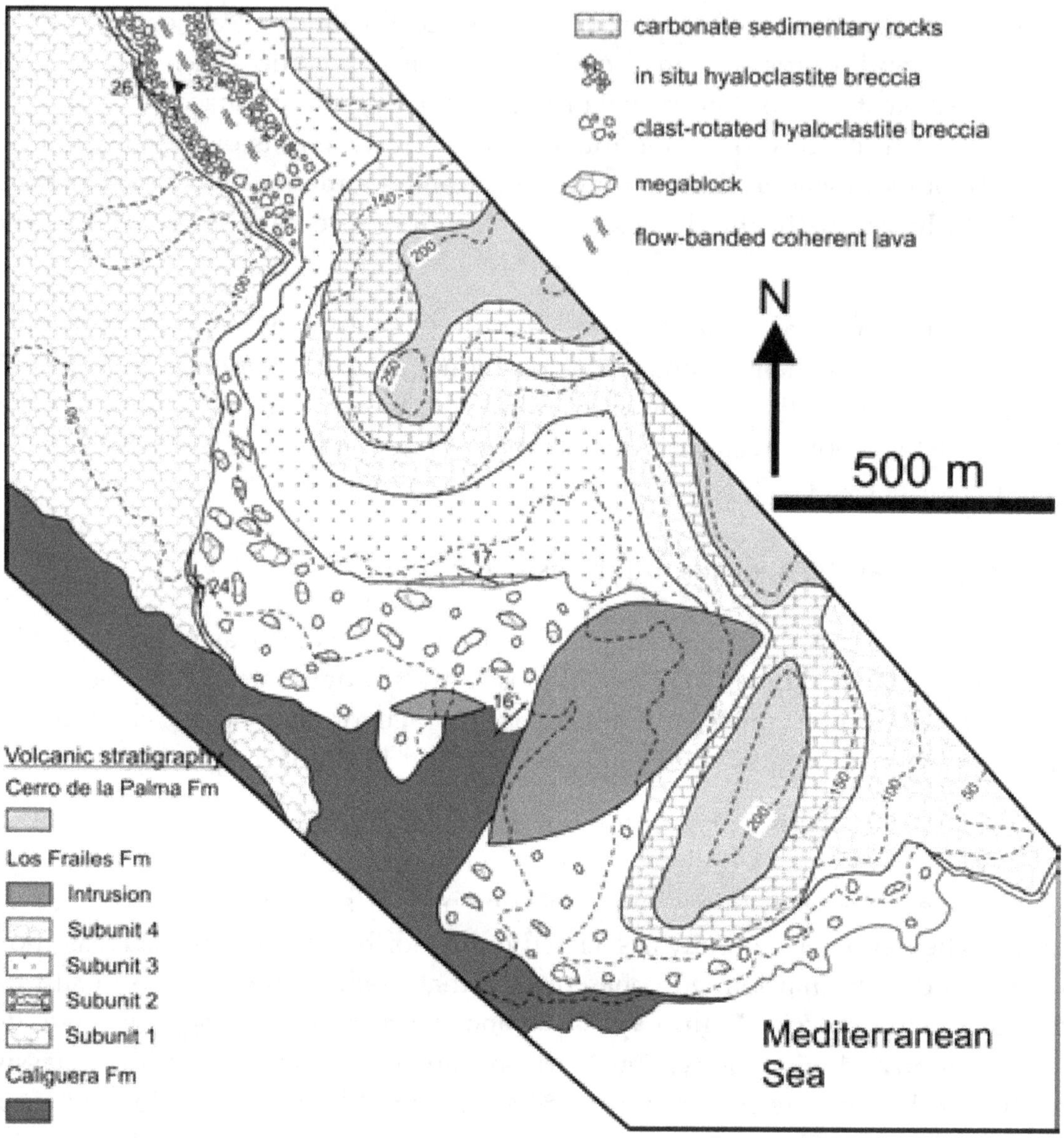

Figure 9. Geologic map of the Los Frailes Formation (see **Figure 4** for location).

The lateral association of the debris-avalanche deposit and lavas within subunit 2, together with their identical composition and mineralogy, indicates that dacite lavas are probably the source area for the volcanic debris of the avalanche deposit. Thickness and areal distribution of the debris-avalanche deposit in **Figure 6** and nearby areas suggests a volume <0.5 km^3 and run-out distance <5 km [8]. These factors together suggest that the debris avalanche is derived from the sector collapse of a submarine lava dome or a dome complex. In this view, megablocks correspond to the coherent core of the lava dome, whereas matrix facies corresponds to the hyaloclastite carapace surrounding the core. In spite of this general provenance for block and matrix facies, block disintegration and clast-to-clast friction during transport are not to be excluded as mechanisms producing block and clast fragmentation [42, 43]. This is particularly evident by the shear deformation and stepped geometry at the base of megablocks.

The debris-avalanche deposit of subunit 4 is thinner and less extended laterally than that of subunit 2. It is massive, monomictic and dacitic in composition and is stratigraphically limited by lower and upper carbonate sedimentary rocks that contain marine fossils. The internal organization of the debris avalanche deposit of subunit 4 is similar to that of subunit 2, consisting of block and mixed facies.

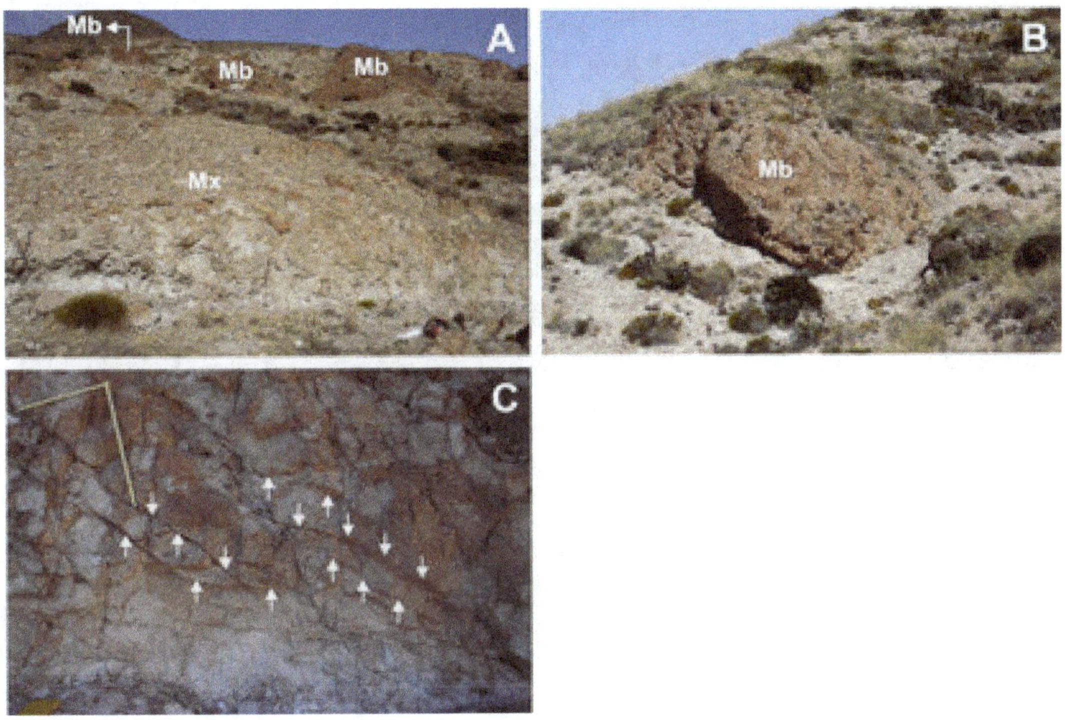

Figure 10. A: panoramic view of the debris-avalanche deposit of subunit 2 of the Los Frailes Formation with megablocks (Mb) and matrix (Mx). **B**: megablock with polyhedral-jointed interior and sharp basal contact. **C**: base of a megablock with anastomosed sigmoid shear planes (arrows).

6.2. The debris avalanche deposit of the Cerro Estorvillas Formation

Inland exposures to the west of Las Negras and to the north of Rodalquilar in the central Cabo de Gata area provide well-preserved profiles of the debris avalanche deposit of the Cerro Estorvillas Formation (**Figure 4**). This unit uncomformably overlies rhyolite bedded pumice breccia of the Cerro de los Guardias Formation and is unconformably overlain by flow banded dacite lavas of the Monte Cinto Formation and by andesite lavas and breccias of the Cerro Negro Formation. The age of the Cerro Estorvillas Formation is stratigraphically constrained between 12.30 Ma and 11.93 Ma (**Figure 3**). The Cerro Estorvillas Formation is dacitic in composition and monomictic and consists of a volcanic succession with a debris avalanche deposit at the base grading upward into bedded pumice breccia that in turn grades upward into bedded and massive breccia (**Figure 11**). To the west of Las Negras, the Cerro Estorvillas Formation also includes dacite lavas.

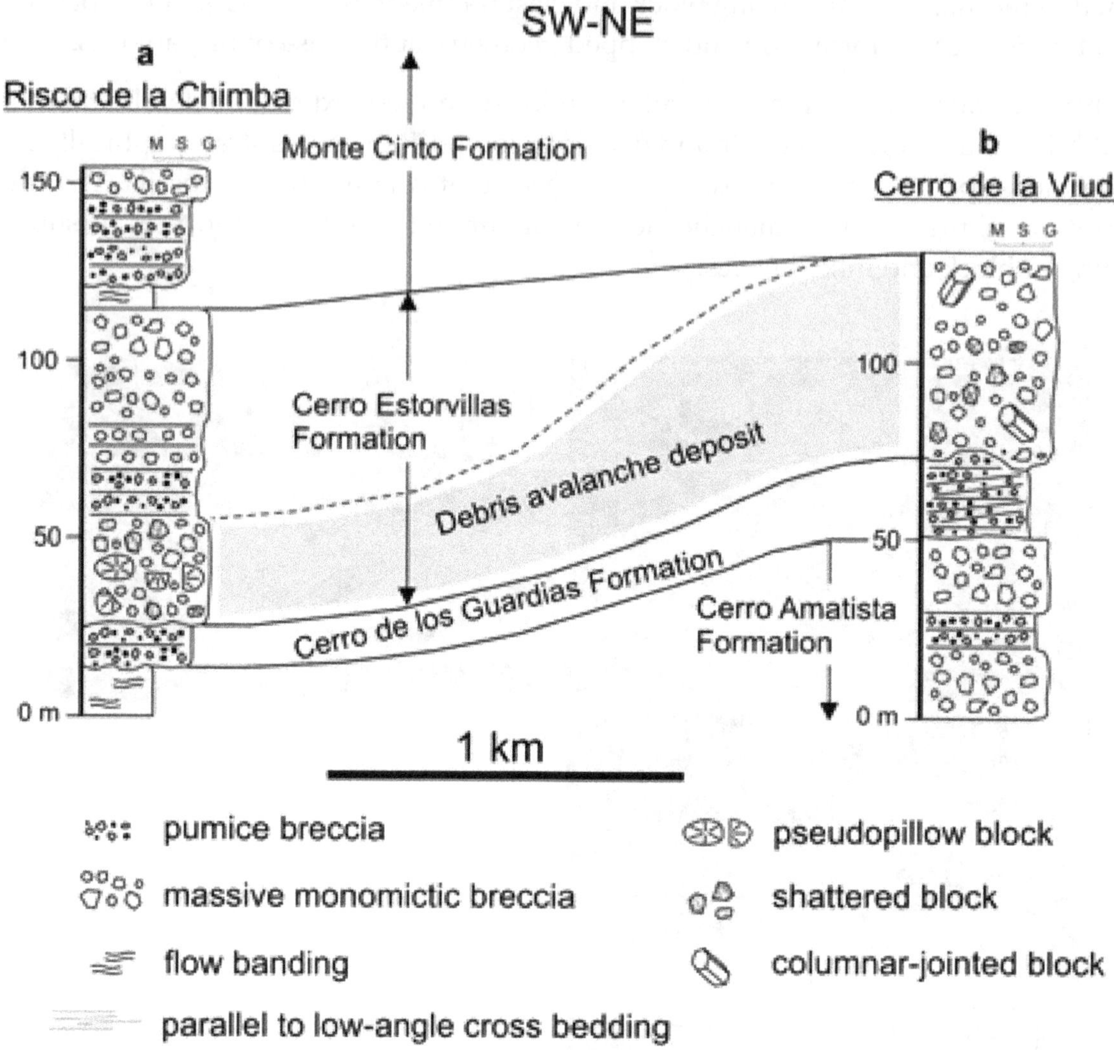

Figure 11. Stratigraphic logs of the Cerro Estorvillas Formation (see **Figure 4** for log location).

The debris avalanche deposit of the Cerro Estorvillas Formation is almost 100 m thick near Las Negras, whereas it is less than 50 m thick farther to the west. This deposit is massive and internally organized in block and mixed facies. Block facies consists of angular clasts up to 6 m across, including meter-size pseudopillow blocks that exhibit radial jointing and meter-size blocks with columnar jointing (**Figure 12A** and **B**). Mixed facies comprises a clast-supported framework of centimeter- to decimeter-size angular clasts of the same composition and mineralogy as the blocks. Blocks are usually shattered with a jigsaw-fit fracture framework. In the lower part of the debris-avalanche deposit, irregular domains with diffuse margins extend laterally for more than 10 m and are less than 5 m thick. These domains consist of well-rounded cobbles of dacite occasionally supported in a matrix of coarse sand that consists of dense rock fragments of dacite and crystals (**Figure 12C**). The contact of the debris avalanche deposit with the underlying Cerro de los Guardias Formation consists of an irregular zone in which clasts and small, usually vein-like, portions of the debris avalanche deposit are mixed with clasts and portions of the underlying pumice breccia deposit (**Figure 12D**). This mixing zone extends up to 2 m upward from the basal contact of the debris-avalanche deposit.

Figure 12. A: columnar-jointed block in the debris-avalanche deposit of the Cerro Estorvillas Formation; **B**: pseudopillow block with radial jointing; **C**: detailed of the cobble domain in the lower part of the debris-avalanche deposit. Note fracture pattern in shattered cobble clasts (arrows); **D**: basal mixed contact of the debris-avalanche deposit of the Cerro Estorvillas Formation to the pumice breccia of the Cerro de los Guardias Formation. Note shattered clasts in the debris-avalanche deposit.

Lack of sorting, massive character, internal organization in block and mixed facies and shattered blocks with jigsaw-fit fractures are usual features of debris avalanche deposits [38, 40, 44, 45]. Pseudopillow radial-jointed blocks may derive from pillows and lava lobes surrounded by hyaloclastite breccia originally deposited in subaqueous settings while the monomictic character indicates a homogenous source for the volcanic debris. Rounded cobbles must have been formed by long-term reworking in highly energetic settings rather than in a nearly instantaneous event like a debris-avalanche. Although marine fossils have not been observed in the cobble domains, rounding of the clasts and the coarse sand matrix fits well with original deposition in a beach environment and subsequent transport within the debris avalanche. Transport and deposition of nearly intact parts of the volcano stratigraphy is a common feature observed in many debris avalanche deposits [39, 41]. Hence, the debris avalanche of the Cerro Estorvillas Formation may have involved the partial collapse of an emerged dacite lava dome likely located to the west of Las Negras, where dacite lavas of the same composition and mineralogy occur. Rhyolite pumice and andesite clasts and irregular portions of pumice breccia from the underlying Cerro de los Guardias Formation at the basal contact of the debris avalanche are interpreted to have been ripped up from the substrate during transport.

7. Tectonic controls on cyclic volcanism

The nearly continuous sedimentation in passive-margins and other types of sedimentary basins allows characterization of basin dynamics by use of the principles of sequence stratigraphy and techniques such as backstripping and decompaction on sedimentary successions. In contrast, the record of volcanism is discontinuous in the geologic history due to its episodic character, whereas volcanic rocks allow a reasonably good determination of absolute ages via radio-isotopic dating. Additionally, in the coherent lithofacies of volcanic successions, compaction effects on accumulated thicknesses are minor and can be likely disregarded.

In Cabo de Gata, volcanic units were emplaced in submarine settings at unknown depth and are interbedded with sedimentary units in which paleobathymetry can be better constrained by analyzing the lithofacies of sedimentary successions. Using the known thickness and ages of volcanic rocks and the better-constrained paleodepth deposition of sedimentary rocks, the paleobathymetric setting of volcanism can be indirectly constrained and the uplift-downdrop history of volcano-sedimentary successions in Cabo de Gata can be roughly assessed. In addition, lateral correlation of the different lithofacies forming subunits of the Los Frailes Formation allows characterization of the eruptive styles and interpretation of the constructive and destructive processes of volcanic edifices during the eruptive periods. This may serve to illustrate the tectonic controls on cyclic volcanism not only of the Los Frailes Formation but also of other units in Cabo de Gata.

Debris-avalanche deposits of subunits 2 and 4 are laterally associated toward the north with coherent lava that grade into hyaloclastite breccias (**Figure 13**). Subunit 1 is also made up of debris-avalanche deposits and coherent lava that grade into hyaloclastite breccia and have a similar dacitic composition and mineralogy to dacite of subunits 2 and 4. Subunit 3 is made up of pumice-rich and glass-shard-rich lithofacies (i.e., massive and bedded pumice breccia, massive pumice tuff and thinly bedded fine tuff). Lithofacies of subunit 3 are fines depleted, have low-angle cross bedding and soft-sediment-deformation structures and show similar dacitic composition and mineralogy to subunits 2 and 4. Although subunit 3 is laterally discontinuous, it constitutes a distinctive stratigraphic marker in the Los Frailes area that allows lateral correlation of subunits within the Los Frailes Formation (**Figure 13**).

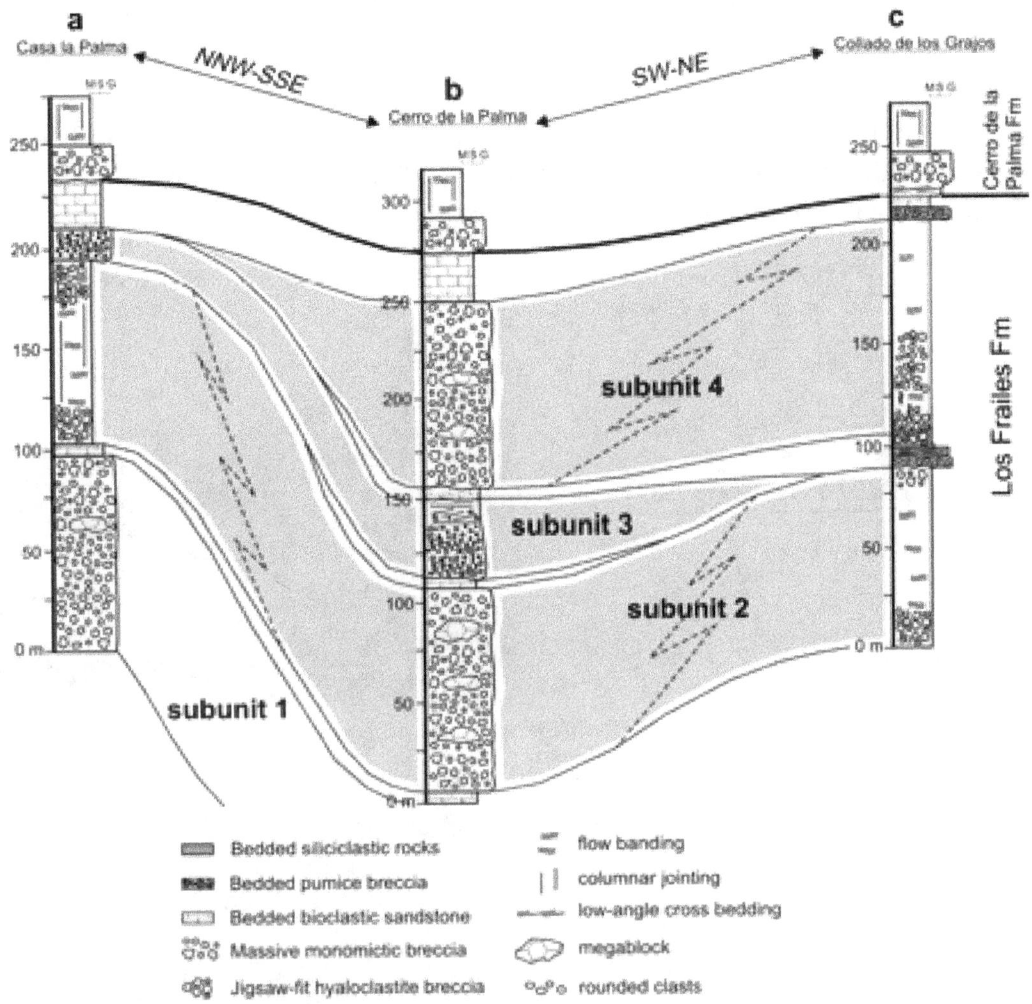

Figure 13. Correlation of volcanic subunits and sedimentary units of the Los Frailes Formation in the Los Frailes area (see **Figure 4** for log location).

At Collado de los Grajos, lava of subunit 2 is overlain by a fining-upward siliciclastic succession that, from base to top, consists of massive conglomerate with well-rounded pebbles of phyllite from the metamorphic basement to the Neogene succession of Cabo de Gata, low-angle cross-bedded coarse sandstone and cross-laminated fine sandstone to siltstone. The siliciclastic succession is overlain by lava of subunit 4, which consists of hyaloclastite breccia grading upward into coherent lava. Fine sandstone and siltstone of the upper siliciclastic succession fluidized and filled the space between clasts of in situ to clast-rotated hyaloclastite breccia of subunit 4. Coherent lava of subunit 4 is capped by a fining- and thinning-upward succession that, from base to top, includes a conglomerate bed, rhodolith-rich facies and cross-bedded sandstone (**Figure 14A**). The conglomerate bed is horizontal, up to 1 m thick and grades upward into a carbonate bed of up to 50 cm thick. The base of the conglomerate bed is irregular on the coherent dacite lava of subunit 4. Conglomerate is composed of well-rounded dacite cobbles with the free space among cobbles filled up with carbonate material infiltrated from the overlying carbonate bed (**Figure 14B**). Dacite cobbles and the underlying coherent lava are remarkably reddish. The upper carbonate bed is massive to crudely bedded and grades upward into low-angle, cross-bedded bioclastic sandstone up to 25 cm thick. The carbonate bed is a rhodolith-rich rudstone that contains angular volcanic clasts, serpulids, bryozoan, bivalve fragments and other bioclasts (**Figure 14C**). Rhodoliths have a concentric algal framework around bioclastic nuclei and occasionally around volcanic clasts. The carbonate succession is overlain by cross-bedded monomict sandstone, massive monomict breccia and coherent lavas of the Cerro de la Palma Formation (**Figure 13**).

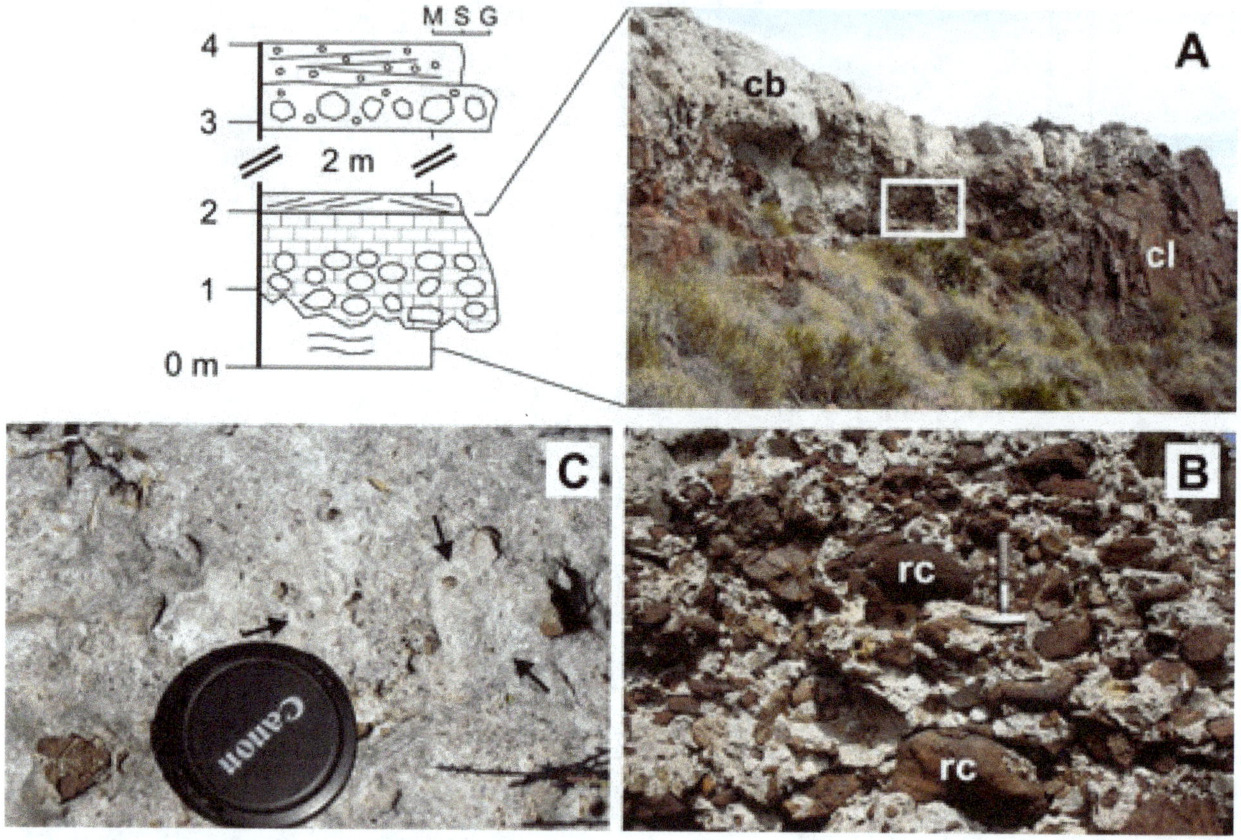

Figure 14. Detailed stratigraphic log of with the sedimentary sequence overlying coherent lava of subunit 4 in the Collado de Los Grajos log (see **Figure 13** for location). **A**: contact of subhorizontal conglomerate bed and carbonate bed (cb) on reddish coherent lava (cl). **B**: conglomerate bed with well-rounded cobbles (rc) and carbonate matrix infilling the the space among cobbles. **C**: top view of the carbonate bed with crustose rhodoliths forming concentric algae framework (arrows).

Fluidized sandstone and siltstone at the base of lava of subunit 4 indicates that the sediment was wet and nonconsolidated at the time of lava emplacement. The reddish color of the uppermost part of coherent lava of subunit 4 and of dacite cobbles of the upper conglomerate bed can be attributed to oxidizing conditions either at lava emplacement or after emplacement. Rhodoliths are encrusting coralline algae usually associated with above wave base environments, although they have been also described in deeper water settings [46, 47]. In terms of modern sequence stratigraphy, rhodolith-rich facies deposited on unconformities have been interpreted as condensed beds, indicating the onset of marine transgression ([48] and references therein). This interpretation fits well with the conglomerate bed on coherent lava of subunit 4 as deposited in a beach environment after lava emplacement and with rhodolith-rich bed and cross-bedded sandstone as a deeper facies in an ongoing marine transgression. Hence, the fining and thinning-upward sedimentary sequence with the conglomerate bed at the base and the cross-bedded sandstone at the top is interpreted to have been deposited above wave base and to reflect deepening water conditions from the shoreface down to infralittoral settings. The depositional model of a Late Holocene clastic prograding wedge in the western Mediterranean has been studied in detail and shows that the mean storm wave base is 20 m below sea level [49]. The present-day configuration of the western Mediterranean was somewhat similar to the paleogeographic configuration in early Tortonian times, in which the Mediterranean Sea was connected to the Atlantic Ocean by narrow passages [50]. Hence, the mean storm wave base during Serravallian-Tortonian times can be reasonably assumed at 20 m below sea level (mbsl) and the exceptional storm wave base at 30 mbsl, both marking the slope break that separates infralittoral and littoral depositional settings from offshore deposition.

Subunits 2, 3, and 4 are separated by sedimentary successions in which silt grain size is minor and clay grain size is absent. These successions contain a variety of shallow-water marine fossils (bryozoan, echinoderms, bivalves, and coralline algae) and have low-angle cross stratification, rounded cobbles and pebbles and other evidence of above wave base settings as shown above. Therefore, the sedimentary units separating the volcanic subunits of the Los Frailes Formation can be collectively interpreted as mixed carbonate-siliciclastic platform-beach facies deposited in infralittoral to littoral settings. The fines-depleted character, soft-sediment deformation structures, low-angle cross-bedding and interbedding with sedimentary rocks from shallow-water settings suggest subaqueous deposition of volcanic subunit 3 of the Los Frailes Formation [8]. Hence, the composite stratigraphic succession obtained from lateral correlation in the Los Frailes area includes thin sedimentary units deposited in infralittoral to littoral settings (0–20 mbsl) and thicker volcanic units emplaced in deeper, likely offshore, settings (>20 mbsl). Based on these paleobathymetyric constraints, the relative sea-level curve can be obtained with hiatuses corresponding to the unconformities at the contacts between volcanic and sedimentary units (**Figure 15**). Fluidization of unconsolidated sediments into overlying lavas suggests that the time encompassed from deposition of sediments to lava emplacement was not very long. Based on the ages of volcanic subunits and on the number of volcanic subunits of the Los Frailes formation, the periodicity of volcanic cycles can be roughly approximated at 10^5 years. Volcanic activity during each volcanic cycle includes constructive and destructive events of volcanic edifices and is assumed to have occurred in a relatively short time span ($<10^4$ years). The Los Frailes area can be regarded as a volcanic field composed of individual, partly overlapping, domes that are active along a time span of about 5×10^5 years with periods of volcanic repose separating dome activity. A similar scenario has been proposed for Surtseyan volcanism of the Eocene-Oligocene Waiareka-Deborah volcanic field in New Zealand but encompassing a time span of several 10^6 years [51]. The accumulated thickness of the stratigraphic succession at Los Frailes is mainly derived from the volcanic activity while the sedimentary succession accumulated during repose periods has much smaller thickness (**Figure 15**). Sea-level oscillations can be explained by the progradation and retrogradation of mixed carbonate-siliciclastic platform-beach wedges after each volcanic cycle and by infilling of offshore settings with volcanic material during volcanic cycles. This dynamics requires tectonic subsidence to accommodate the thickness of volcanic material accumulated during volcanic cycles in infralittoral to littoral sediments. Similar to other Neogene basins of the Betic-Rif Orogen [52–54], we propose that tectonic subsidence in Cabo de Gata is controlled by uplift-downdrop displacements along major strike-slip faults that bound the Almeria-Níjar basin (i.e., the Carboneras fault) and/or by vertical displacements along minor faults associated with these major faults.

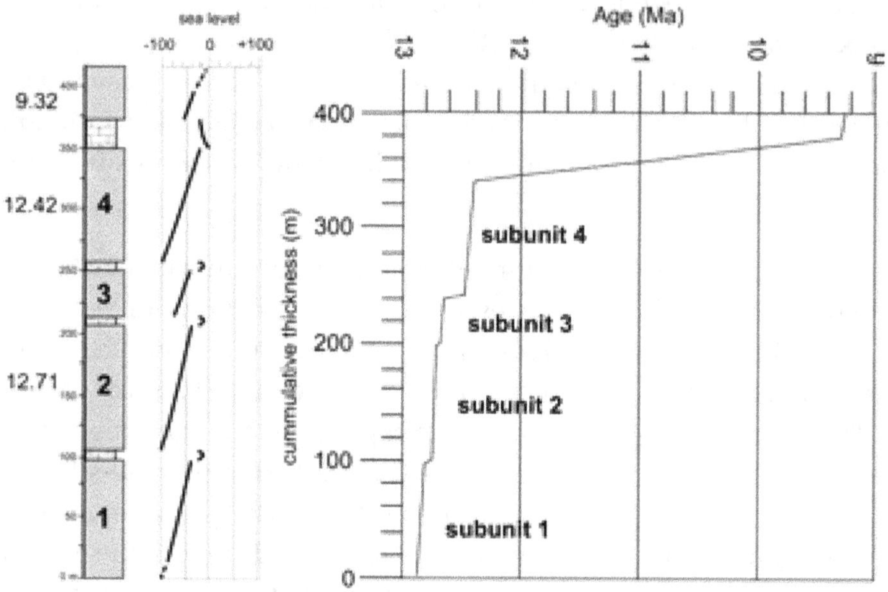

Figure 15. Composite stratigraphic log of the Los Frailes area based on the lateral correlation of **Figure 13** with sea-level curve based on the bathymetric constrains from sedimentary units and volcanic units (1–4) and cumulative thickness versus age.

8. Conclusion

Volcanism in Cabo de Gata is cyclic, with eruptive periods being characterized by both construction and partial dismantling of volcanic edifices, whereas noneruptive periods are characterized by volcanic repose, sedimentation of carbonate and siliciclastic rocks and erosion of volcanic edifices and deposits. The El Barronal Formation, the Los Frailes Formation and many other units in Cabo de Gata, which are divided into volcanic subunits separated by sedimentary units, are the geological record of this cyclic volcanism and volcanic hiatus history. Volcanic activity in Cabo de Gata includes effusive and explosive eruptions and syn-eruptive partial collapse of volcanic edifices yielding avalanches of volcanic debris deposited in submarine settings. The characteristic eruption model in Cabo de Gata likely began with explosive eruptions that produced juvenile and lithic fragments and deposition of pyroclastic density currents and ended with the effusion of lavas. Volcanic debris-avalanche deposits of the Los Frailes Formation and the Cerro Estorvillas Formation resulted from the sector collapse of submarine to emergent lava domes and were deposited in offshore settings. They constitute rare examples in the geological record of submarine volcanic debris avalanches that can be studied by direct access on exposed outcrops. Based on depositional setting of sedimentary rocks and on thickness and ages of volcanic rocks, it is suggested that the volcanic pile produced during each volcanic cycle in infralittoral to littoral settings is accommodated by tectonic subsidence along major faults controlling the Almeria-Níjar basin.

Acknowledgements

We thank Miguel Garcés for comments and suggestions on a first version of the manuscript. This research has been partly funded by project 2014SGR1595 and by grant PRX14/00303 to Carles Soriano.

Author details

Carles Soriano[1*], Ray A.F. Cas[2*], Nancy R. Riggs[3*] and Guido Giordano[4*]

*Address all correspondence to: csoriano@ictja.csic.es, ray.cas@nau.edu, nancy.riggs@nau.edu and guido.giordano@uniroma3.it

1 Institute of Earth Sciences "Jaume Almera", Superior Council for Scientific Research c/ Lluís Solé Sabarís s/n, Barcelona, Spain

2 School of Earth, Atmosphere and Environment, Monash University, Monash University, Calyton Campus, Australia

3 Geology Program, School of Earth Sciences and Environmental Sustainability, Northern Arizona University, Flagstaff, USA

4 Department of Science, Geological Sciences Section, Roma Tre University, Largo S. Leonardo Murialdo, Roma, Italy

References

[1] Fuster JM, Aguilar MJ, García A. Las sucesiones volcánicas en la zona del Pozo de los Frailes dentro del vulcanismo cenozoico del Cabo de Gata (Almería). Estudios Geológicos. 1965; XXI: 199–222.

[2] Páez Carrión A, Sánchez Soria P. Vulcanología del Cabo de Gata, entre San José y Vela Blanca. Estudios Geológicos. 1965; XXI: 223–246.

[3] León C. Las formaciones volcánicas del Cerro de los Lobos (Almería, SE. de España). Estudios Geológicos. 1967; XXIII: 15–28.

[4] Di Battistini G, Toscani L, Iaccarino S, Villa IM. K/Ar ages and the geological setting of calc-alkaline volcanic rocks from Sierra de Gata, SE Spain. Neues Jahrbuch für Mineralogie, Monatshefte. 1987; H 8: 369–383.

[5] Cunningham CG, Arribas A Jr, Rytuba JJ, Arribas A. Mineralized and unmineralized calderas in Spain; Part 1, evolution of the Los Frailes caldera. Mineralium Deposita 1990; v. 25 (supp.): S21–S28.

[6] Arribas A Jr, Cunningham CG, Rytuba JJ, Rye RO, Kelly WC, Podwysocki MH, McKee EH, Tosdal RM. Geology, Geochronology, Fluid inclusions, and Isotope Geochemistry of the Rodalquilar Gold Alunite Deposit Spain. Economic Geology.1995; 90: 795–822.

[7] Cas RAF, Giordano G. Submarine volcanism: a review of the constrains, processes and products and relevance to the Cabo de Gata volcanic succession. Italian Journal of Geosciences. 2014; 133: 362–377.

[8] Soriano C, Riggs N, Giordano G, Porreca M, Conticelli S. Cyclic growth and mass wasting of submarine Los Frailes lava flow and dome complex in Cabo de Gata, SE Spain. Journal of Volcanology and Geothermal Research. 2012; 231–232: 72–86.

[9] Soriano C, Giordano G, Cas R, Riggs N, Porreca M. Facies architecture, emplacement mechanisms and eruption style of the submarine andesite El Barronal complex, Cabo de Gata, SE Spain. Journal of Volcanology and Geothermal Research. 2013; 264: 210–222.

[10] Soriano C, Giordano G, Riggs N, Porreca M, Bonamico A, Iosimi D, Cifelli F, Mattei M, De Benedetti A, Guarnieri L, Marchionni S. Geologic map, volcanic stratigraphy and structure of the Cabo de Gata volcanic zone, Betic-Rif Orogen, SE Spain. Italian Journal of Geosciences. 2014; 133: 325–340.

[11] Porreca M, Cifelli F, Soriano C, Giordano G, Romano C, Conticelli S, Mattei M. Hyaloclastite fragmentation below the glass transition: an example from El Barronal submarine volcanic complex (Spain). Geology. 2014; 42: 87–90.

[12] Porreca M, Cifelli F, Soriano C, Giordano G, Mattei M. Magma flow within dykes in submarine hyaloclastite environments: an AMS study of the Miocene Cabo de Gata volcanic units. In: Ort MH, Porreca M, Geissman JW, editors. The Use of Palaeomagnetism and Rock Magnetism to Understand Volcanic Processes. Geological Society, London, Special Publication, 396; 2015. p. 133–157.

[13] van Otterloo J, Cas RAF, Scutter C. The fracture behaviour of volcanic glass and relevance to quench fragmentation during formation of hyaloclastite and phreatomagmatism. Earth Science Reviews. 2015; 151: 79–116.

[14] Sharqawy M, Lienhard JH, Zubair SM. Thermophysical properties of seawater: a review of existing correlations and data. Desalination and water treatment. 2010; 16: 354–380.

[15] Höskuldsson A, Sparks RSJ. Thermodynamics and fluid dynamics of effusive subglacial eruptions. Bulletin of Volcanology. 1997; 59: 219–230.

[16] Zimanowski B, Buttner R. Phreatomagmatic explosions in subaqueous volcanism. In: White JDL, Smellie JL, Clague DA, editors. Submarine Explosive Volcanism. American Geophysical Union, Monograph, 140; 2003. pp. 51–60.

[17] Wohletz KH. Water/magma interaction: physical considerations for the deep submarine environment. In: White JDL, Smellie JL, Clague DA, editors. Submarine Explosive Volcanism. American Geophysical Union, Monograph, 140; 2003. pp. 25–49.

[18] Doblas M, López-Ruiz J, Cebriá JM. Cenozoic evolution of the Alboran Domain: a review of the tectonomagmatic models. In: Beccaluva L, Bianchini G, Wilson M, editors Cenozoic Volcanism in the Mediterranean area. Geological Society of America, Special Paper, 418; 2007. pp. 303–320.

[19] Lustrino M, Duggen S, Rosenberg C. The central-western Mediterranean: anomalous igneous activity in an anomalous collisional tectonic setting. Earth Science Reviews. 2011; 104: 1–40.

[20] Vergés J, Fernàndez M. Tethys-Atlantic interaction along the Iberia-Africa plate boundary: The Betic-Rif orogenic system. Tectonophysics. 2012; 579: 144–172.

[21] Conticelli S, Guarnieri L, Farinelli A, Mattei M, Avanzinelli R, Bianchini G, Boari E, Tommasini S, Tiepolo M, Prelevic D, Venturelli G. Trace elements and Sr-Nd-Pb isotopes of K-rich, shoshonitic, and calc-alkaline magmatism of the Western Mediterranean Region: genesis of ultrapotassic to calc-alkaline magmatic associations in a post-collisional geodynamic setting. Lithos. 2009; 107: 68–92.

[22] Mattei M, Riggs N, Conticelli S, Giordano G, Cifelli F, Soriano C, Jicha B, Marchionni S, Guarnieri L, Tommasini S, Jasim A, Franciosi L, Porreca M. Geochronology geochemistry and geodynamics of the Cabo de Gata volcanic zone, South Eastern Spain. Italian Journal of Geosciences. 2014; 133: 341–361.

[23] Faccenna C, Piromallo C, Crespo-Blanc A, Jolivet L, Rossetti F. Lateral slab deformation and the origin of the western Mediterranean arcs. Tectonics. 2004; 23: TC1012.

[24] Comas MC, Platt JP, Soto JI, Watts AB. The origin and tectonic history of the Alboran basin; insights from Leg 161 results. In: Zahn R, Comas MC, Klaus A, editors. Proceedings of the Ocean Drilling Program, Scientific Results, 161; 1999. pp. 555–580.

[25] McClusky HL, Conner ME, Smith LD, Holloway MV, Conover TA, Beasley DE. Mapping of the lateral flow field in typical subchannels of a support grid with vanes. Journal of Fluids Engineering. 2003; 125(6): 987–996.

[26] Iribarren L, Verges J, Camurri F, Fullea J, Fernandez M. The structure of the Atlantic–Mediterranean transition zone from the Alboran Sea to the Horseshoe Abyssal Plain (Iberia-Africa plate boundary). Marine Geology. 2007; 243: 97–119.

[27] Bell JW, Amelung F, King GCP. Preliminary late quaternary slip history of the Carboneras fault, Southeastern Spain. Journal of Geodynamics. 1997; 24: 51–66.

[28] Scotney P, Burgess R, Rutter EH. 40Ar/39Ar age of the Cabo de Gata volcanic series and displacements on the Carboneras fault zone, SE Spain. Journal of the Geological Society, London. 2000; 157: 1003–1008.

[29] Rutter EH, Burgess R, Faulkner DR. Constraints on the movement history of the Carboneras Fault Zone (SE Spain) from stratigraphy and ^{40}Ar-^{39}Ar dating of Neogene volcanic rocks. In: Llana-Fúnez S, Marcos A, Bastida F, editors. Deformation Structures and Processes within the Continental Crust. Geological Society, London, Special Publications, 394; 2014. pp. 79–99.

[30] Pedrera A, Marín-Lechado C, Galindo-Zaldívar J, Rodríguez-Fernández LR, Ruiz-Constán A. Fault and fold interaction during development of the Neogene-Quaternary Almería-Níjar basin (SE Betic Cordilleras). In: Moratti G, Chalouan A, editors Tectonics of the Western Mediterranean and North Africa. Geological Society, London, Special Publication, 262; 2006. pp. 217–230.

[31] Borehole dataset of the Instituto Geologico y Minero of Spain. Available from www.igme.es

[32] Brachert TC, Hultzsch N, Knoerich AC, Krautworst UMR, Stückrad OM. Climatic signature in shallow-water carbonates: high-resolution stratigraphic markers in carbonate build-ups (Late Miocene, southern Spain). Palaeogeography, Palaeoclimatology, Palaeoecology. 2001; 175: 211–237.

[33] Ballesteros M, Rivera J, Muñoz A, Muñoz-Martín A, Acosta J, Carbo A, Uchupi E. Aboran basin southern Spain-Part II. Neogen tectonic implications for the orogenic float model. Marine and Petroleum Geology. 2008; 25: 75–101.

[34] (IGME) Instituto Geológico y Minero de España. Mapa Geológico de España 1:50000, El Pozo de los Frailes 1060. Servicio de Publicaciones del Ministerio de Industria y Energía, Madrid. 1983.

[35] Duggen S, Hoernle K, van den Bogaard P, Harris C. Magmatic evolution of the Alboran region: The role of subduction in forming the western Mediterranean and causing the Messinian Salinity Crisis. Earth and Planetary Science Letters. 2004; 218: 91–108.

[36] Montgomery P, Farr MR, Franseen EK, Goldstein RH. Constraining controls of carbonate sequences with high-resolution chronostratigraphy: Upper Miocene, Cabo de Gata region, SE Spain. Palaeogeography, Palaeoclimatology, Palaeoecology. 2001; 176: 11–45.

[37] Trofimovs J, Cas RAF, Davis BK. An Archean submarine volcanic debris avalanche deposit, Yilgarn Craton, western Australia, with komatiite, basalt and dacite megablocks. The product of dome collapse. Journal of Volcanology and Geothermal Research. 2004; 138: 111–126.

[38] Crandell DR, Miller CD, Glicken HX, Christiansen RL, Newhall CG. Catastrophic debris avalanche from ancestral Mount Sashta, California. Geology. 1984; 12: 143–146.

[39] Glicken HX. Sedimentary architecture of large volcanic-debris avalanches. In: Smith GA., Fisher RV, editors. Sedimentation in volcanic settings. SEPM Tulsa Special Publication, 45; 1991. pp. 99–106.

[40] Palmer BA, Alloway BV, Neall VE. Volcanic-debris-avalanche deposits in New Zealand-Lithofacies organization in unconfined, wet-avalanche flows. In: Smith GA., Fisher RV, editors. Sedimentation in Volcanic settings. SEPM Tulsa Special Publication, 45; 1991. pp. 89–98.

[41] Stoopes GR, Sheridan MF. Giant debris avalanche from the Colima Volcanic Complex, México, implications for long runout landslides (<100°km) and hazard assessment. Geology. 1992; 20: 299–302.

[42] Roverato M, Cronin S, Procter J, Capra L. Textural features as indicators of debris avalanche transport and emplacement, Taranaki volcano. Geological Society of America bulletin. 2014. DOI:10.1130/B30946.1.

[43] Voight B, Komorowski JC, Belusov AB, Belusova M, Boudon G, Francis PW, Franz V, Heinrich P, Sparks RSJ, Young SR. The 26 December (Boxing Day) 1997 sector collapse and debris avalanche at Soufrière Hills Volcano, Montserrat. In: Druitt TH, Kokelaar BP editors. The eruption of Soufrière Hills Volcano, Montserrat, from 1995–1999. Geological Society London Memoir 21; 2002. pp. 363–407.

[44] Siebert L. Large volcanic debris avalanche: characteristics of source areas, deposits and associated eruptions. Journal of Volcanology and Geothermal Research. 1984; 22: 163–197.

[45] Ui T, Glicken HX. Internal structural characteristics of debris avalanche from Mount Sahsta, California, U.S.A. Bulletin of Volcanology. 1986; 48: 189–194.

[46] Flügel E. Microfacies of Carbonate Rocks. Springer Berlin Heidelberg; 2004. 976 p.

[47] Brandano M, Ronca S. Depositional processes of the mixed carbonate-siliciclastic rhodolith beds of the Miocene Saint-Florent Basin, northern Corsica. Facies. 2014; 60: 73–90.

[48] Nalin R, Nelson CS, Basso D, Massari F. Rhodolith-bearing limestones as transgressive marker beds: fossil and modern examples from North Island, New Zealand. Sedimentology. 2008; 55: 249–274.

[49] Hernández-Molina FJ, Fernández-Salas LM, Lobo F, Somoza L, Díaz-del-Río V, Alveirinho Dias JM. The infralittoral prograding wedge: a new large-scale progradational sedimentary body in shallow marine environments. Geo-Marine Letters. 2000; 20: 109–117.

[50] Martín JM, Braga JC, Aguirre J, Puga-Bernabéu A. History and evolution of the North-Betic Strait (Prebetic Zone, Betic Cordillera): A narrow, early Tortonian, tidal-domi-

nated, Atlantic-Mediterranean marine passage. Sedimentary Geology. 2009; 216: 80–90.

[51] Moorhouse BL, White JDL, Scott, J. Cape Wanbrow: A stack of Surtseyan-style volcanoes built over millions of years in the Waiareka-Deborah volcanic field, New Zealand. Journal of Volcanology and Geothermal Research. 2015; 298: 27–46.

[52] Soria J, Viseras C, Fernández J. Late Miocene-Pleistocene tectono-sedimentary evolution and subsidence history of the central Betic Cordillera (Spain): a case study of the Guadix intramontane basin. Geological Magazine. 1998; 135: 565–574.

[53] Garcés M, Krijgsman W, Agustí J. Chronostratigraphic framework and evolution of the Fortuna basin (Eastern Betics) since the Late Miocene. Basin Research. 2001; 13: 199–216.

[54] Rodríguez-Fernández J, Sanz de Galdeano C. Late orogenic intramontane basin development: the Granada basin, Betics (southern Spain). Basin Research. 2006; 18: 85–102.

5

Ordovician and Carboniferous Volcanism/ Plutonism in Central Inner Mongolia, China and Paleozoic Evolution of the Central Asian Orogenic Belt

Yuruo Shi

Abstract

Adakite was originally proposed as a genetic term to define intermediate to high silica, high Sr/Y and La/Yb volcanic and plutonic rocks derived from melting of young, subducted lithosphere. However, most volcanic rocks in modern island arcs and continental arcs are probably derived from melting in the mantle wedge. Trace element chemistry with high Sr/Y ratios is a distinguishing characteristic of adakites. Ordovician and Carboniferous volcanic/plutonic rocks with high Sr/Y ratios occur in Central Inner Mongolia, which is situated on the southern margin of the Central Asian Orogenic Belt (CAOB). The samples are mostly granodiorite, tonalite and quartz-diorite in composition with intermediate to high-silica, high Na_2O (3.08–4.26 wt.%), low K_2O (0.89–2.86 wt.%) and high Na_2O/K_2O and Sr/Y ratios. Their chondrite-normalized REE patterns are characterized by LREE enrichment. In mantle-normalized multi-element variation diagrams, they show typical negative Nb anomalies, and all samples display positive $\varepsilon_{Hf}(t)$ and $\varepsilon_{Nd}(t)$ values, and low I_{Sr}. The Ordovician rocks, however, show higher Sr/Y and La/Yb ratios than the Carboniferous samples, implying that the older granitoids represent adakitic granitoids, and the Carboniferous granitoids are typical subduction-related arc granitoids but also with adakite-like compositions. The results are compatible with the view that the Central Asian Orogenic Belt (CAOB) in Inner Mongolia evolved through operation of several subduction systems with different polarities: an early-middle Paleozoic subduction and accretion system along the northern margin of the North China Craton and the southern margin of the Mongolian terrane, and late Paleozoic northward subduction along the northern orogen and exhumation of a high-pressure metamorphic terrane on the northern margin of the North China Craton.

Keywords: Adakitic, Ordovician and Carboniferous, Geochemistry, Hf-in-zircon isotopes, Central Inner Mongolia, CAOB

1. Introduction

It is generally agreed that the Solonker suture zone represents the southernmost termination of the Central Asian Orogenic belt (CAOB; [1–5]). However, there are a lot of controversies about the timing of the amalgamation of the Central Asian Orogenic belt with continental blocks to the south [1–9]. It is still debated whether the CAOB evolved through subduction and accretion of a single, long-lasting, subduction system [10] or through several subduction systems with different polarities and through collision/accretion of arcs and microcontinents [11–15].

Adakite was originally proposed as a genetic term to define intermediate to high-silica, high Sr/Y and La/Yb volcanic and plutonic rocks derived from melting of young, subducted lithosphere [16]. However, most volcanic rocks in modern island arcs and continental arcs are probably derived from melting in the mantle wedge [17]. Trace element chemistry with high Sr/Y ratios is a distinguishing characteristic of adakites [16, 18]. Ordovician and Carboniferous volcanic/plutonic rocks with high Sr/Y ratios occur in Central Inner Mongolia, which is situated on the southern margin of the Central Asian Orogenic Belt (CAOB, [19]). Early Paleozoic [6–9, 20–22] and Late Paleozoic [2–4, 23] arc volcanism/plutonism as part of trench-island arc-basin systems occurred along the southern margin of South Mongolian microcontinent and the northern margin of North China Craton, suggesting concurrent two-way subduction towards opposing continental margins. The chapter focuses on early and late Paleozoic volcanic/plutonic rocks with high Sr/Y ratios in Central Inner Mongolia, and contributes geochemical data to the evolution of the CAOB.

2. Geotectonic situation

Central Asian Orogenic Belt (CAOB, [19]) is a giant accretionary orogen [15], bounded by the Siberian, Tarim and North China Craton ([19, 24]; **Figure 1**), and reflects a complex evolution from the late Mesoproterozoic to late Palaeozoic [1, 6, 8, 14, 26, 27].

In Central Inner Mongolia and adjacent southern Mongolia, the Solonker suture zone can be traced for ca. 1000 km by dismembered ophiolite fragments (**Figure 1**) and represents a major paleo-plate boundary in Central Asia that stretches northeastwards for more than 2500 km in Mongolia and China [28]. It has been variably interpreted as the southernmost limit of the Altaids ([10]) or the southernmost termination of the CAOB [1]. The Solonker suture zone separates two continental blocks (**Figure 1**) [3]. The Northern Block consists of the Southern Mongolia (or Hutag Uul) block (gneissic granite, 1784 ± 7 Ma, Shi et al., unpublished data) and the Northern Orogen, which includes metamorphic complex (an orthogneiss has a zircon age of 437 ± 3 Ma, [29]), an ophiolitic mélange with blueschist, a near-trench granitoid (ca. 498–461 Ma) and a juvenile arc (ca. 484–469 Ma, [3]). The Southern Block comprises the southern orogen and the northern margin of the North China Craton.

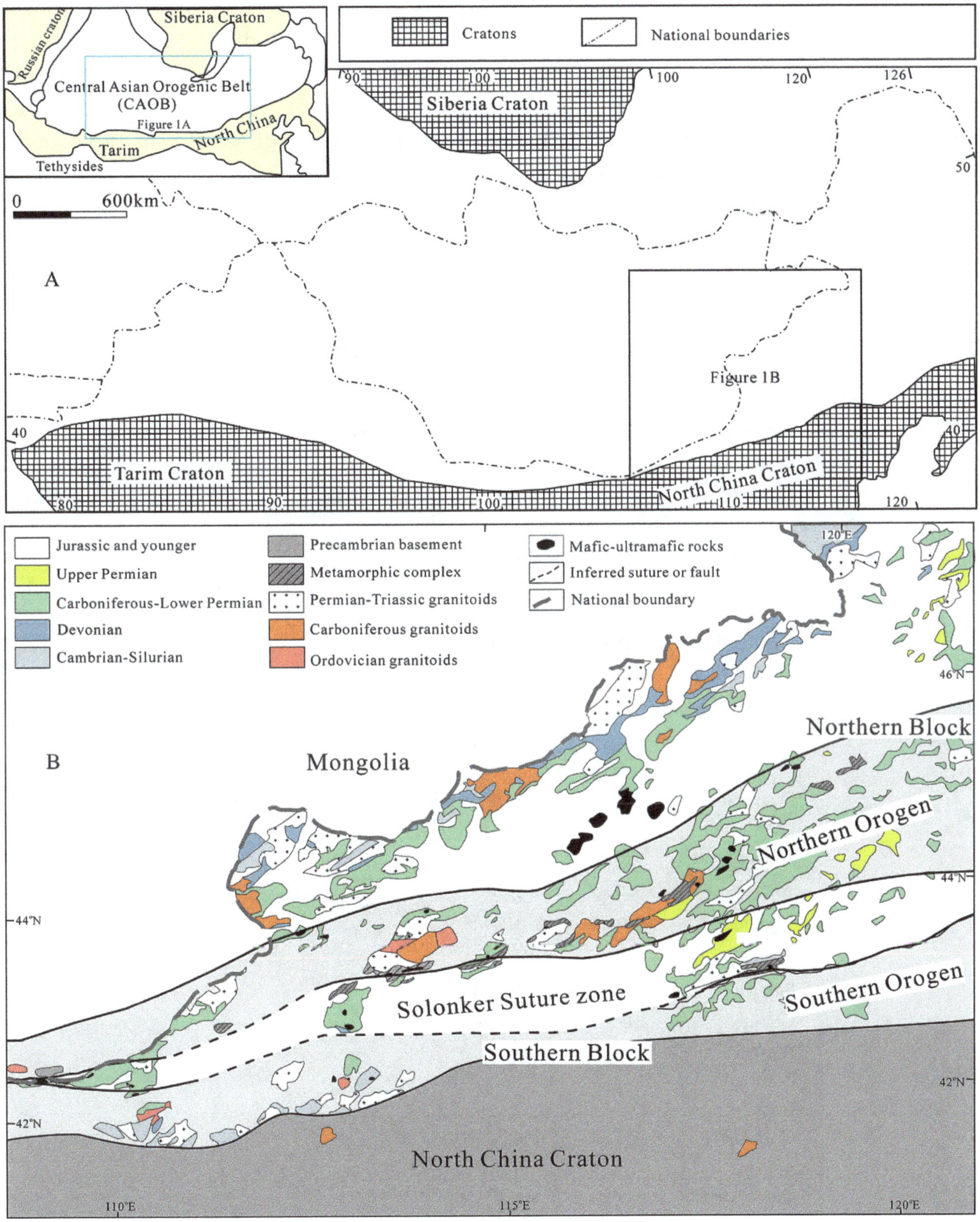

Figure 1. Geological sketch map of the southeastern CAOB (the inset map of **Figure 1A** compiled after [19]; **Figure 1B** after [3, 25]). In **Figure 1B**, the Solonker suture zone represents the tectonic boundary between the northern and the southern continental blocks [3].

Paleozoic volcanic rocks and granitoids are widely distributed along the margin of the Solonker suture zone. Ordovician granitoids (quartz-diorite, granodiorite, diorite, tonalite, and trondjemite; **Table 1** and **Figure 2**) occur in the northern and southern orogen [7, 8, 20, 21, 42, 43]; **Figure 1**), whereas Carboniferous volcanic rocks and granitoids (quartz-diorite, granodiorite, tonalite, and granite; **Table 1** and **Figure 2**) are mainly distributed in the northern orogen ([2, 23, 30, 31, 35, 37, 38, 40, 41]; **Figure 1**), and scattered along the northern margin of the North China Craton [44, 45]. The geochemical data of representative rocks are listed in **Table 2**.

Unit	Episode	Lithology	Zircon age (Ma)	Method	$\varepsilon_{Hf(t)}$ (Zircon)	$\varepsilon_{Nd(t)}$ (Whole rock)	Initial $^{87}Sr/^{86}Sr$ (whole rock)	Reference
Northern Orogen	Ordo vician	Quartz diorite	490 ± 8	SHRIMP				[23]
		Tonalite	479 ± 8	SHRIMP		+1.5	0.7053	[7]
		Quartz diorite	475 ± 6	SHRIMP				[7]
		Granodiorite	472 ± 3	SHRIMP	+7.4 to +10.7	+2.2	0.7060	This study
		Tonalite	464 ± 8	SHRIMP		+1.4	0.7053	[7]
	Carboni ferous	Tonalite	329 ± 3	SHRIMP		+5.1	0.7043	This study
		Quartz diorite	325 ± 3	SHRIMP				[30]
		Quartz diorite	323 ± 4	SHRIMP				[31]
		Quartz diorite	322 ± 3	SHRIMP				[30]
		Monzogranite	322 ± 1	LA-ICP-MS	+10.6 to +14.0			[32]
		Quartz diorite	320 ± 3	SHRIMP	+8.1 to +12.3	+2.1	0.7051	This study
		Granodiorite	320 ± 8	SHRIMP		+1.0	0.7055	This study
		Andesite	320 ± 7	SHRIMP				[33]
		Granite	319 ± 4	LA-ICP-MS				[34]
		Granodiorite	319 ± 3	SHRIMP				[35]
		Basalt	318 ± 3	LA-ICP-MS				[4]
		Granite	317 ± 2	LA-ICP-MS				[36]
		Garnet bearing granite	316 ± 3	SHRIMP				[29]
		Granodiorite	316 ± 1	LA-ICP-MS	+3.0 to +12.6			[32]
		Quartz diorite	315 ± 4	SHRIMP				[31]
		Basalt	315 ± 4	LA-ICP-MS				[4]
		Monzonitic granite	314 ± 2	LA-ICP-MS				[37]
		Quartz diorite	313 ± 5	SHRIMP				[31]
		Granodiorite	312 ± 1	LA-ICP-MS				[38]

Unit	Episode	Lithology	Zircon age (Ma)	Method	$\varepsilon_{Hf(t)}$ (Zircon)	$\varepsilon_{Nd(t)}$ (Whole rock)	Initial $^{87}Sr/^{86}Sr$ (whole rock)	Reference
		Monzonitic diorite	312 ± 4	SHRIMP				[39]
		Diorite	311 ± 2	SHRIMP				[35]
		Gabbroic diorite	310 ± 5	SHRIMP	+5.4 to +11.5	+2.5	0.7052	[2]
		Quartz diorite	310 ± 2	SHRIMP				[35]
		Volcanic rock	310 ± 1	LA-ICP-MS				[38]
		Quartz diorite	309 ± 8	SHRIMP		−0.2	0.7056	[23]
		Volcanic rock	309 ± 2	LA-ICP-MS				[40]
		Monzonitic granite	308 ± 2	LA-ICP-MS				[36]
		Monzonitic granite	307 ± 2	SHRIMP				[41]
		Volcanic rock	307 ± 6	LA-ICP-MS				[40]
		Rhyolite	303 ± 6	SHRIMP				[33]
Mongolia Hutag Uul		Gneissic granite	1784 ± 7	SHRIMP				Shi et al., unpub lished
		Granodiorite	454 ± 10	SHRIMP				Shi et al., unpub lished
Southern Orogen	Ordo vician	Tonalite	491 ± 8	SHRIMP		+5.2	0.7047	[8]
		Diorite	472					[42]
		Dacite	459 ± 8	SHRIMP				[43]
		Dacite	458 ± 3	SHRIMP		+7.1	0.7058	[8]
		Quartz diorite	454 ± 4	SHRIMP		+2.0	0.7056	[8]
		Diorite	452 ± 3	SHRIMP				[8]
		Trondjemite	451 ± 7	SHRIMP				[43]
		Granodiorite	450					[42]
Northern margin of NCC	Carboni ferous	Biotite K-feldspar granite	342 ± 5	SHRIMP				[44]
		Quartz diorite	324 ± 6	SHRIMP				[45]
		Quartz diorite	311 ± 2	SHRIMP				[45]
		Granodiorite	310 ± 5	SHRIMP				[45]
		Quartz diorite	302 ± 4	SHRIMP				[45]
Ophiolitic block								
Erlianhot-Hegenshan		Gabbro	354 ± 7	SHRIMP		+9.8	0.7043	[3]
		Gabbro	298 ± 9	SHRIMP		+8.1	0.7037	[25]
Jiaoqier-		Gabbro	483 ± 2	SHRIMP				[8]

Unit	Episode	Lithology	Zircon age (Ma)	Method	$\varepsilon_{Hf(t)}$ (Zircon)	$\varepsilon_{Nd(t)}$ (Whole rock)	Initial $^{87}Sr/^{86}Sr$ (whole rock)	Reference
Xilinhot								
Solonker-Linxi		Trondjemite	324 ± 3	SHRIMP		+8.4	0.7039	[3]
		Plagiogranite	288 ± 6	SHRIMP		+7.8	0.7039	[3]
		Gabbro	284 ± 4	SHRIMP		+6.8	0.7043	[3]
Wenduer miao-Xar Moron		Gabbro	480 ± 3	SHRIMP		+9.2	0.7059	[8]

Table 1. Summary of zircon ages, Hf isotopic data and whole-rock Sr-Nd isotopic data.

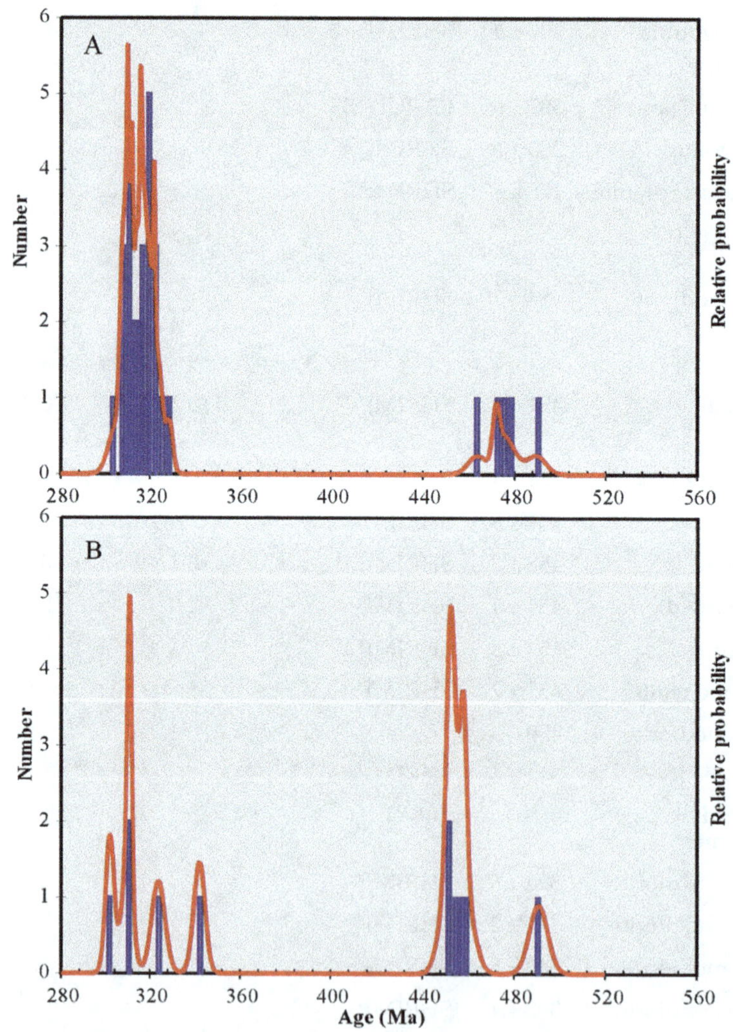

Figure 2. Cumulative plot for zircon U-Pb ages of Ordovician and Carboniferous rocks from Central Inner Mongolia (data and references are in **Table 1**). A for rocks from the Northern Block, which consists of the Southern Mongolia (or Hutag Uul) block and the northern orogen; and B for rocks from the Southern Block, which is composed of the northern margin of North China Craton and the southern orogen [3].

Sample	MS02-7	MB1-3	MS3-5	MB1-6	MB1-1	MB1-5	MB1-2	MB1-4
Lithology	Tonalite	Granodiorite	Tonalite	Tonalite	Granodiorite	Quartz-diorite	Granite	Granite
Age (Ma)	479 ± 8	472 ± 3	464 ± 8	329 ± 3	ca. 320	320 ± 3	297 ± 2	--
SiO_2	61.13	67.37	61.62	61.98	66.47	54.96	75.19	71.94
TiO_2	0.42	0.25	0.41	0.59	0.43	0.68	0.18	0.17
Al_2O_3	17.05	16.31	16.56	16.22	15.63	18.80	13.76	15.68
TFe_2O_3	5.88	3.68	5.62	5.82	4.17	7.98	1.92	1.30
MnO	0.14	0.08	0.14	0.08	0.06	0.12	0.02	0.02
MgO	2.34	1.14	2.27	2.95	1.69	3.65	0.70	0.47
CaO	5.69	3.91	5.78	4.93	3.43	6.25	0.39	1.54
Na_2O	3.56	4.26	3.08	3.22	3.37	3.14	3.92	5.47
K_2O	1.34	1.37	1.74	1.49	2.86	0.89	2.62	2.89
P_2O_5	0.19	0.12	0.18	0.16	0.14	0.22	0.09	0.11
LOI	1.84	1.40	2.78	2.48	1.81	2.95	1.32	0.73
TOTAL	99.58	99.89	100.18	99.92	100.06	99.64	100.11	100.32
Na_2O/K_2O	2.66	3.11	1.77	2.16	1.18	3.53	1.50	1.89
Sc	12.8	5.60	13.4	17.1	9.5	20.0	2.69	0.60
V	115	64	107	123	83	151	33.9	25.1
Cr	20.44	5.0	79.5	45	21	26	8.8	8.1
Co	12.1	6.0	10.6	17.3	10.7	21.8	3.23	2.80
Ni	10.3	3.6	18	29.1	11.4	19.8	7.0	3.9
Cu	5.4	9.1	6.2	38.7	10.3	51.4	23.0	15.8
Zn	51.5	39.2	49.5	55.9	42.7	87.8	18.3	30.9
Ga	16.5	17.4	16.3	16.9	16.1	19.4	12.7	18.1
Ge	1.38	1.48	1.42	1.45	1.17	1.28	1.21	0.78
Rb	32.26	51.1	42.09	69.08	96.9	24.05	99.6	66.5
Sr	649	711	604	304	373	473	198	581
Zr	84.8	78.3	81.9	149.4	171	52.6	75.1	104.4

Sample	MS02-7	MB1-3	MS3-5	MB1-6	MB1-1	MB1-5	MB1-2	MB1-4
Lithology	Tonalite	Granodiorite	Tonalite	Tonalite	Granodiorite	Quartz-diorite	Granite	Granite
Age (Ma)	479 ± 8	472 ± 3	464 ± 8	329 ± 3	ca. 320	320 ± 3	297 ± 2	--
Nb	3.64	4.73	3.2	5.23	6.20	4.96	6.12	1.96
Cs	0.55	0.794	0.8	1.25	1.12	1.54	1.82	1.27
Ba	685.0	471.8	862.4	241.8	687.1	173.8	487.8	511.8
Hf	2.43	2.09	2.43	3.59	4.50	1.33	2.18	2.88
Ta	0.24	0.26	0.23	0.45	0.55	0.25	0.61	0.19
Th	3.78	10.46	2.76	5.83	11.26	0.49	11.83	2.31
U	1	1.48	1.05	1.277	2.16	0.264	0.58	1.11
La	10.92	25.5	7.35	14.67	19.3	9.45	8.91	5.41
Ce	22.7	49.3	17.01	29.3	37.9	20.9	28.4	14.3
Pr	2.85	4.58	2.11	3.78	4.23	2.63	1.97	1.25
Nd	11.85	15.8	9.03	16.2	16.2	11.6	7.12	5.11
Sm	2.67	2.19	2.32	3.63	3.09	2.61	1.33	1.00
Eu	0.84	0.61	0.77	1.00	0.81	0.88	0.32	0.38
Gd	2.57	2.09	2.26	3.84	2.81	2.72	1.45	0.93
Tb	0.42	0.23	0.38	0.64	0.40	0.42	0.22	0.11
Dy	2.39	1.32	2.26	3.77	2.31	2.54	1.46	0.72
Ho	0.52	0.26	0.51	0.87	0.48	0.53	0.33	0.12
Er	1.52	0.78	1.39	2.32	1.29	1.41	0.95	0.35
Tm	0.25	0.12	0.23	0.39	0.20	0.22	0.16	0.038
Yb	1.64	0.95	1.61	2.61	1.42	1.50	1.18	0.37
Lu	0.27	0.15	0.28	0.44	0.24	0.24	0.18	0.028
Y	16.2	9.1	14.5	27.0	14.1	16.9	11.1	4.68
La/Yb	6.7	26.8	4.6	5.6	13.6	6.3	7.6	14.6
Sr/Y	40	78	42	11	27	28	18	124

Table 2. Major oxide (wt.%) and trace element (ppm) composition of representative samples.

Figure 3 shows the photographs of field occurrences and photomicrographs of some representative samples. **Figure 3A** was taken from Central Inner Mongolia to show the beautiful landscape; **Figure 3B** shows the Carboniferous volcanic rocks which are located in the Southern Block.

Granodiorite sample MB1-3 (**Figure 3C** and **3D**), collected from Baiyinbaolidao, southern Sonidzuoqi, which is located in the Northern Block, is medium-grained, foliated and consists of plagioclase (45–50 vol.%), quartz (20–25%), K-feldspar (10–15), biotite (5–10%), hornblende (1–5%), accessory zircon, apatite and sphene. Plagioclase is partially epidotized, sericitized and biotite grains are chloritized.

Figure 3. Photographs to show field occurrences and photomicrographs of some representative samples.

Tonalite sample MB1-6 (**Figure 3E** and **3F**), which is also collected from Baiyinbaolidao, Southern Sonidzuoqi, is medium-grained and consists of plagioclase (60–65%), quartz (20–25%), hornblende (10–15%) and biotite (1–5%) with trace amounts of zircon, apatite and sphene. Plagioclase is partially epidotized, and biotite grains are chloritized.

3. Petrogenesis of the Ordovician and Carboniferous volcanic rocks and granitoids

The Ordovician granitoid samples have intermediate to high-silica (61.13–67.37 wt.%), high Al_2O_3 (mostly >15 %), higher Na_2O than K_2O ($Na_2O > K_2O$, $Na_2O/K_2O = 1.77–3.11$), low MgO (<3%), low HREE (**Figure 4**), depleted HFSE (**Figure 5**), Y and Yb (Y < 18 ppm, Yb < 1.9 ppm), high Sr (604–711 ppm), Sr/Y mostly >40 (40.1–78.1) (**Table 2; Figure 6**) and low I_{Sr} with positive $\varepsilon_{Nd}(t)$ isotope ratios (**Table 1; Figure 7**). The Ordovician granitoid samples therefore represent adakitic compositions ([16, 48]; **Table 3**).

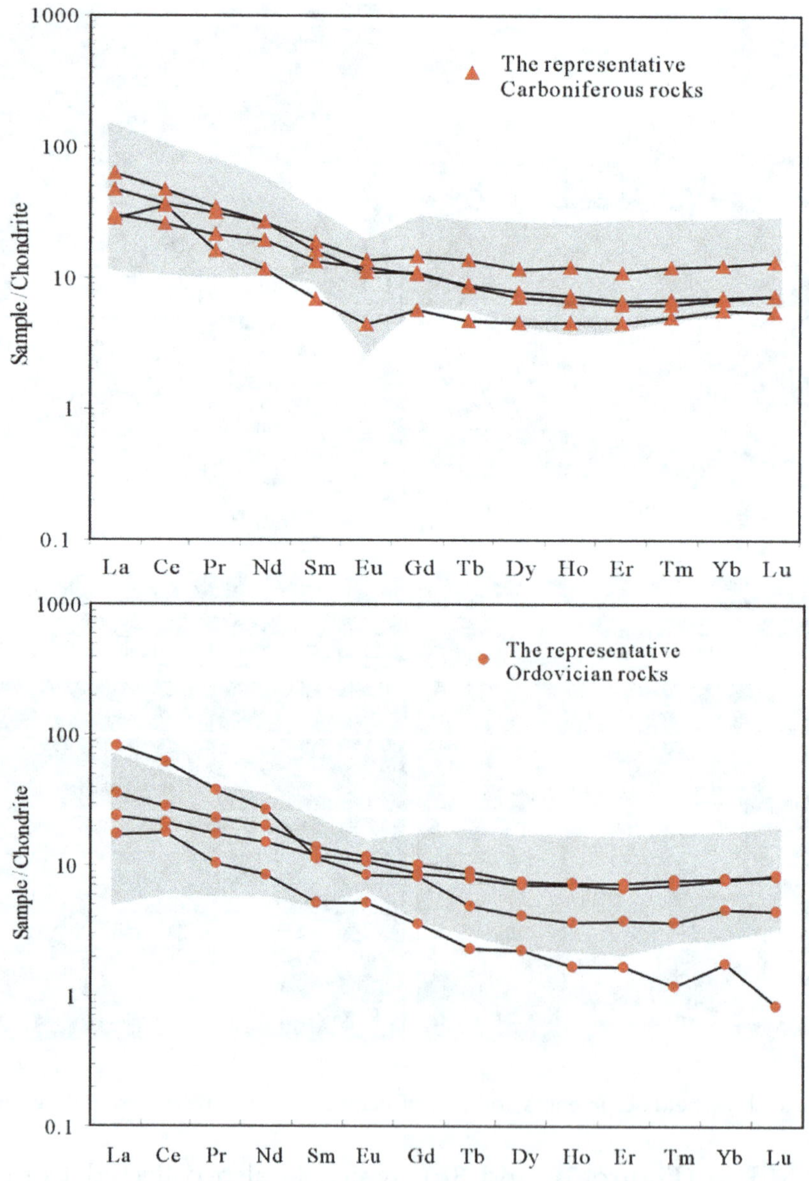

Figure 4. Chondrite (CHON)-normalized REE patterns for representative samples (grey fields show data from [7, 8, 43] for Ordovician granitoids; and from [23, 30, 32, 34, 36, 37] for Carboniferous granitoids). Chondrite values are from [46].

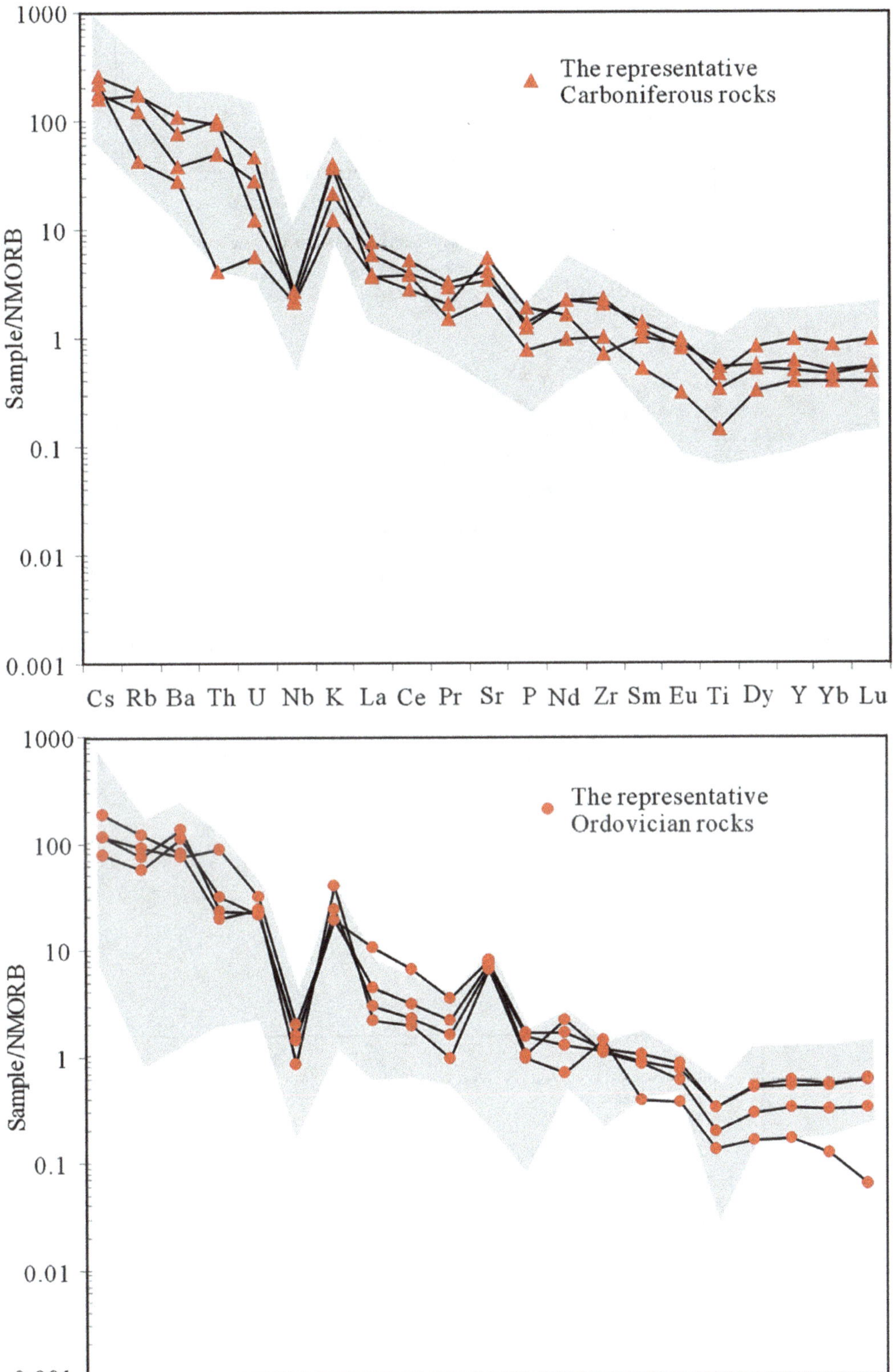

Figure 5. N-MORB-normalized trace element variation diagrams for representative samples (grey fields show data from [7, 8, 43] for Ordovician granitoids; and from [23, 30, 32, 34, 36, 37] for Carboniferous granitoids). N-MORB values are from [47].

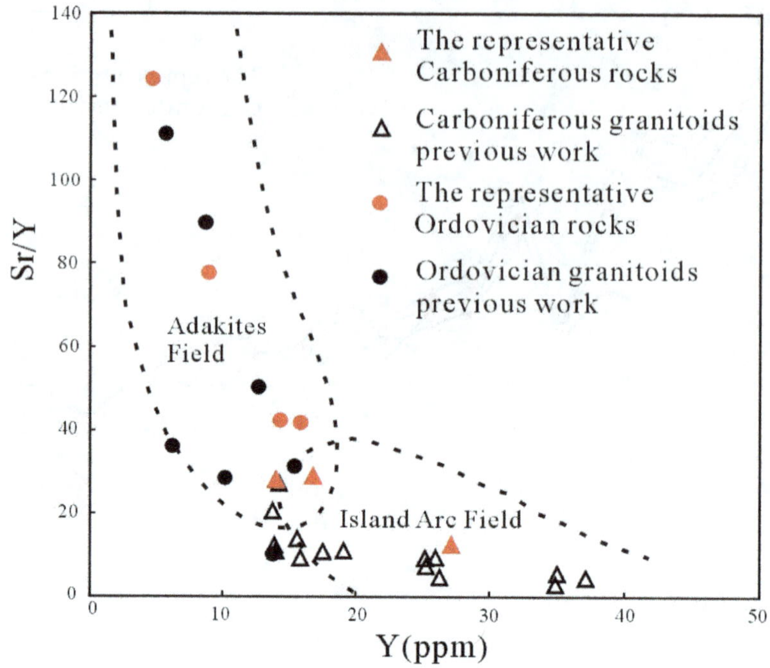

Figure 6. Y vs. Sr/Y plot showing adakitic rocks (after [18]) (data from [7, 8, 43] for Ordovician rocks; and from [23, 30, 32, 34, 36, 37] for Carboniferous rocks).

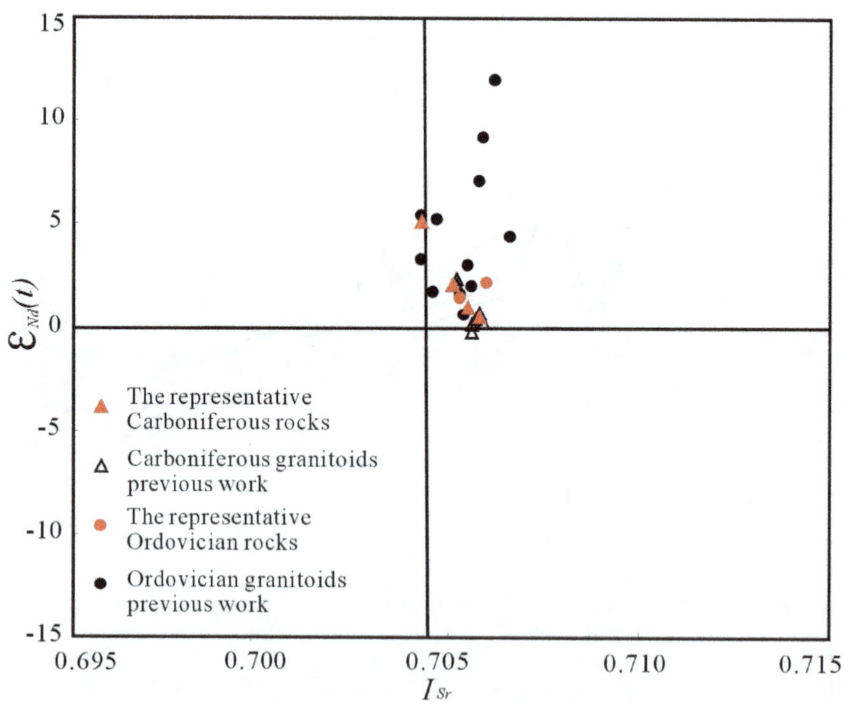

Figure 7. I_{Sr} vs. $\varepsilon_{Nd(t)}$ for some typical Ordovician and Carboniferous rocks with high Sr/Y ratio from Central Inner Mongolia (data from [8, 23]).

	Adakites			MS02-7 Tonalite	MB1-3 Granodiorite	MS3-5 Tonalite	Oceanic arc granites			Active continental margin arc granites	MB1-6 Tonalite	MB1-1 Granodiorite	MB1-5 Quartz-diorite
	Adakites[a]	Cook island[b]	Cerro Pampa[c]				Oman[d]	Little Port[d]	Jamaica[d]	Chile[d]			
SiO_2 (%)	≥56.0	61.4	62.6	61.13	67.37	61.62	70.1	69.5	68.4	74.5	61.98	66.47	54.96
Al_2O_3 (%)	≥15.0	18.4	17.3	17.05	15.68	16.56	12.0	14.60	14.44	12.52	16.22	15.63	18.80
Na_2O/K_2O	>1.00	7.75	3.82	2.66	3.11	1.77	15.75	4.37	1.18	0.65	2.16	1.18	3.53
MgO (%)	<3			2.34	1.14	2.27					2.95	1.69	3.65
Y (μg/g)	≤18.00	6		16.2	9.1	14.5	44	19	10	30	27.0	14.1	16.9
Yb (μg/g)	≤1.90	0.85	0.72	1.64	0.15	1.61	4.54		1.37	3.12	2.61	1.42	1.50
Sr (μg/g)	>400	1910	1886	649	711	604	200	274	210	93	304	373	473
Sr/Y	>20	319		40	78	42	4.6	14.4	21.0	3.1	11	26	28
Sr anomaly	Positive	Positive		Positive	Positive	Positive					Positive	Positive	Positive
Eu nomaly	Positive or weakly negative			Weakly negative	Negative	Positive					Negative	Negative	Weakly positive
Age (Ma)	<25 Ma	<24 Ma	ca. 12 Ma	479 ± 8	472 ± 3	464 ± 8					329 ± 3	ca. 320	320 ± 3

[a] [16, 49].
[b] Cook island adakites [50].
[c] Cerro Pampa adakites [51].
[d] [52].

Table 3. The comparison of geochemical characteristics between the rocks from Central Inner Mongolia, the typical adakitic and arc rocks.

The genesis of adakites is extensively debated, and there are four proposed origins, namely partial melting of young subducted lithosphere [16], melting of newly underplated lower continental crust [53], differentiation of a parental basaltic magma [54, 55] and melting of foundered mafic lower continental crust [56]. High-Al, high Na_2O and calc-alkaline adakites are generally interpreted to have formed due to the melting of subducted oceanic crust and are different from high-K, high total alkali (Na_2O + K_2O) and low Al_2O_3 adakites that form through melting of thickened basaltic lower continental crust [16, 51, 53, 57–60].

The Inner Mongolian Ordovician granitoids of this study have depleted HREE, Nb, positive Sr anomalies, low Y and Yb contents and positive to weakly negative Eu anomalies. These characteristics are consistent with the loss of plagioclase and the presence of garnet as residual phases, probably related to partial melting of the source material under eclogite-facies conditions [61, 62]. The petrology and geochemistry of the Ordovician adakitic granitoids indicate a contribution from melting of subducted oceanic crust in their formation rather than melting of thickened basaltic lower continental crust.

The Carboniferous samples in this area have intermediate to high-silica (54.96–66.47 wt.%), high Al_2O_3 (15.63–18.80 %), higher Na_2O than K_2O ($Na_2O > K_2O$, Na_2O/K_2O = 1.18–3.53), low HREE (**Table 2**; **Figure 4**), and with low I_{Sr} (0.7043–0.7060), positive $\varepsilon_{Nd}(t)$ (+1.0 to +5.1) and $\varepsilon_{Hf}(t)$ (+8.1 to +12.3) isotope ratios (**Table 1**; **Figures 7** and **8**). However, most of them have lower Sr and Sr/Y ratio than those of Ordovician adakitic granitoids in this area (**Table 2**; **Figure 6**), which are typical subduction-related arc granitoids [52, 63, 64] although still with adakite-like compositions [16, 48].

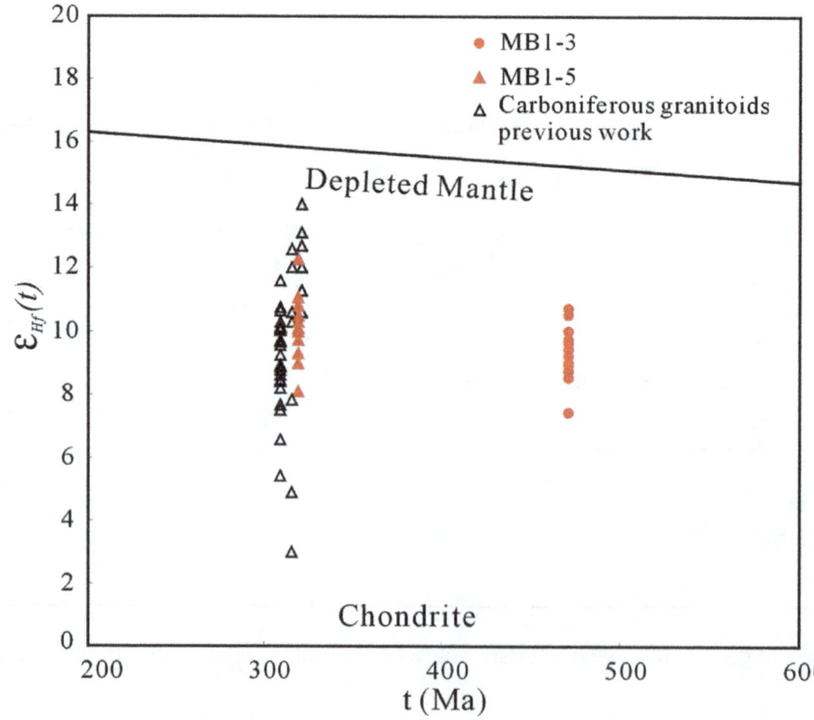

Figure 8. U-Pb age vs. $\varepsilon_{Hf(t)}$ for zircons from (data from [2], and [32] for Carboniferous granitoids).

4. Geodynamic significance of the Ordovician and Carboniferous volcanic rocks and granitoids

A subduction-accretion complex usually forms along a convergent plate boundary where an oceanic plate subducts beneath another oceanic or continental plate [65]. Early Paleozoic arc plutonism as part of trench-island arc-basin systems ([6, 8, 21, 22]; **Table 2**) occurred in the southern orogen, along the northern margin of the North China Craton, and late Silurian molasse deposits unconformably overlie these rocks [6, 66]. Coeval adakitic plutonism is emplaced in the northern orogen, along the southern margin of the Mongolian terrane [20]. Silurian high-pressure metamorphic rocks [67] and Silurian syncollisional magmatism in the northern orogen along the Solonker suture [68] were also reported. All these features indicate an early-middle Paleozoic subduction and accretion system along the northern margin of the North China Craton and the southern margin of the Mongolian terrane. After demise of the ocean in the southern orogen, caused by subduction of a ridge crest and by ridge collision with supra-subduction zone ophiolite in the Silurian [8], the southern orogen became tectonically consolidated and turned into a post-orogenic setting [69].

There has been some debate about whether the Carboniferous calc-alkaline granitoids formed in a subduction zone [23, 30] or in a late- to post-orogenic setting [31]. Carboniferous calc-alkaline plutonic rocks (ca. 328–308 Ma) in the northern orogen were suggested by [2, 23, 30] as subduction genesis, which can be related to the northward subduction of Asian ocean slab. Bao et al. [31], however, thought these Carboniferous granitoids formed in a Late Paleozoic rift area because of Permian bimodal volcanic rocks. These Carboniferous granitoids include variably foliated gabbro, diorite, quartz diorite, granodiorite, tonalite and granite [23, 30], which belong to low-K tholeiitic and calk-alkaline series, and are enriched in large ion lithophile elements (LILE) and depleted in high field strength elements (HFSE) [2, 23, 30], low I_{Sr}, positive $\varepsilon_{Nd}(t)$ and $\varepsilon_{Hf}(t)$ isotope ratios ([23]) showing subduction-related arc granitoids characteristics [52, 63, 64].

Additionally, a subduction-accretion complex was identified from previously defined late Carboniferous and early Permian strata in the Daqing pasture, southern Xiwuqi, Inner Mongolia [4]. In addition to this subduction-accretion complex, most magmatic rocks are considered to have formed in a subduction setting [23, 30], and the spatial configuration of both geological units indicates that the subduction polarity was from south to north [4] along the northern orogen.

Carboniferous granitoids on the northern margin of North China craton also have the composition of tholeiitic and calk-alkaline island-arc rocks and adakitic compositions [45], however, low negative whole-rock $\varepsilon_{Nd}(t)$ and zircon $\varepsilon_{Hf}(t)$ isotope ratios indicate that they were derived mainly from anatectic melting of the ancient lower crust with some involvement of mantle materials [70]. The Carboniferous plutons were interpreted as subduction-related and emplaced in an Andean-style continental-margin arc [70].

On the northern margin of the North China craton, however, Carboniferous eclogites are exposed at least 200 km south of the Solonker suture zone and have tholeiitic protoliths (MORB and IAT), and eclogite-facies metamorphism reflects deep subduction of oceanic lithosphere [71]. The granitoids (330–298 Ma) of this area were emplaced and deformed during, and/or shortly after eclogite-facies metamorphism (ca. 331–319 Ma) [71]. This close temporal relationship indicates that magmatism closely followed the exhumation of the high-pressure metamorphic terrane [3].

5. A possible model for the discrete evolution of CAOB

The southeastern CAOB was formed by the concurrent two-way subduction of Paleo-Central Asian Ocean towards opposing continental margins in the early Paleozoic (**Figure 9A**). In the south is an arc-trench complex, which can be regarded as an analogue of the Izu-Bonin-Mariana arc [72], and in the north a product of ridge-trench interaction [8]. In the late Paleozoic, however, Andean-type orogenesis was induced by subduction of Central Asian Ocean beneath either the northern (e.g. [4]) or southern (e.g. [45]) continental blocks (**Figure 9B**). Plutonic magmatism [45] was accompanied by exhumation of a high-pressure metamorphic terrane [71] in the south; and a subduction-accretion complex [4], together with most arc-related magmatic rocks [23, 30] was formed along the northern orogen.

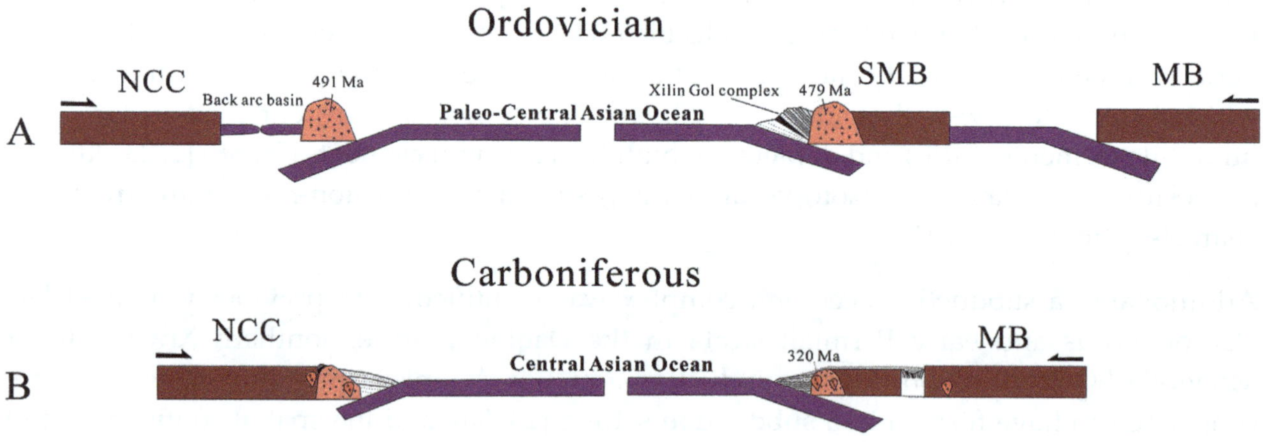

Figure 9. A possible model for Ordovician and Carboniferous evolution of Central Inner Mongolia. Abbreviation: NCC, North China Craton; SMB, South Mongolia Block; MB, Mongolia Block.

6. Modern equivalent

6.1. Cook Island and Cerro Pampa adakites

Cenozoic andesitic to dacitic rocks collected from Cerro Pampa [51] and andesites from Cook Island [50] have intermediate to high-silica, high Al_2O_3, higher Na_2O than K_2O, low HREE, depleted HFSE, Y and Yb, high Sr, and high Sr/Y ratios (**Table 3**), and low I_{Sr} with positive ε_{Nd} isotope ratios. The samples, therefore, represent adakites [50, 51]. Cerro Pampa adakitic magmas formed in response to melting of hot slab that was subducting beneath South America [51], and similar petrogenesis for the Austral Volcanic Zone adakites [50]. Ordovician adakitic rocks from Central Inner Mongolia show similar petrogenesis and geotectonic setting with the Cenozoic adakites from Cook Island [50], Cerro Pampa [51] and Aleutian arc [16].

6.2. Oman and Chile volcanic arc granites

Volcanic arc granites from Oman and Chile have high-silica, intermediate Al_2O_3, low HREE [52] (**Table 3**), and with low Sr and Sr/Y ratios than the adakites (**Table 3**), which are typical subduction-related arc granitoids derived from melting in the mantle wedge. Most Carboniferous volcanic rocks and granitoids present similar petrogenesis and geotectonic setting with the Cenozoic subduction-related arc granitoids.

7. Conclusions

1. The Ordovician and Carboniferous volcanic rocks and granitoids are mostly intermediate to high-silica, high Na_2O/K_2O ratio, high Sr/Y ratios. They are characterized by LREE enrichment and exhibit typical negative Nb anomalies. All samples show positive $\varepsilon_{Hf}(t)$, $\varepsilon_{Nd}(t)$ values and low I_{Sr}.

2. The Ordovician rocks show higher Sr/Y ratio than the Carboniferous rocks, suggesting that the former represent adakitic rocks and the latter are typical subduction-related arc rocks with adakite-like compositions.

3. The Central Asian Orogenic Belt evolved through several subduction systems with different polarities in Central Inner Mongolia, namely an early-middle Paleozoic subduction and accretion system along the northern margin of the North China Craton and the southern margin of the Mongolian terrane, and late Paleozoic northward subduction along the northern orogen and exhumation of a high-pressure metamorphic terrane on the northern margin of the North China Craton.

Acknowledgements

The editorial patience and the comments of Dr. Karoly Nemeth are appreciated. This study was financially supported by the National Natural Science Foundation of China (Grant nos. 40703012) and Geological Survey of China (Grant nos. 1212011121075, 12120114020901, 12120114064301, and 1212011120332).

Author details

Yuruo Shi

Address all correspondence to: shiyuruo@bjshrimp.cn

Beijing SHRIMP Center, Institute of Geology, Chinese Academy of Geological Sciences, Beijing, China

References

[1] Xiao W.J., Windley B.F., Hao J., Zhai M.G., 2003. Accretion leading to collision and the Permian Solonker suture, Inner Mongolia, China: termination of the Central Asian orogenic belt. Tectonics 22, 8–20.

[2] Chen B., Jahn B.M., Tian W., 2009. Evolution of the Solonker suture zone: constraints from zircon U-Pb ages, Hf isotopic ratios and whole-rock Nd–Sr isotope compositions of subduction- and collision-related magmas and forearc sediments. Journal of Asian Earth Sciences 34, 245–257.

[3] Jian P., Liu D.Y., Kröner A., Windley B.F., Shi Y.R., Zhang W., ZhangF.Q., MiaoL.C., ZhangL.Q., Tomurhuu D., 2010. Evolution of a Permian intraoceanic arc–trench system in the Solonker suture zone, Central Asian Orogenic Belt, China and Mongolia. Lithos 118(1), 169–190.

[4] Liu J.F., Li J.Y., Chi X.G., Qu J.F., Hu Z.C., Fang S., Zhang Z., 2013. A late-Carboniferous to early early-Permian subduction–accretion complex in Daqing pasture, southeastern Inner Mongolia: evidence of northward subduction beneath the Siberian paleoplate southern margin. Lithos 177, 285–296.

[5] Shi Y.R., Liu C., Deng J.F., Jian P., 2014. Geochronological frame of granitoids from Central Inner Mongolia and its tectonomagmatic evolution. Acta Petrologica Sinica 30 (11), 3155–3171 (in Chinese with English abstract).

[6] Tang K.D., 1990. Tectonic development of Paleozoic foldbelts at the north margin of the Sino-Korean craton. Tectonics 9, 249–260.

[7] Shi Y.R., Liu D.Y., Zhang Q., Jian P., Zhang F.Q., Miao L.C., ShiG.H., ZhangL.Q., Tao H., 2004. SHRIMP dating of diorites and granites in southern Suzuoqi, Inner Mongolia. Acta Geologica Sinica 78(6), 789–799. (in Chinese with English abstract).

[8] Jian P., Liu D.Y., Kröner A., Windley B.F., Shi Y.R., Zhang F.Q., Shi G.H., Miao L.C., Zhang W., Zhang Q., 2008. Time scale of an early to mid-Paleozoic orogenic cycle of the long-lived Central Asian Orogenic Belt, Inner Mongolia of China: Implications for continental growth. Lithos 101, 233–259.

[9] Xu B., Zhao P., Wang Y.Y., Liao W., Luo Z.W., Bao Q.Z., Zhou Y.H., 2015. The pre-Devonian tectonic framework of Xing'an-Mongolia orogenic belt (XMOB) in north China. Journal of Asian Earth Sciences 97, 183–196.

[10] Sengör A.M.C., Natal'in B.A., Burtman V.S., 1993. Evolution of the Altaid tectonic collage and Paleozoic crustal growth in Eurasia. Nature 364, 299–307.

[11] Coleman R.G., 1989. Continental growth of Northwest China. Tectonics 8, 621–635.

[12] Mossakovskii A.A., Ruzhentsev S.V., Samygin S.G., Kheraskova T.N., 1993. Central Asian Foldbelt: geodynamic evolution and formation history. Geotektonika 6, 3–32 (in Russian).

[13] Kröner A., Windley B.F., Badarch G., Tomurtogoo O., Hegner E., Jahn B.M., Gruschka S., Khain E.V., Demoux A., Wingate M.T.D., 2007, Accretionary growth and crust formation in the Central Asian Orogenic Belt and comparison with the Arabian-Nubian-Shield. Geological Society of America, Memorials 200, 181–209.

[14] Kröner A., Kovach V., Belousova E., Hegner E., Armstrong R., Dolgopolova A. Seltmann R., Alexeiev D.V., Hoffmann J.E., Wong J., Sun M., Cai K., Wang T., Tong Y., Wilde S.A., Degtyarev K.E., Rytsk E., 2014. Reassessment of continental growth during the accretionary history of the Central Asian Orogenic Belt. Gondwana Research 25, 103–125.

[15] Windley B.F., Alexeiev D., Xiao W.J., Kröner A., Badarch G., 2007. Tectonic models for accretion of the Central Asian Orogenic Belt. Journal of the Geological Society 164 (1), 31–47.

[16] Defant M.J., Drummond M.S. 1990. Derivation of some modern arc magmas by melting of young subducted lithosphere. Nature 347, 662–665.

[17] Gill J.B., 1981. Orogenic Andesites and Plate Tectonics. Springer, Berlin.

[18] Defant M.J., Drummond M.S., 1993. Mount St. Helens: potential example of the partial melting of the subducted lithosphere in a volcanic arc. Geology 21(6), 547–550.

[19] Jahn B.M., Wu F.Y., Chen B., 2000. Granitoids of the Central Asian Orogenic Belt and continental growth in the Phanerozoic. Transactions Royal Society of Edinburgh: Earth Sciences 91, 181–193.

[20] Shi Y.R., Liu D.Y., Zhang Q., Jian P., Zhang F.Q., Miao L.C., Shi G.H., Zhang L.Q., Tao H., 2005a. The petrogenesis and SHRIMP dating of the Baiyinbaolidao adakitic rocks in south Suzuqi, Inner Mongolia. Acta Petrologica Sinica 21, 143–150 (in Chinese with English abstract).

[21] Zhang W., Jian P., 2008. SHRIMP dating of Early Paleozoic granites from North Damaoqi, Inner Mongolia. Acta Geologica Sinica 82, 778–787 (in Chinese with English abstract).

[22] Zhang W., Jian P., Kröner A., Shi Y.R., 2013. Magmatic and metamorphic development of an early to mid-Paleozoic continental margin arc in the southernmost Central Asian Orogenic Belt, Inner Mongolia, China. Journal of Asian Earth Sciences 72, 63–74.

[23] Chen B., Jahn B.M., Wilde S., Xu B., 2000. Two contrasting Paleozoic magmatic belts in northern Inner Mongolia, China: petrogenesis and tectonic implications. Tectonophysics 328, 157–182

[24] Badarch G., Cunningham W.D., Windley B.F., 2002. A new terrane subdivision for Mongolia: implications for the Phanerozoic crustal growth of Central Asia. Journal of Asian Earth Sciences 21 (1), 87–110.

[25] Miao L.C., Fan W.M., Liu D.Y., Zhang F.Q., Sh,i Y.R., Guo, F., 2008. Geochronology and geochemistry of the Hegenshan ophiolitic complex: Implications for late-stage tectonic evolution of the Inner Mongolia-Daxinganling Orogenic Belt, China. Journal of Asian Earth Sciences 32: 348–370.

[26] Dobretsov N.L., Berzin N.A., Buslov M.M., 1995. Opening and tectonic evolution of the Paleo-Asian Ocean. International Geology Review 37 (4), 335–360.

[27] Jian P., Shi Y.R., Zhang F.Q., Miao L.C., Zhang L.Q., Kröner A., 2007. Geological excursion to Inner Mongolia, China, to study the accretionary evolution of the southern margin of the Central Asian Orogenic Belt. Structural and tectonic correlation across the Central Asia Orogenic Collage: Implications for continental growth and intracontinental deformation (Abstracts and Excursion guidebook). The Third International workshop and field excursions for IGC Project 480: 49–72.

[28] Li J.Y., 2006. Permian geodynamic setting of Northeast China and adjacent regions: closure of the Paleo- Asian Ocean and subduction of the Paleo-Pacific Plate. Journal of Asian Earth Sciences 26, 207–224.

[29] Shi G.H., Liu D.Y., Zhang F.Q., Jian P., Miao L.C., Shi Y.R., Tao H., 2003. Zircon SHRIMP U-Pb geochronology and significance of the Xilinhot metamorphic complex, Inner Mongolia, China. Chinese Science Bulletin 48(20), 2187–2192 (in Chinese).

[30] Liu J.F., Chi X.G., Zhang X.Z., Ma Z.H., Zhao Z., Wang T.F., Hu Z.C., Zhao X.Y., 2009. Geochemical characteristic of Carboniferous quartz-diorite in the southern Xiwuqi area, Inner Mongolia and its tectonic significance. Acta Geologica Sinica 83 (3), 365–376 (in Chinese with English abstract).

[31] Bao Q.Z., Zhang C.J., Wu Z.L., Wang H., Li W., Sang J.H., Liu Y.S., 2007. SHRIMP U-Pb zircon geochronology of a Carboniferous quartz-diorite in Baiyingaole area, Inner Mongolia and its implications. Journal of Jilin University (Earth Science Edition) 37 (1), 15–23 (in Chinese with English abstract).

[32] Hu C.S., Li W.B., Xu C., Zhong R.C., Zhu F., 2015. Geochemistry and zircon U-Pb-Hf isotopes of the granitoids of Baolidao and Halatu plutons in Sonidzuoqi area, Inner

Mongolia: Implications for petrogenesis and geodynamic setting. Journal of Asian Earth Sciences 97, 294–306.

[33] Xin H.T., Teng X.J., Cheng Y.H., 2011. Stratigraphic Subdivision and Isotope Geochronology Study on the Baoligaomiao Formation in the East Ujimqin County, Inner Mongolia. Geological Survey and Research 34, 1–8 (in Chinese with English abstract).

[34] Liang Y.W., Yu C.L., Shen G.Z., Sun Q.R., Li J.W., Yang Y.C., She H.Q., Zhang B., Tan G., 2013. Geochemical characteristics of granites in the Suonaga Pb-Zn-Ag deposit of Dong Ujimqin Banner, Inner Mongolia, and their tectonic and ore-forming implications. Geology in China 40, 767–779 (in Chinese with English abstract).

[35] Xue H.M., Guo L.J., Hou Z.Q., Tong Y., Pan X.F., Zhou X.W., 2010. SHRIMP zircon U-Pb ages of the middle Neopaleozoic unmetamorphosed magmatic rocks in the southwestern slope of the DaHinggan Mountains, Inner Mongolia. Acta Petrologica et Mineralogica 29, 811–823 (in Chinese with English abstract).

[36] Xu L.Q., Ju W.X., Liu C., He H.Y., Li M.Y., 2012. Sr-Yb classification and genesis of late Carboniferous granites in Arenshaobu area of Erenhot, Inner Mongolia. Geological Bulletin of China 31, 1410–1419 (in Chinese with English abstract).

[37] He F.B., Xu J.X., Gu X.D., Cheng X.B., Wei B., Li Z., Liang Y.N., Wang Z.L., Huang Q., 2013. Ages, origin and geological implication of the Amuguleng composite granite in East Ujimqin Banner, Inner Mongolia. Geological Review 59, 1150–1164 (in Chinese with English abstract).

[38] Li K., Zhang Z.C., Feng Z.S., Li J.F., Tang W.H., Luo Z.W., C.Y., 2015. Two-phase magmatic events during late Paleozoic in the north of the Central Inner Mongolia-Da Hinggan orogenic belt and its tectonic significance. Acta Geologica Sinica 89, 272–288 (in Chinese with English abstract).

[39] Yun F., Nie F.J., Jiang S.H., Liu Y., Zhang W.Y., 2011. Zircon SHRIMP U-Pb age of Monuogechin monzodiorite of Inner Mongolia and its geological significance. Mineral Deposits 30, 504–510 (in Chinese with English abstract).

[40] Li K., Zhang Z.C., Feng Z.S., Li J.F., Tang W.H., Luo Z.W., 2014. Zircon SHRIMP U-Pb dating and its geological significance of the Late-Carboniferous to Early-Permian volcanic rocks in Bayanwula area, the central of Inner Mongolia. Acta Petrologica Sinica 30, 2041–2054 (in Chinese with English abstract).

[41] Cheng Y.H., Teng X.J., Xin H.T., Yang J.Q., Ji S.P., Zhang Y., Li Y.F., 2012. SHRIMP zircon U-Pb dating of granites in Mahonondor area, East Ujimqin Banner, Inner Mongolia. Acta Petrologica et Mineralogica 31, 323–334 (in Chinese with English abstract).

[42] Xu L.Q., Deng J.F., Chen Z.Y., Tao J.X. 2003. The identification of Ordovician adakites and its signification in northern Damao,Inner Mongolia. Geoscience 17, 428–434 (in Chinese with English abstract).

[43] Liu D.Y., Jian P., Zhang Q., Zhang F.Q., Shi Y.R., Shi G.H., Zhang F.Q., Tao H., 2003. SHRIMP dating of adakites in the Tulingkai ophiolite,Inner Mongolia: evidence for the early Paleozoic subduction. Acta Geologica Sinica 77, 318–327 (in Chinese with English abstract).

[44] Zhang C., Liu S.W., Han B.F., He G.Q. Huang B.L., 2007a. SHRIMP U-Pb dating of Dashigou biotite-K-felspar granites in Shangdu, Inner Mongolia, and its significance. Acta Petrologica Sinica 23(3), 591–596 (in Chinese with English abstract).

[45] Zhang S.H., Zhao Y., Song B., Yang Z.Y., Hu J.M., Wu H., 2007b. Carboniferous granitic plutons from the northern margin of the North China block: Implications for a Late Paleozoic active continental margin. Journal of the Geological Society London 164, 451–463.

[46] Boynton W.V., 1984. Geochemistry of the rare earth elements: meteorite studies. In: Henderson P. (Ed.), Rare Earth Element Geochemistry. Elsevier, 63–114.

[47] Sun S.S., McDonough W.F., 1989. Chemical and isotope systematics of oceanic basalts: implications for mantle composition and processes. Geological Society, London, Special Publications 42, 313–345.

[48] Martin H., 1986. Effect of steeper Archean geothermal gradient on geochemistry of subduction-zone magmas. Geology 14(9), 753–756.

[49] Drummond M.S., Defant M.J., 1990. A model for trondhjemite-tonalite-dacite genesis and crustal growth via slab melting: Archean to modern comparisons. Journal of Geophysical Research: Solid Earth (1978–2012), 95(B13), 21503–21521.

[50] Stern C.R., Kilian R., 1996. Role of the subducted slab, mantle wedge and continental crust in the generation of adakites from the Andean Austral Volcanic Zone. Contributions to Mineralogy and Petrology 123(3), 263–281.

[51] Kay S.M., Ramos V.A., Marquez M., 1993. Evidence in Cerro Pampa volcanic rocks of slab melting prior to ridge trench collision in southern South America. Journal of Geology 101, 703–714.

[52] Pearce J.A., Harris N.B.W., Tindle A.G., 1984. Trace-element discrimination diagrams for the tectonic interpretation of granitic rocks. Journal of Petrology 25, 956–983.

[53] Atherton M.P., Petford N., 1993. Generation of sodium-rich magmas from newly underplated basaltic crust. Nature 362, 144–146.

[54] Castillo P.R., Janney P.E., Solidum R.U., 1999. Petrology and geochemistry of Camiguin Island, southern Philippines: insights to the source of adakites and other lavas in a complex arc setting. Contributions to Mineralogy and Petrology 134, 33–51.

[55] Castillo P.R., 2006. An overview of adakite petrogenesis. Chinese Science Bulletin 51, 257–268.

[56] Gao S., Rudnick R.L., Yuan H.L., Liu X.M., Liu Y.S., Xu W.L., LingW.L., AyersJ., WangX.C., Wang Q.H., 2004. Recycling lower continental crust in the North China craton. Nature 432, 892–897.

[57] Peacock S.M., Rushmer T., Thompson A.B., 1994. Partial melting of subducting oceanic crust. Earth and Planetary Science Letters 121, 227–244.

[58] Drummond M.S., Defant M.J., Kepezhinskas P.K., 1996. Petrogenesis of slab derived trondhjemite-tonalite-dacite/adakite magmas. Geological Society of America Special Papers 315, 205–215.

[59] Defant M.J., Xu J.F., Kepezhinskas P., Wang Q., Zhang Q., Xiao L., 2002. Adakites: some variations on a theme. Acta Petrologica Sinica 18, 129–142.

[60] Kay S.M., Mpodozis C., 2002. Magmatism as a probe to the Neogene shallowing of the Nazca plate beneath the modern Chilean flat-slab. Journal of South American Earth Sciences 15, 39–57.

[61] Rapp R.P., Watson E.B., Miller C.F., 1991. Partial melting of amphibolite/eclogite and the origin of Archean trondhjemites and tonalites. Precambrian Research 51, 1–25.

[62] Rapp R.P., Watson E.B., 1995. Dehydration melting of metabasalt at 8–32 kbar: implications for continental growth and crust-mantle recycling. Journal of Petrology 36, 891–931.

[63] Harris N.B.W., Pearce J.A., Tindle A.G., 1986. Geochemical characteristics of collision zone magmatism. In: Coward M.P., Ries A.C. (Eds.), Collision Tectonics: Geological Society Special Publications 19, 67–81.

[64] Pearce J.A., 1996. Sources and settings of granitic rocks. Episodes 19, 120–125.

[65] Isozaki Y., 1996. Anatomy and genesis of a subduction-related orogen: A new view of geotectonic subdivision and evolution of the Japanese Islands. Island Arc 5, 289–320.

[66] Zhang Y., Tang K., 1989. Pre-Jurassic tectonic evolution of intercontinental region and the suture zone between the North China and Siberian platforms. Journal of South East Asian Earth Sciences 3, 47–55.

[67] Yan Z.Y., Tang K.D., Bai J.W., Mo Y.C., 1989. High pressure metamorphic rocks and their tectonic environment in northeastern China. Journal of Southeast Asian Earth Sciences 3, 303–313.

[68] Shi Y.R., Liu D.Y., Jian P., Zhang Q., Zhang F.Q., Miao L.C., Shi G.H., Zhang L.Q., Tao H., 2005b. Zircon SHRIMP dating of K-rich granites in Sonid Zuoqi. Central Inner Mongolia. Geological Bulletin of China 25, 424–428 (in Chinese with English abstract).

[69] Shi Y.R., Liu D.Y., Miao L.C., Zhang F.Q., Jian P., Zhang W., HouK.J., Xu J.Y., 2010. Devonian A-type granitic magmatism on the northern margin of the North China Craton: SHRIMP U-Pb zircon dating and Hf-isotopes of the Hongshan granite at Chifeng, Inner Mongolia, China. Gondwana Research 17(4), 632–641.

[70] Zhang S.H., Zhao Y., Song B., Hu J.M., Liu S.W., Yang Y.H., 2009. Contrasting Late Carboniferous and Late Permian-Middle Triassic intrusive suites from the northern margin of the North China craton: Geochronology, petrogenesis, and tectonic implications. Geological Society of America Bulletin 121(1–2), 181–200.

[71] Ni Z.Y., Zhai M.G., Wang R.M., Tong Y., 2006. Late Paleozoic retrograded eclogites from within the northern margin of the North China Craton: evidence for subduction of the Paleo-Asian ocean. Gondwana Research 9, 209–224.

[72] Stern R.J., Bloomer S.H., 1992. Subduction zone infancy: examples from the Eocene Izu-Bonin-Mariana and Jurassic California arcs, Geological Society of America Bulletin 104, 1621–1636.

6

Effect of Volcano-Polluted Seawater on the Corrosion Behaviour of Different Alloys

Rita Sánchez-Tovar, Ramón Fernández-Domene, Rafael Leiva-García, Clara Escrivà-Cerdán, Bianca Lucas-Granados and José García-Antón

Abstract

During a subsea volcano eruption, gases and thermal water emissions are released. This might change the behaviour of the materials that are in contact with the seawater caused by the decrease of the pH value. For this reason, the materials for marine applications are selected to maintain the integrity of the structure and to be corrosion resistant. In spite of this, corrosion can cause great damage to marine steel infrastructures such as bridges, wharfs, platforms and pipeline systems. These corrosion problems could be aggravated if the medium is altered, due to volcano emissions, since the resistance of the surface film is influenced by the environmental conditions.

Electrochemical techniques are useful to evaluate the corrosion behaviour of different metals and alloys. However, literature regarding the influence of underwater-volcano-polluted seawater on the passivation/corrosion behaviour of stainless steels is scarce. According to this, the objective of the chapter is the evaluation of the influence of an underwater volcano eruption on the corrosion behaviour of different materials that might be used in seawater environments. Electrochemical techniques are applied to evaluate the performance of the different materials when they are immersed in polluted volcano seawater. The results are relevant in the evaluation of natural hazard risk assessments involving critical infrastructures.

Keywords: submarine volcano, corrosion, water chemistry, electrochemistry, stainless steels

1. Introduction

The aim of natural hazard risk assessment is the evaluation of the extent and nature of risk in a particular area by evaluating potential hazards that together could harm people, property and services. Therefore, an effective natural hazard risk assessment requires the characterisation of both hazards and vulnerabilities of exposed elements [1]. Disasters such as the Eyjafjallajökull eruption in Iceland (2010), or the Tōhoku earthquake and tsunami in Japan (2011), suggest the need for effective natural hazard risk management and sustainable development.

Threats and issues with volcanoes are important, around 600 million people are living in areas that could be affected by volcano eruptions [2]. Therefore, the assessment of the volcano impacts on the different areas together with volcanic monitoring can help to prevent and minimise losses associated with these natural disasters [3]. Among the issues related with volcanoes that can be assessed the effects on critical infrastructures are important. Critical infrastructure is defined as a network of man-made systems and processes that function collaboratively to produce and distribute essential goods and services [4] which are heavily relied upon by society for daily function. Critical infrastructures include electrical supply networks, water and wastewater distribution systems or terrestrial transportation systems. Changes in the chemistry of water sources in the area, tephra falls, lars and lava flows are some of the volcano hazards that can affect to the critical infrastructures. One of the hazards that can affect the critical infrastructures in both short and long term is the corrosion of structural materials.

According to the American Society for Testing and Materials, corrosion can be defined as 'The chemical or electrochemical reaction between a material, usually a metal, and its environment that produces a deterioration of the material and its properties [5]'. Corrosion has serious economical, technological and safety consequences to society. From the economical point of view, it has to be taken into consideration all the costs related with over dimensioning during the design, inspection, maintenance and replacement of the equipment. Several studies performed in different countries have tried to determine the cost of corrosion [6, 7]. The most throughout of these studies was the one conducted in the United States in 1976 which found that the overall annual cost of metallic corrosion to the U.S. economy was $70 billion, or 4.2% of the gross national product [8]. Besides, any loss of mechanical properties in the metallic materials may lead to an accident with human or environmental consequences. Therefore, any risk-based inspection methodology has to consider the effect of corrosion on infrastructures, and if facilities are installed in areas with volcanic activity, the hazard effects of volcanoes have to be taken into account.

Corrosion is a well-established phenomenon in volcanically active areas [9, 10]. Hydrogen sulphide, gas, SO_2, CO_2 and other pollutants such as selenium, mercury and arsenic react with oxygen and the atmospheric moisture to produce volcanic smog and acid rain. Thus, those changes in the environment related with volcano eruptions can accelerate the corrosion of the surrounding infrastructures [11]. Specifically, widely distributed volcanic ash derived from explosive volcanic eruption creates short- and long-term hazards to infrastructures including exposed building materials such as metal roofing [12, 13]. The metal corrosion related with volcanoes has been attributed to volcanic ashes in several studies [14]. In order to study the effect of volcanoes on corrosion, Hawthorn et al. [10] installed a corrosion test site at the Kilauea volcano in Volcanoes National park of the island of Hawai in 2006. They monitored during 1 year the atmospheric conditions, rainwater chemistry and chlorides deposition. They also exposed some aluminium coupons to the volcano environment. After 1 year of experiments, they observed that the corrosion rate was seven times higher in the volcanic environment than in an industrial site in Oahu [10]. Oze and collaborators also checked the effect of the volcanic ashes on the corrosion of metal roof materials [15]. Increased corrosion of metal roofing from chemically reactive volcanic ash following ash deposition post eruption is a major concern due to the decreasing on the functionality and stability of roofs. They tried to obtain some quantitative information about corrosion rates by coating different materials with volcanic ashes; they could not observe clear trends in the corrosion rates with the performed experiments; however, they observed that in field situations, the corrosion was increased by the presence of volcanic ashes. Therefore, they concluded that volcanic corrosion could be a major problem and the experiments performed had to be improved in order to have into account all the circumstances that can be found in volcanic environments [15]. Therefore, the corrosion induced by the presence of volcanic activity is a problem that has to be taken into consideration during the natural hazards assessments.

However, not too much data have been published on the effect of subaquatic volcanoes on the corrosion of offshore structural materials [16, 17]. Subaquatic volcanoes are normal features on certain regions of the ocean floor. Some may be currently active and, in shallow water, show their actions by blasting steam and debris over the surface of the sea. Many others form at great depths where the weight of the water above them results in confining pressure preventing the violent release of steam and gases. They are estimated to account for 75% of annual magma output. The majority are located near areas of tectonic plate movement (ocean ridges). Submarine volcanoes inject large amounts of material of variable size, texture and chemical composition into the oceans [18]. The state of a submarine eruption evolves from ongoing magmatic activity to highly evolved hydrothermal systems. Some of the features observed during submarine volcanic eruptions are as follows: lava flows, bubbles of lava, lava debris, ash, pumice, steam-blast eruptions, hydrothermal vents, plumes of fine materials and dissolved gas, explosions and discolouration of seawater to light blue and reddish brown. Subaerial volcanos are well characterised; however, there is a lack of study of those in submarine settling [19, 20]. Direct observations of submarine eruptions have been scarce, as they normally occur in remote locations. The first registered underwater eruption was that of the NW Rota in the Southwest Pacific Ocean [21] in 2006. More recently, the dynamic of the Monowai underwater volcano has been described [22]. In all the cases, the carbonate system is modified and the pH reduced as a consequence of the emission of magmatic gas and other compounds [23].

Santiana-Casiano et al. [24] studied the eruption of the island of El Hierro (Canary Islands, Spain) in 2011. This was the first submarine eruption to be monitored from the initial unrest. The eruption started in October 11, 2011, first evidence of eruption was observed in the form of a discolouration of the surface water in the area, going from light green to pale blue to a brown colour [24]. Those changes may be related to the discharge of high-temperature hydrothermal fluids as well as gases and volcanic particles [25]. Bubbling took place, and the presence of rock fragments was observed floating on the surface of the ocean during the initial steps of the eruption. During the strongest episodes of the eruption, an elevated quantity of dissolved inorganic carbon and a large decrease in the pH value of the surrounding water were observed. The lowest values of pH were concentrated in a layer of 75–100 m of depth; however, due to the intensity of the ejected gases, low values of pH reached the surface water. They also observed at some point during the eruption that the water chemistry of the entire southeast of the island was affected by the presence of the volcano. In April 2012, when the eruption stopped, the system evolved to a hydrothermal system and the pH of the area got to normal values and only some peaks of pH were observed around the area of the volcano [24]. Regarding the dissolved CO_2, the concentration in the southeast of the island reached levels of 11,000–19,000 πatm, which was an important change in the chemistry of the water. The same rise was observed for sulphur species and oxidants. Therefore, in the event of the eruption of a submarine volcano and the subsequent hydrothermal activity, a change in the chemistry and properties of the seawater in the area should be expected (presence of CO_2, H_2S, SO_2, Fe^{2+} and drop in the local pH). Not only submarine eruptions can change the local chemical-physical properties of the water but also passive emissions of gasses and acidic compounds to the environment in islands with volcanic activity can modify the water chemistry. Oceans are a significant store of CO_2. They exchange it with the atmosphere and provide a reservoir for CO_2. Therefore, the emissions of CO_2 and other gasses to the environment due to the volcanic activity can decrease the local pH in harbours of volcanic islands.

Those changes can strongly affect the corrosion activity of seawater, raising the probability of corrosion of the metallic materials of the shore and offshore facilities. This corrosion could be generalised, in the case of non-passive materials or localised (pit formation) in the case of passive materials (stainless steels). Corrosion can cause great damage to marine steel infrastructures such as bridges, wharfs, platforms, ships and pipeline systems [26]. These corrosion problems will be aggravated due to the alteration on the chemistry of the water as a consequence of the volcano emissions [27]. Therefore, an adequate selection of the materials and a proper hazard assessment of the volcano effects are very important to predict how some of the critical infrastructures will be affected. Austenitic stainless steel grades 1.4401 (AISI 316) and its derivatives are suitable for shore and offshore service, splash zone applications and temporary immersion in seawater. Duplex stainless steels (e.g. 1.4462/ASTM S31803) may be used in estuaries where the chloride content of the water is lower than that of the open sea. Additionally, superduplex stainless steels (e.g. 1.4410/ASTM S32750) may also be used in direct and continuous contact with seawater (e.g. in offshore oil platforms or desalination plants).

There are some studies dealing with the passivation/corrosion processes of stainless steels in natural seawater [28, 29]. Corrosion studies can be performed in different ways; there are direct measurements (weight loss) and indirect measurements (electrochemical techniques). Electrochemical techniques are useful to evaluate the corrosion behaviour of different metals and alloys in a short period of time providing useful information about the corrosion mechanisms taking place.

According to this, the objective of this chapter will be to evaluate the influence of an underwater volcano eruption on the corrosion behaviour of different materials that might be used in seawater environments. In this way, the electrochemical behaviour of different stainless steels, that is AISI 316 SS and duplex SS, when immersed in solutions which have been polluted by volcano-seawater (collected from the volcano that erupted in the Atlantic Ocean near the island of El Hierro) have been studied. To achieve this objective, different electrochemical techniques have been used, notably open-circuit potential (OCP) measurements, polarization tests and electrochemical impedance spectroscopy (EIS). All these techniques would make possible to achieve a comprehensive view of the corrosion behaviour of stainless steels in volcano seawater. These results may provide and insight in the corrosion effects of volcanic eruptions that could be used in the preparation of natural hazard risk assessments related with volcanoes. This approach has been presented for submarine volcanoes, but the same methodology could be applied to the assessment of corrosion risks due to subaerial volcanoes. In this book chapter, a whole section describing in detail the electrochemical techniques used in the characterisation of the studied stainless steels, has been included, so the readership could better understand the obtained results. These results are used to demonstrate, using electrochemical techniques, that volcano eruptions can negatively affect the lifetime of very important metallic infrastructures, such as harbours, and to establish a link between electrochemistry of materials and volcanology.

2. Electrochemical techniques

Since corrosion occurs via electrochemical reactions, electrochemical techniques are useful for the study of the corrosion processes. In these electrochemical techniques, a metal sample is used to model the metal in a corroding environment. The metal sample is immersed in a solution similar to the corroding environment of the metal that is being studied. Additional electrodes are introduced to the system to make an electrochemical cell, which allows measuring the different parameters to study the system. In order to evaluate the effect of volcano-polluted seawater on the corrosion behaviour of different alloys, several electrochemical techniques have been used and they are explained in this section. These techniques are open-circuit potential (OCP), polarization tests and electrochemical impedance spectroscopy (EIS).

2.1. Open-Circuit Potential (OCP)

The open-circuit potential (OCP) or corrosion potential is a parameter that indicates the thermodynamically tendency of a material to the electrochemical oxidation in a corrosive medium. According to the ASTM G-15 Standard [5], the OCP is the measured potential between the working electrode and the reference electrode at which there is no current in the circuit. In order to measure the OCP value of a metal, the studied material must be immersed in the solution and after a certain period it stabilizes around a stationary value. When the metal is immersed in the solution, the corrosion starts to occur and electrochemical reactions characteristic of the metal-solution interface take place at the surface of the metal (oxidation formation of a passive layer or immunity), which can produce variations of the OCP value during the stabilization [5, 30, 31].

2.2. Polarization tests

Polarization tests of a metal-electrolyte solution are used in order to characterize the metal by its current-potential relationship. Experimentally, the polarization tests involve changing the potential of the working electrode and register the current as a function of the potential obtaining potentiodynamic curves. Since the current density (the current divided into the area) can experiment high variations, the potentiodynamic curves represent the logarithm of the current density vs the potential [32]. The potential starts at a value lower than the corrosion potential (cathodic region) and increases in the anodic direction. In order to study the repassivation behaviour of a material, the direction of the cyclic potentiodynamic curve can be changed backward to cathodic potentials after achieving a certain current value (once the pitting potential was reached). **Figure 1** shows as an example, a cyclic potentiodynamic curve, with the corresponding corrosion parameters that can be obtained from the curve.

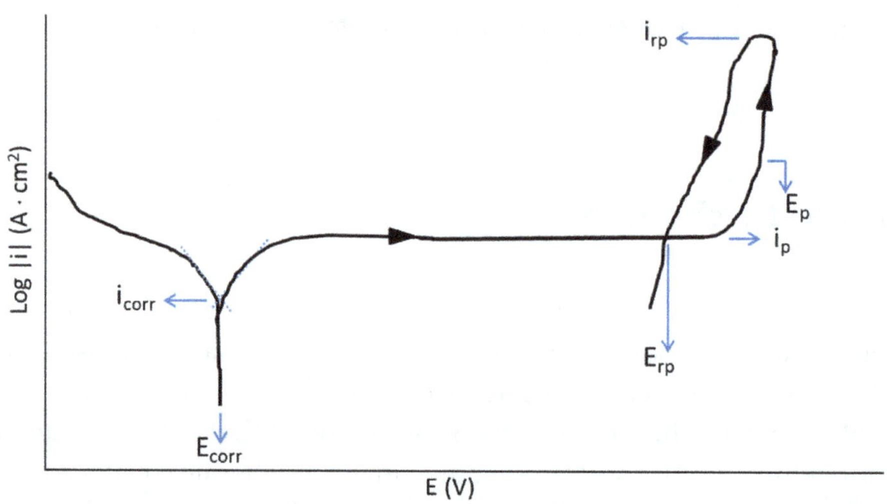

Figure 1. Cyclic potentiodynamic curve with the corrosion parameters. i_{corr}: corrosion current density, E_{corr}: corrosion potential, i_p: current density of passivation, E_p: pitting potential, i_{rp}: repassivation current density, E_{rp}: repassivation potential.

When the metal is immersed in the corrosive medium, reduction and oxidation processes occur in its surface. Generally, the metal oxidizes (corrodes) and the medium is reduced both anodic and cathodic currents occur in its surface. If the anodic and cathodic currents are equals in magnitude (i.e. the rate of oxidation and reduction are the same), there is no net current to be measured. The potential at which the current is non-measurable is called corrosion potential (E_{corr}).

At potentials more negative than the corrosion potential ($E < E_{corr}$), the cathodic current predominates whereas the anodic current is insignificant, so the curve is in the cathodic region. In contrast, if the potential is more positive than the corrosion potential ($E > E_{corr}$), the predominant current is the anodic one and the curve is in the anodic region [32, 33].

In order to calculate the current density (i), the Butler-Volmer equation [Eq. (1)] can be useful as it considers the corrosion of a system with cathodic and anodic reaction.

$$i = i_{corr} \cdot [\exp(\frac{2.303 \cdot (E - E_{corr})}{b_a})] - \exp(-\frac{2.303 \cdot (E - E_{corr})}{b_c}) \tag{1}$$

In Eq. (1), i is the external current density, i_{corr} is the corrosion current density, E is the applied potential to polarize the system, E_{corr} is the corrosion potential, b_a is the anodic Tafel slope and b_c is the cathodic Tafel slope. If the applied potential is equal to the corrosion potential then, $i = i_{corr}$, but if E is different from E_{corr}, Eq. (1) gives the Tafel law [Eq. (2)]:

$$E = a \pm b \cdot \log|i| \tag{2}$$

where a is a constant and b is b_a or b_c according to the region of the curve (anodic or cathodic, respectively) [32–34].

In the potentiodynamic curves, for applied potentials more positive than the ones of the cathodic region, in the materials that can be passivated, there is a region where the current density becomes constant (passivation current density, i_p). This region is called passivation region, and it is characterized because a passive layer (metal oxide) is formed on the surface of the metal and its chemical reactivity is reduced.

As the applied potential continues increasing, the current density could experiment an abrupt increase and the passive layer could locally breaks by pitting corrosion if the medium contains aggressive ions. The potential at which the pitting corrosion starts is called pitting potential (E_p, when the current density reaches 100 μA/cm^2). Finally, if the potential scan is reversed, the current density decreases and eventually the backward curve intersects the forward one: this final region is called repassivation region because the metal recovers the passive state (the potential and current density of the repassivation regions are E_{rp} and i_{rp}, respectively).

2.3. Electrochemical impedance spectroscopy

Electrochemical impedance spectroscopy is a non-destructive technique that is used to study the metal/electrolyte interface. It consists of applying an electrical disturbance (potential) to an electrical circuit in a range of frequencies and measure the current response of the system to each frequency. The applied disturbance of potential can be expressed as a sinusoidal signal as Eq. (3) shows.

$$E = E_0 \cdot \sin(\omega \cdot t) \tag{3}$$

where ω is the angular frequency (2π times the conventional frequency in Hz). On the other hand, the current response can also be expressed as a sinusoidal signal that is desphased certain angle (ϕ), as Eq. (4) shows.

$$I = I_0 \cdot \sin(\omega \cdot t + \phi) \tag{4}$$

The impedance is the quotient of the applied potential and the measured current and it follows the Ohm law; therefore, it can be expressed as Eq. (5) shows.

$$Z = \frac{E}{I} = \frac{E_0 \cdot \sin(\omega \cdot t)}{I_0 \cdot \sin(\omega \cdot t + \phi)} = Z_0 \cdot \frac{\sin(\omega \cdot t)}{\sin(\omega \cdot t + \phi)} \tag{5}$$

where Z is the complex impedance (which in some electrochemical systems depends on applied potential frequency). The complex impedance can be expressed in terms of a real (Z') and an imaginary component (jZ'') as Eq. (6) shows.

$$Z(\omega) = Z_{\text{Re}} - jZ_{Im} \equiv Z' - jZ'' \tag{6}$$

Plotting the results of the impedances in a correct frequency range, it is possible to extract conclusions about the physical and chemical properties of the materials and the electrochemical systems. Two diagrams are typically extracted from the EIS measurements: Nyquist and Bode diagrams. In the Nyquist diagrams, the real and imaginary components of the impedance at different frequencies are represented. The Bode diagrams represent the log $|Z|$ vs log f (Bode-modulus), and ϕ vs log f (Bode-phase). **Figure 2** shows an example of a Nyquist diagram with the equivalent circuit and an example of a Bode diagram [35, 36].

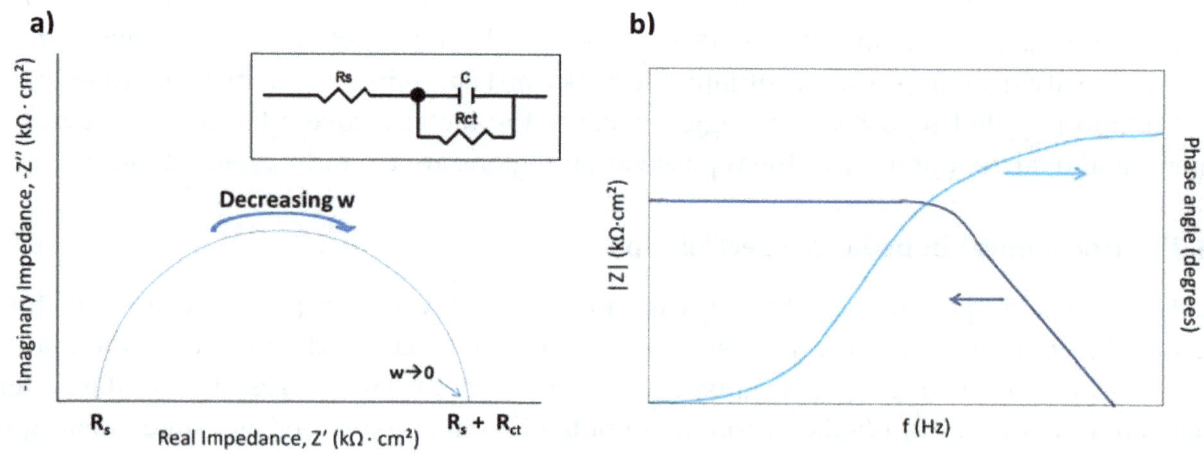

Figure 2. Nyquist diagram with the equivalent circuit (**a**) and Bode-modulus and Bode-phase plots (**b**). R_S is the solution resistance and R_{ct} the charge transfer resistance.

In order to solve the mathematical difficulties in the data treatment of the EIS, the results can be adjusted to an equivalent electrical circuit (**Figure 2**). These circuits try to represent the electrical characteristics of the model and are composed by basic elements such as resistances, capacitances and inductances.

3. Experimental procedure

In order to evaluate the influence of an underwater volcano eruption on the corrosion behaviour of different materials, a practical example is detailed in this chapter. In this way, the electrochemical behaviour of different stainless steels, that is AISI 316 stainless steel (SS) and duplex SS, immersed in volcano-seawater (collected from a volcano that erupted in the Atlantic Ocean near the island of El Hierro, Canary Islands, Spain) have been studied. To achieve this objective, different electrochemical techniques have been used. That is OCP measurements, polarization tests and electrochemical impedance spectroscopy (EIS). All these techniques would make possible to achieve a comprehensive view of the corrosion behaviour of stainless steels in volcano seawater.

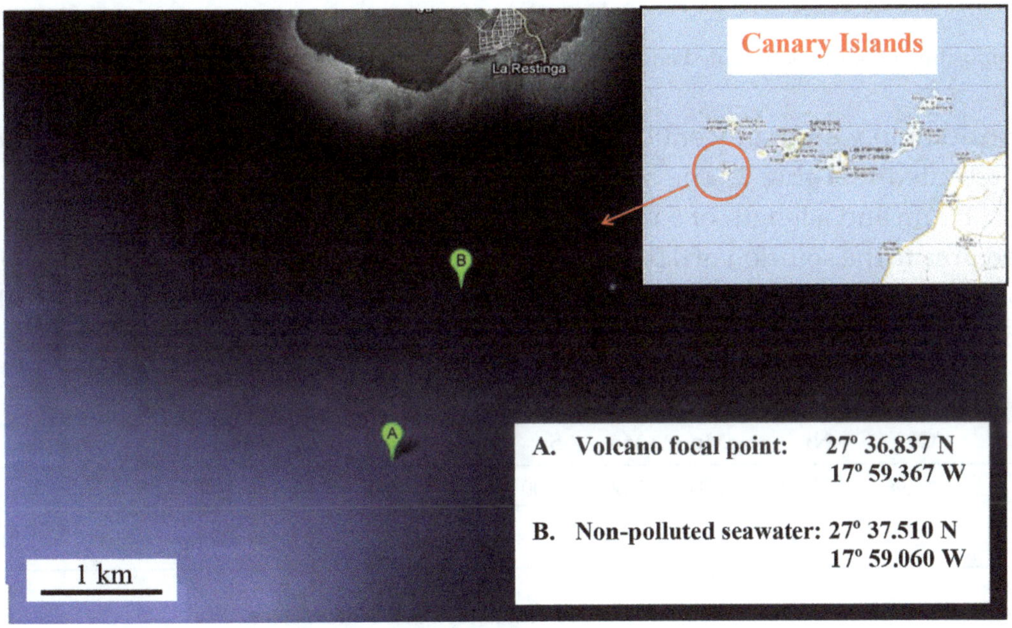

Figure 3. Map showing the zones near the island of El Hierro where the non-polluted and polluted seawater samples were taken.

3.1. Materials, solution and electrochemical cell

Experiments were performed in non-polluted (NPS) and polluted seawater (PS). **Figure 3** shows the coordinates where the seawater samples were taken on December 19, 2011. The polluted water was obtained on the focus of the volcano. The non-polluted samples were taken at enough distance to avoid the influence of the pollutants spread by the volcano, following the suggestions of the Spanish National Geographic Institute. Additionally, **Table 1** presents the main parameters of the non-polluted and polluted by the "El Hierro" volcano eruption seawaters. Prior to each experiment the pH and conductivity of the non-polluted and polluted seawater was measured. As shown in **Table 1**, the most substantial difference between both seawaters is pH.

Value	Unpolluted seawater	Polluted seawater
pH	8.02	6.23
Conductivity (μS cm^{-1})	65,850	66,400
Sodium (mg l^{-1})	11,080	11,868
Potassium (mg l^{-1})	438	447
Calcium (mg l^{-1})	558	400
Magnesium (mg l^{-1})	1486	1320
Sulphates (mg l^{-1})	2400	2600
Chlorides (mg l^{-1})	20,926	21,200
Bicarbonates (mg l^{-1})	159	220
Carbonates (mg l^{-1})	5.53	0.00

Table 1. Composition of both non-polluted and polluted seawater samples, given by the Spanish Institute of Geodesy.

The composition in weight percent of the AISI 316 SS and duplex SS (Alloy 900) used in this work is presented in **Table 2**. The shape of the electrodes was cylindrical, and they have a diameter of 8 mm and a length of 5 mm. They were coated in polytetrafluoroethylene, having an exposed area to the solution of 0.5 cm^2. The samples were wet abraded with different silicon carbide papers (from 500 to 4000), and they were finally cleaned with ethanol, distilled water and dried in air.

	Cr	Ni	Mn	Mo	S	Si	P	C	Fe	Cu	N	Ti
AISI 316	16.96	10.17	1.34	2.30	0.00	0.37	0.03	0.08	Bal.	–	–	–
Alloy 900	22.34	4.85	1.59	2.69	–	0.35	0.02	0.03	67.80	0.13	0.20	0.01

Table 2. Chemical composition in weight % of AISI 316 and Alloy 900 stainless steels.

An electrochemical vertical cell made of glass was used to carry out the tests. A configuration of three electrodes was used to perform the electrochemical measurements, that is the stainless steel was the working electrode, a platinum wire was the counter electrode and a silver/silver chloride (Ag/AgCl 3M KCl) was used as reference electrode. All these electrodes were connected to a pontentiostat (Autolab PGSTAT302N). The electrochemical tests were repeated at least three times to verify reproducibility of the data obtained in the non-polluted and polluted seawater. The tests were carried out at 25°C.

3.2. OCP measurement and polarization tests

Before each polarization measurement, the open-circuit potential (OCP) was measured for 1 h in the test solution. The average value of the potentials recorded during the last 300 s was the value of the OCP. Afterwards, the potential was shifted to −0.25 $V_{Ag/AgCl}$ with respect to the OCP, in order to start the polarization measurements. Then, the potential was scanned in the anodic direction until a value of current density of 10 mA/cm^2 was obtained. In this moment, the potential scan was reversed. The scan rate used to perform the polarization measurements was 0.5 mV s^{-1}.

3.3. EIS and capacitance measurements

EIS measurements were performed after forming a stable passive film on the surface of the samples at a constant applied potential (−0.1 $V_{Ag/AgCl}$) for 1 h. EIS measurements were conducted at that formation potential in the frequency range of 100 kHz–10 mHz, with a signal amplitude of 10 mV.

4. Results and discussion

4.1. OCP

The OCP value was calculated as the mean value of the last 5 min of the complete OCP register, which lasted 1 h. The OCP values obtained for the SSs in the different media (polluted and non-polluted seawater) are shown in **Table 3**. In relation to **Table 3**, it can be pointed out that there is no significant influence of the composition of the seawater on the OCP values. On the other hand, the OCP values obtained for Alloy 900 in the non-polluted and polluted seawaters are slightly more negative than those determined for the AISI 316 SS.

Seawater	AISI 316	Alloy 900
Non-polluted	−90 ± 7	−161 ± 2
Polluted	−116 ± 4	−155 ± 8

Table 3. Open-circuit potential values for AISI 316 SS and Alloy 900 in both non-polluted and polluted seawater.

4.2. Polarization tests

One of the reproducible cyclic polarization curves obtained for the SS in non-polluted and polluted seawater are presented in **Figure 4**. From the polarization curves shown in **Figure 4**, several corrosion and repassivation parameters can be calculated. These parameters, all presented in **Table 4**, are the corrosion potential (E_{corr}), the corrosion current density (i_{corr}), the passivation current density (i_p), pitting potential (E_p) and the repassivation potential (E_{rp}). The magnitude order of the corrosion potential is the same regardless of the studied media; however, there is an important influence of the seawater on the passivation current densities, which were obtained when the current was stable within the passivation range. Corrosion resistance of these stainless steels is directly associated with a passive film mainly formed by chromium III oxide [37]. It is important to point out that **Table 4** only shows the i_p values as a measure of the corrosion rates of passive alloys (this parameter has been selected to quantify the corrosion rate instead of i_{corr} since the SSs used in this work are passivable in the studied media). In relation to the passivation current densities, they importantly increase in polluted

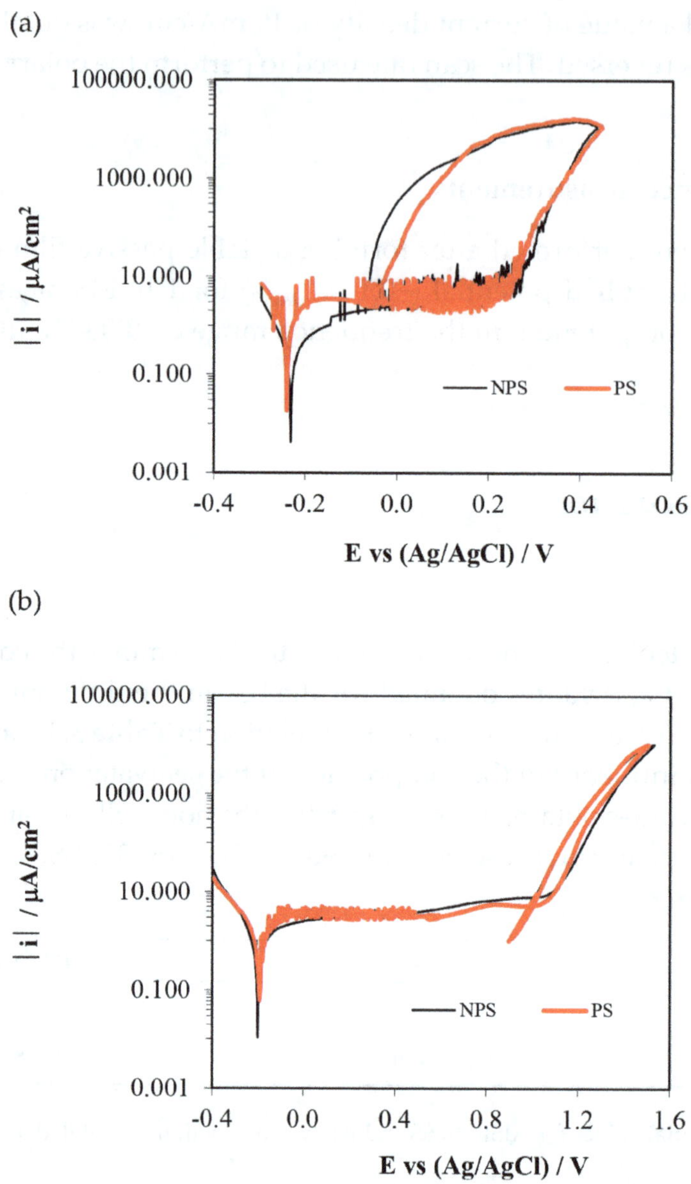

Figure 4. Potentiodynamic polarization cyclic curves for AISI 316 SS (**a**) and Alloy 900 (**b**) in both non-polluted (NPS) and polluted seawater (PS) solutions at 25°C.

seawater. This fact could be associated to the lower pH which presented this kind of medium. Besides, Alloy 900 SS has higher i_p values in comparison with the ones obtained for the AISI 316 SS. However, several oscillations in the current density register of the AISI 316 SS in both seawaters and in the register of the Alloy 900 in polluted seawater can be clearly observed in **Figure 4**. These current oscillations indicate that polluted seawater increase the susceptibility of the stainless steels to metastable pitting [38]. In order to evaluate the susceptibility to pitting corrosion of the SS the pitting potential value was evaluated. This parameter is the potential at which the current density reaches 100 μA cm^{-2} and refers to the susceptibility to local breakdown and stable pit initiation [39]. The stainless steels are susceptible to pitting corrosion in both media since chlorides (present in the seawater) are very aggressive ions which are responsible for the breakdown of the passive films formed on these SSs [39, 40]. **Table 4** shows that the E_p values for the Alloy 900 are more positive in comparison to AISI 316. This is due to the fact that Duplex SSs, such as Alloy 900, possess a very high resistance to localized corrosion due to their high nitrogen, chromium and molybdenum contents (see **Table 2**). On the other hand, **Table 4** also shows that the values of the pitting potential are similar despite the studied media, but they are to some extent more positive in the non-polluted seawater.

	E_{corr} vs (Ag/AgCl)/mV	i_p/μA cm^{-2}	E_p vs (Ag/AgCl)/mV	E_{rp} vs (Ag/AgCl)/mV
Seawater	AISI 316			
Non-polluted	−230 ± 8	3.89 ± 0.32	309 ± 10	−69 ± 11
Polluted	−239 ± 5	4.66 ± 0.11	302 ± 7	−55 ± 15
	Alloy 900			
Non-polluted	−203 ± 11	4.35 ± 0.44	1220 ± 47	990 ± 22
Polluted	−197 ± 6	5.02 ± 0.09	1200 ± 29	990 ± 30

Table 4. Electrochemical parameters for AISI 316 SS and Alloy 900 in both non-polluted and polluted seawater, obtained from cyclic polarization curves.

The repassivation potentials were obtained at the crossing point between the backward and forward scans of the polarization curves. This parameter refers to the limit below which the material remains passive and active pits can repassivate. According to this, an E_{rp} value which is more negative than the corresponding E_{corr} indicates that the SS is not able to repassivate the pits. Contrary to this, an E_{rp} value which is more positive than the corrosion potential indicates that the SSs can regenerate an eventual breakdown of the passive film. **Table 4** shows that the repassivation potentials for the studied SS are, in all the studied media, more positive than the corresponding corrosion potential; therefore, the SSs can repassivate the pits.

4.3. EIS measurements at an applied anodic potential

In order to characterise the electrochemical behaviour of the passive film/electrolyte interface for the two stainless steels under study, EIS measurements were performed at an applied potential of −0.1 $V_{Ag/AgCl}$ in both non-polluted and polluted seawater, after a passive film was formed on the electrodes' surface at the same potential. **Figure 5** shows the EIS spectra of AISI 316 and Alloy 900 in the form of Nyquist and Bode plots. In all cases, EIS diagrams exhibit a typical passive state shape characterized by a semicircular shape and high impedance values in Nyquist plots, and by phase angle values close to 90° in Bode-phase plots, suggesting that a stable passive film is formed on all the electrodes [41, 42]. In general, higher impedance values are observed for passive films formed in non-polluted seawater (NPS). The highest impedances are obtained for the duplex stainless steel.

The equivalent electric circuit used to interpret EIS spectra of anodically formed passive films is shown in **Figure 6**. In this equivalent circuit, R_S represents the electrolyte resistance, CPE_1 is related to the capacitance of the passive film/electrolyte interface, R_1 corresponds to the charge transfer resistance at that interface and Z_{WS} is a Warburg impedance, which has been used to interpret the transport of vacancies within the passive film, in the frame of the Point Defect Model [42–44]. The Warburg impedance used here models dimensional diffusion through a layer of finite thickness with absorbing boundary condition [45, 46]:

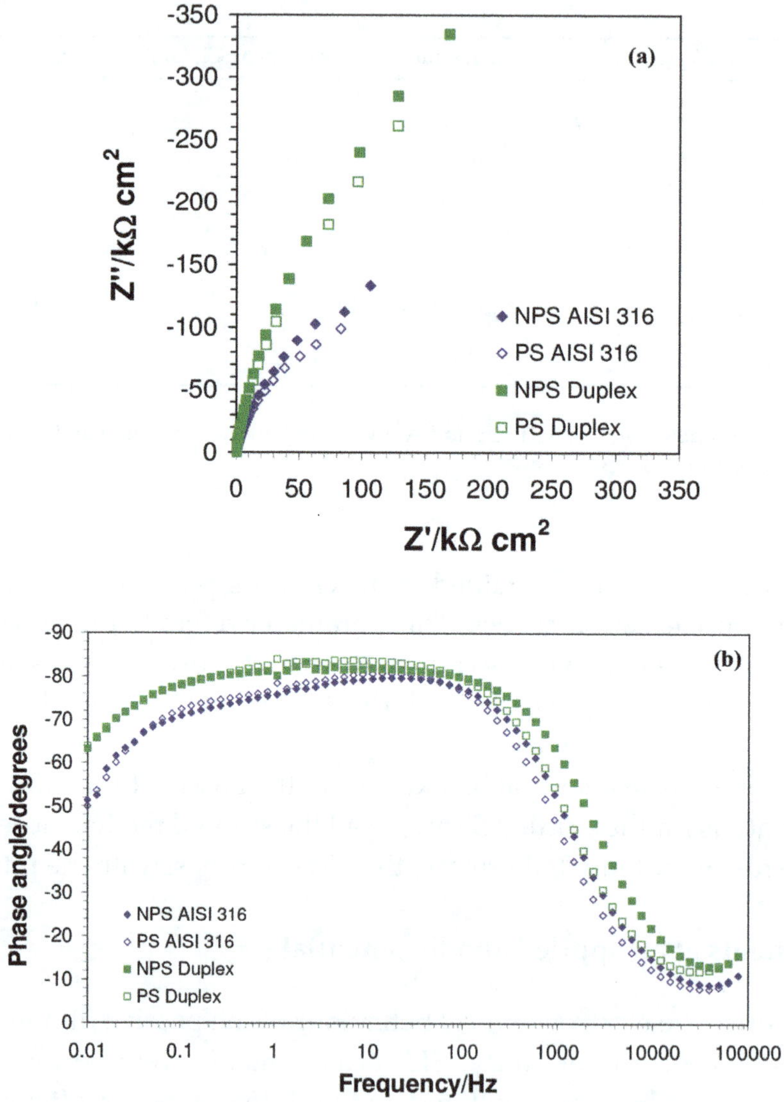

Figure 5. Nyquist (**a**) and Bode-phase (**b**) plots for AISI 316 and Alloy 900 in both non-polluted (NPS) and polluted seawater (PS) at 25°C, at a film formation potential of −0.1 $V_{Ag/AgCl}$.

$$Z_{WS} = \frac{R_W \tanh(B\sqrt{j\omega})}{B\sqrt{j\omega}} \tag{7}$$

where $B = \delta/(D)^{1/2}$, δ is the diffusion layer thickness, D is the diffusion coefficient of the diffusing species, R_W is the Warburg resistance, ω is the angular frequency, and j is the imaginary unit.

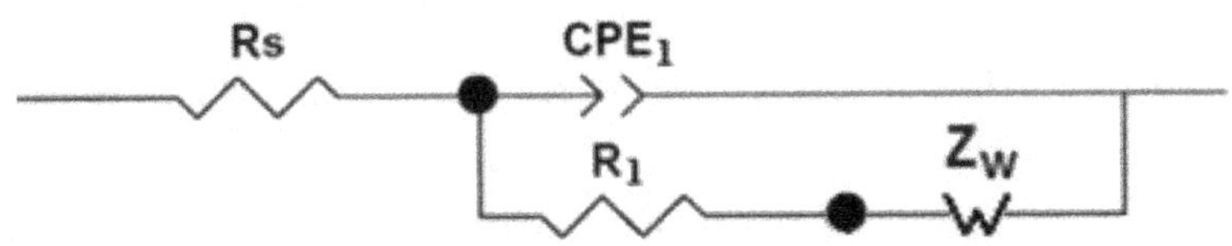

Figure 6. Equivalent circuit used to interpret the EIS data.

A constant-phase element (CPE) representing a shift from the ideal capacitor has been used instead of the capacitance itself. The impedance of a constant-phase element is defined as:

$$Q = Z_{CPE} = [C(j\omega)^n]^{-1} \tag{8}$$

where n, defined as a CPE power, is an adjustable parameter that lies between −1 and 1. For $n = 1$ the CPE describes an ideal capacitor, and for $0.5 < n < 1$ the CPE describes a frequency dispersion of time constants due to local heterogeneities in the dielectric material.

The CPE used in the equivalent electric circuit of **Figure 6** has been converted into a pure capacitance (C) through the following equation [47, 48]:

$$C = Q^{1/n}(R_S^{-1} + R_1^{-1})^{(n-1)/n} \tag{9}$$

where $Q = Z_{CPE}$ [Eq. (9)].

The parameters of the equivalent circuit obtained from the fitting procedure are shown in **Table 5**. The values of χ^2 are of the order of 10^{-3}, indicating that the fittings are good from a mathematical point of view.

		R_S/Ω cm²	R_1/k Ω cm²	C_1/μF cm⁻²	n_1	R_W/k Ω cm²	χ^2 (×10⁻³)
AISI 316	Non-polluted seawater	5.4 ± 0.1	26 ± 4	16.2 ± 1.1	0.90 ± 0.01	295 ± 14	3.8
	Polluted seawater	4.8 ± 0.2	26 ± 3	24.8 ± 1.7	0.90 ± 0.01	207 ± 18	3.7
Alloy 900	Non-polluted seawater	4.7 ± 0.3	535 ± 30	12.4 ± 1.3	0.91 ± 0.01	753 ± 53	3.0
	Polluted seawater	4.9 ± 0.3	320 ± 11	18.3 ± 1.2	0.92 ± 0.01	544 ± 18	2.7

Table 5. Equivalent circuit parameters obtained by fitting the experimental results of EIS, for AISI 316 and Alloy 900 in both non-polluted and polluted seawater, at −0.1 $V_{Ag/AgCl}$.

It can be observed from **Table 5** that the charge transfer resistance at the interface, R_1, is equal or lower, in general, for passive films formed in the polluted seawater. The Warburg resistance, R_W, follows the same tendency, decreasing in the polluted electrolyte. These results indicate an enhancement of charge transfer processes at the passive film/electrolyte interface and of point defects transport through the passive film for the samples immersed in the polluted seawater, reflecting the poorer protective properties of passive films formed in this electrolyte.

On the other hand, the capacitance of the passive film/electrolyte interface, C_1, is higher in the polluted seawater. The interfacial capacitance has a strong influence on electrochemical processes in passive systems. This is due to the fact that its characteristics establish the potential drop at the semiconducting passive film/electrolyte interface [42, 49, 50]. Therefore, an increase in C_1 suggests, along with the decrease observed in R_1 and R_W, that passive films formed in polluted seawater are less protective. The values of the exponent n_1 (lower than unity in all the cases) support the introduction of CPEs in the equivalent circuit rather than pure capacitors.

Concerning the comparison between the protective properties of passive films formed on the two stainless steels under study, R_1 values are significantly lower in the case of AISI 316 (more than 10 times lower than in Alloy 900), and R_W is the lowest for this stainless steel, as well. Besides, C_1 values for AISI 316 are the highest, regardless of the type of seawater (**Table 5**). On the other hand, the highest resistances (R_1 and R_W) and the lowest interfacial capacitance values (C_1) are obtained for the duplex stainless steel (Alloy 900), denoting that passive films formed on Alloy 900 are the most resistant to corrosion in both seawater media.

5. Conclusion

The electrochemical techniques used in this work have been proved to be useful in order to study the corrosion behaviour of stainless steels in volcano polluted seawater. From polarization curves, it has been demonstrated that Alloy 900 is more resistant to pitting corrosion than AISI 316. Moreover, the decrease of the charge transfer resistance, R_1, and the Warburg resistance, R_W, along with the increase of the interfacial capacitance, C_1, indicate worse protective properties of passive films formed in the polluted seawater, especially for AISI 316 SS.

Comparing both stainless steels, it can be concluded that passive films formed on Alloy 900 present better protective properties than those of AISI 316 SS.

Acknowledgements

We wish to express our gratitude to the Ministerio de Ciencia e Innovación (Project CTQ2009-07518), to Prof. Dr. Aurora Santos López, Dr. Rosario Lunar Hernández and Dr. Jose Arnoso Sampedro (Instituto de Geociencias (CSIC, UCM)), Dr. Carmen López Moreno (Head of the Volcano Monitoring Unit, Spanish National Geographic Institute), Humberto Gutiérrez García (Head of the Civil Protection and Emergency Management Service of the General Directorate of Security and Emergencies of the Canary Islands) and Alejandro Ramos Fernández (Coordinador Multisectorial del Centro Coordinador de Emergencias 1-1-2 del Gobierno de Canarias) for their help in supplying both non-polluted and polluted seawater samples.

Author details

Rita Sánchez-Tovar[1], Ramón Fernández-Domene[1], Rafael Leiva-García[2], Clara Escrivà-Cerdán[1], Bianca Lucas-Granados[1] and José García-Antón[1*]

*Address all correspondence to: jgarciaa@iqn.upv.es

1 Chemical and Nuclear Engineering Department, Polytechnic University of Valencia, Electrochemical Engineering and Corrosion (IEC), Valencia, Spain

2 School of Materials, University of Manchester, Manchester, UK

References

[1] UNISDR 2009. *UNISDR Terminology on Disaster Risk reduction*, United Nations, Geneva, Switzerland (2009).

[2] M.R. Auker, R.S.J. Sparks, L. Siebert, H.C. Crossweller, J. Ewert. *A statistical analysis of the global volcanic fatalities record*. Journal of Applied Volcanology 2013, 2: 1–24. doi: 10.1186/s13617-014-0014-6

[3] R.S.J. Sparks, W.P. Aspinall, H.S. Crossweller, T.K. Hincks. *Risk and uncertainty assessment of volcanic hazards*. In: J. Rougier. R.S.J. Sparks, L. Hill (Eds). *Risk and uncertainty assessment of for natural hazards*. Cambridge University Press, Cambridge (2013), p. 588.

[4] S.M. Rinaldi, J.P. Peerenboom, T.K. Kelly. *Identifying understanding and analyzing critical infrastructure interdependencies*. IEEE Control Systems 2001, 21: 11–25. doi: 10.1109/37.969131

[5] ASTM G15-89A, Standard Terminology Relating to Corrosion and Corrosion testing. ASTM, Philadelphia (1990).

[6] T.P. Hoar. *Report of the Committee on Corrosion and Protection—A Survey of Corrosion Protection in the United Kingdom*. HMSO, London (1971).

[7] C.G. Okamoto *Report of the Committee on Corrosion and Protection—A Survey of the Cost of Corrosion to Japan*, Japan Society of Corrosion Engineering and Japan Association of Corrosion Control (1977).

[8] *Research Needs Report Energy Convention Research Task force on Energy Conversion Research*, Am. Society of Mechanical Engineering, New York (1976).

[9] M. Watanabe, A. Hokazono, T. Handa, T. Ichino, N. Kuwani. *Corrosion of copper and silver plates by volcanic gases*. Corrosion Science 2006, 48: 3759–3776. doi: 10.1016/j.corsci.2005.12.009

[10] G.A. Hawthorn, N. Mullet, R. Srinivasan, L.H. Hihara. *Corrosion testing and atmospheric monitoring in an active volcanic environment*. Proceedings of the 2007 Tri-service conference, US Department of Defence (2007).

[11] R.J. Blong. *Volcanic hazards: a sourcebook on the effects of eruptions*. Academic Press, North Ryde, NSW (1984). p. 424.

[12] S. Izumo, H. Sueyoshi, K. Kitamura, Y Ohzono. *Corrosion of metal in volcanic environments*. Corrosion Engineering 1990, 39: 271–281.

[13] R.J. Blong. *Building damage in Rabaul, Papua, New Guinea*. Bulletin of Volcanology 2003, 65: 43–54. doi: 10.1007/s00445-002-0238-x

[14] J. Becker, R. Smith, D. Johnston, A. Munro. *Effects of the 1995–1996 Ruapehu eruptions on communities in central North Island, New Zealand, and people's perceptions of volcanic hazards after the event*. Australian Journal of Disaster Trauma Studies 2001, 1: 23.

[15] C. Oze, J. Cole, A. Scott, T. Wilson, G. Wilson, S. Gaw, S. Hampton, C. Doyle, Z. Li. *Corrosion of metal roof materials related to volcanic ash interactions*. Natural Hazards 2014, 71: 785–802. doi: 10.1007/s11069-013-0943-0

[16] R.M. Fernández-Domene, R. Sánchez-Tovar, C. Escrivà-Cerdán, R. Leiva-García, J. García-Antón. *Study of passive films formed on AISI 316L stainless steel in non-polluted and underwater-volcano-polluted seawater*. Corrosion 2014, 70: 390–401. doi: 10.5006/0934

[17] R.M. Fernández-Domene, R. Sánchez-Tovar, C. Escrivá-Cerdán, R. Leiva-García, J. García-Antón. *Comparison of the effect of non-polluted and underwater-volcano-polluted seawater on the corrosion resistance of different stainless steels*. Materials and Corrosion. 2015, 66: 1279–1289. doi: 10.1002/maco.201408172

[18] V.M. Mantas, A.J.S.C. Pereira, P.V. Morais. *Plumes of discolored water of volcanic origin and possible implications for algal communities. The case of the Home Reef eruption of 2006 ga, Southwest Pacific Ocean)*. Remote Sensing of Environmental 2011, 115: 1341–1352.

[19] J.A. Resing, E.T. Baker, J.E. Lupton, K. Nakamura. *Chemistry of hydrothermal plumes above submarine volcanoes of the Mariana Arc*. Geochemistry Geophysics Geosystems 2009, 10: Q02009. doi: 0.1029/2008GC002141

[20] C.E.J. de Ronde, E.T. Baker, G.J. Massoth, J.E. Lupton, I.C. Wright, R.J. Sparks, S.C. Bannister, M.E. Reyners, S.L. Walker, R.R. Greene, J. Ishibashi, K. Faure, J.A. Resing, G.T. Lebon. *Submarine hydrothermal activity along the mid-Kermadec Arc, New Zealand: large-scale effects on venting*. Geochemistry Geophysics Geosystems 2007, 8: Q07007.

[21] R.W. Embley. *Extensive and diverse submarine volcanism and hydrothermal activity in the NE Lau Basin*. Eos, Transactions, AGU (Fall Meeting Supplement) 90, abstr. V51D–1719 (2009).

[22] A.B. Watts. *Rapid rates of growth and collapse of Monowai submarine volcano in the Kermadec Arc*. Nature Geoscience 2012, 5: 510–515. doi: 10.1038/ngeo1473

[23] H. Guillou, J.C. Carracedo, F. Pérez Torrado, E. Rodriguez Badiol. *K-Ar ages and magnetic stratigraphy of a hotspot-induced, fast grown oceanic island: El Hierro, Canary Islands*. Journal of Volcanology and Geothermal Research 1996, 73: 141–155. doi: 10.1016/0377-0273(96)00021-2

[24] J.M. Santana-Casiano, M. Gonzalez-Davila, E. Fraile-Nuez, D. de Armas, A.G. Gonzalez, J.F. Domınguez-Yanes, J. Escanez. *The natural ocean acidification and fertilization event caused by the submarine eruption of El Hierro*. Scientific Reports Nature 2013, 3: Article number: 1140. doi: 10.1038/srep01140

[25] K. Nogami, M. Yoshida, J. Osaka. *Chemical composition of discolored seawater around Satsuma-Iwojima, Kagoshima, Japan*. Bulletin Volcanology Society of Japan 1993, 38: 71–77.

[26] J. Duan, S. Wu, X. Zhang, G. Huang, M. Du, B. Hou. *Corrosion of carbon steel influenced by anaerobic biofilm in natural seawater*. Electrochimica Acta 2008, 54: 22–28. doi: 10.1016/j.electacta.2008.04.085

[27] L. Freire, M.J. Carmezim, M.G.S. Ferreira, M.F. Montemor. *The passive behaviour of AISI 316 in alkaline media and the effect of pH: a combined electrochemical and analytical study*. Electrochimica Acta 2010, 55: 6174–6181. doi: 10.1016/j.electacta.2009.10.026.

[28] A. Al Odwani, M. Al-Tabtabaei, A. Abdel-Nabi. *Performance of high chromium stainless steels and titanium alloys in Arabian Gulf seawater*. Desalination 1998, 120: 73–81. doi: 10.1016/S0011-9164(98)00203-3

[29] S.A. Al-Fozan, A.U. Malik. *Effect of seawater level on corrosion behavior of different alloys*. Desalination 2008, 228: 61–67. doi: 10.1016/j.desal.2007.08.007

[30] C.J. Brabec, A. Cravino, D. Meissner, N. Serdar Sariciftci, T. Fromherz, M.T. Rispens, L. Sanchez, J.C. Hummelen. *Origin of the open circuit voltage of plastic solar cells*. Advanced Functional Materials 2001, 11: 374–380. doi: 10.1002/1616-3028(200110)11

[31] G. Berthomé, B. Malki, B. Baroux. *Pitting transients analysis of stainless steels at the open circuit potential*. Corrosion Science 2006, 48: 2432–2441. doi: 10.1016/j.corsci.2005.09.012.

[32] X.L. Zhang, Zh.H. Jiang, Zh. P. Yao, Y. Song, Zh. D. Wu. *Effects of scan rate on the potentiodynamic polarization curve obtained to determine the Tafel slopes and corrosion current density*. Corrosion Science 2009, 51: 581–587. doi: 10.1016/j.corsci.2008.12.005.

[33] E. McCafferty. *Validation of corrosion rates measured by the Tafel extrapolation method*. Corrosion Science 2005, 47: 3202–3215. doi: 10.1016/j.corsci.2005.05.046.

[34] O.A. Petrii, R.R. Nazmutdinov, M.D. Bronshtein, G.A. Tsirlina. *Life of the Tafel equation: current understanding and prospects for the second century*. Electrochimica Acta 2007, 52: 3493–3504. doi: 10.1016/j.electacta.2006.10.014.

[35] D.V. Ribeiro, J.C.C. Abrantes. *Application of electrochemical impedance spectrosopy (EIS) to monitor the corrosion of reinforced concrete: a new approach*. Construction and Building Materials 2016, 111: 98–104. doi:10.1016/j.conbuildmat.2016.02.047.

[36] R.L.C. Naumann. Functional Polymer Films: 2 Volume Set. Chapter 25. *Electrochemical Impedance Spectroscopy (EIS).* Wiley Online Library, Germany. doi: 10.1002/9783527638482.ch25.

[37] N.E. Hakiki, S. Boudin, B. Rondot, M. Da Cunha Belo. *The electronic structure of passive films formed on stainless steels.* Corrosion Science 1995, 37:1809–1822. doi: 10.1016/0010-938X(95)00084-W

[38] R.C. Newman. *Understanding the corrosion of stainless steel.* Corrosion 2001, 57: 1030–1041. doi: doi: 10.5006/1.3281676

[39] M. Kaneko, H.S. Isaacs. *Pitting of stainless steel in bromide, chloride and bromide/chloride solutions.* Corrosion Science 2000, 42: 67–78. doi: 10.1016/S0010-938X(99)00056-6

[40] M. Kaneko, H.S. Isaacs. *Effects of molybdenum on the pitting of ferritic- and austenitic-stainless steels in bromide and chloride solutions.* Corrosion Science 2002, 44: 1825–1834. doi: 10.1016/S0010-938X(02)00003-3.

[41] J. Pan, D. Thierry, C. Leygraf. *Electrochemical impedance spectroscopy study of the passive oxide film on titanium for implant application.* Electrochimica Acta. 1996, 41: 1143–1153. doi: 10.1016/0013-4686(95)00465-3.

[42] R.M. Fernández-Domene, E. Blasco-Tamarit, D.M. García-García, J. García-Antón. *Passive and transpassive behaviour of alloy 31 in a heavy brine libr solution.* Electrochimica Acta 2013, 95: 1–11. doi: 10.1016/j.electacta.2013.02.024.

[43] D.D. Macdonald, R.Y. Liang, B.G. Pound. *An electrochemical impedance study of the passive film on single crystal Ni(111) in phosphate solutions.* Journal of Electrochemical Society 1987, 134: 2981–2986. doi: 10.1149/1.2100326

[44] M. Sánchez, J. Gregori, M.C. Alonso, J.J. García-Jareño, F. Vicente, *Anodic growth of passive layers on steel rebars in an alkaline medium simulating the concrete pores.* Electrochimica Acta 2006, 52: 47–53. doi: 10.1016/j.electacta.2006.03.071.

[45] J. Bisquert, G. Garcia-Belmonte, F. Fabregat-Santiago, P.R. Bueno. *Theoretical models for ac impedance of finite diffusion layers exhibiting low frequency dispersion.* Journal of Electroanalytical Chemistry 1999, 475: 152–163. doi: 10.1016/S0022-0728(99)00346-0

[46] Z. Grubac, Z. Petrovic, J. Katic, M. Metikos-Hukovic, R. Babic. *The electrochemical behaviour of nanocrystalline nickel: a comparison with polycrystalline nickel under the same experimental condition.* Journal of Electroanalytical Chemistry 2010, 645: 87–93. doi: 10.1016/j.jelechem.2010.04.018.

[47] G.J. Brug, A.L.G. van den Eeden, M. Sluyters-Rehbach, J.H. Sluyters. *The analysis of electrode impedances complicated by the presence of a constant phase element.* Journal of Electroanalytical Chemistry 1984, 176: 275–295. doi: 10.1016/S0022-0728(84)80324-1.

[48] B. Hirschorn, M.E. Orazem, B. Tribollet, V. Vivier, I. Frateur, M. Musiani. *Determination of effective capacitance and film thickness from constant-phase-element parameters.* Electrochimica Acta 2010, 55: 6218–6227. doi: 10.1016/j.electacta.2009.10.065.

[49] J.O. Bockris, K. Uosaki, H. Kita. *Interfacial electron transfer as a significant step in photo-electrochemical reactions on some semiconductors.* Journal of Applied Physics 1981, 52: 808–810. doi: 10.1063/1.328847

[50] K. Uosaki, H. Kita. *Effects of the Helmholtz Layer capacitance on the potential distribution at semiconductor/electrolyte interface and the linearity of the Mott-Schottky plot.* Journal of Electrochemical Society 1983, 130: 895–897. doi:10.1149/1.211985

7

How Polygenetic are Monogenetic Volcanoes: Case Studies of Some Complex Maar-Diatreme Volcanoes

Boris Chako Tchamabé, Gabor Kereszturi,
Karoly Németh and Gerardo Carrasco-Núñez

Abstract

The increasing number of field investigations and various controlled benchtop and large-scale experiments have permitted the evaluation of a large number of processes involved in the formation of maar-diatreme volcanoes, the second most common type of small-volume subaerial volcanoes on Earth. A maar-diatreme volcano is recognized by a volcanic crater that is cut into country rocks and surrounded by a low-height ejecta rim composed of pyroclastic deposits of few meters to up to 200 m thick above the syn-eruptive surface level. The craters vary from 0.1 km to up to 5 km wide and vary in depth from a few dozen meters to up to 300 m deep. Their irregular morphology reflects the simple or complex volcanic and cratering processes involved in their formation. The simplicity or complexity of the crater or the entire maar itself is usually observed in the stratigraphy of the surrounding ejecta rings. The latter are composed of sequences of successive alternating and contrastingly bedded phreatomagmatic-derived dilute pyroclastic density currents (PDC) and fallout depositions, with occasional interbedded Strombolian-derived spatter materials or scoria fall units, exemplifying the changes in the eruptive styles during the formation of the volcano. The entire stratigraphic sequence might be preserved as a single eruptive package (small or very thick) in which there is no stratigraphic gap or significant discordance indicative of a potential break during the eruption. A maar with a single eruptive deposit is quantified as monogenetic maar, meaning that it was formed by a single eruptive vent from which only a small and ephemeral magma erupted over a short period of time. The stratigraphy may also display several packages of deposits separated either by contrasting discordance surfaces or paleosoils, which reflect multiple phases or episodes of eruptions within the same maar. Such maars are characterized as complex polycyclic maars if the length of time between the eruptive events is relatively short (days to years). For greater length of time (thousands to millions of years), the complex maar will be quantified as polygenetic. These common depositional breaks interpreted as signs of temporal interruption of the eruptions for various timescales also indicate deep magma system processes; hence magmas of different types might erupt during the formation of both simple and complex maars. The feeding dikes can interact with groundwater and form closely distributed small craters. The latter can coalesce to form a final crater with various shapes depending on the distance between them. This observation indicates the significant role of the magmatic plumbing system on the formation and growth of complex and polygenetic maar-diatreme volcanoes.

Keywords: maar-diatreme, polygenetic, monogenetic volcanism, complex maars, dike injection

1. Introduction

Monogenetic volcanoes represent the most common type of subaerial volcanoes not only on Earth but in the solar system [1–3]. These volcanoes, including maars, tuff rings, tuff cones and scoria cones, and sometimes short lava flows, are generally believed to form by a single short-lived eruption probably during a brief period of time (e.g., hours to days). The eruptions are typically fed by small volume of magma of any type, producing simple and small-volume volcanoes that predominantly clustered in lowland volcanic fields at the footprints of polygenetic volcanoes (e.g., [4–7]). The term polygenetic is often used to refer to volcanoes constructed by multiple eruptive events over a large timespan and characterized by complex volcanic and geochemical evolution such as stratovolcanoes (e.g., [8]). The same applies in this study. Most of the small volcanoes are generally characterized by simple and easy to understand volcanic sequences. However, the reality is different and that is what this book chapter tries to demonstrate. For instance, increasing investigations in several volcanic fields have shown that small volcanoes can exhibit, in special cases, a contrasting stratigraphy consisting of tephra units and other lava flows sometimes of different geochemical compositions (e.g., [8–10]). The eruptive units are sometimes deposited periodically with short or even prolonged inactive periods between eruptive events, indicating an evolution that cannot be explained with a single eruption (e.g., [11–13]). Additionally, such intricate trend of complexity is even more obvious with maar volcanoes that are the results of explosive magma-water interactions (molten-fuel-coolant interaction), a process causing fragmentation of both the magma and country rocks close to the surface (e.g., [14–19]. Their eruptive sequences are, therefore, heavily influenced by the magma internal (physicochemical) attributes as well as the environmental parameters that are largely expressed by the nature of the geologic substrate and the availability of external water to cause explosions. Maars are, therefore, probably the most complex and suitable bunch of small-volume volcanoes where complex sedimentary succession of their rim (and underground) architecture is expected. In this chapter, we intend to explore this diversity through some recent studies and own research. Examples will include recently documented small volcanoes showing complex stratigraphy, compound volcano edifice, and complex eruptive history, in various volcanic fields, including the Western Australian volcanic field (e.g., [10, 20]), the Coli Albani volcanic complex in Italy [13, 21], the Cameroon volcanic line [11, 22, 23], the Eiffel volcanic field [24, 25], the Trans-Mexican Volcanic Belt [26–28], the Central Volcanic Range of Costa Rica [12], the Bakony-Balaton Highland Volcanic Field in Hungary [29–31], as well as Auckland volcanic field in New Zealand [32–34].

This chapter focuses on phreatomagmatic-derived volcanoes, especially maar-diatremes, for which an important number of field investigations and various types of controlled benchtop and large-scale experiments have permitted to constrain processes involved in their formation, such as the mechanism of explosion (e.g., [35–39]), the quantification and control of magma-water interactions during the fragmentation process in regard to the potential maximum energy release that such process can provide (e.g., [35, 40]), the cratering process (e.g., [17–19]), the resulting volcanic facies preserved in a volcanic edifice and depositional processes associated with the distal regions of such volcanoes in the inter-volcano region (e.g., [29, 41–47]), as well as the characterization of the changes in the eruptive styles in the course of their formation (e.g., [48–52]), and the geochemical processes associated with their formation from the melt extraction and deep fractionation prior to eruptions to the processes influencing the rising magma batches (e.g., [8, 21, 52–55]). The reader can therefore refer to [56–58] for detailed review on scoria cones. The main purpose of this chapter is to highlight the main features that could help to understand the formation of complex maars and how we can recognize and discriminate The latter from simple maar volcanoes. Thus, we will emphasize where poly-activity have been identified such as the Purrumbete Maar, Australia (e.g., [20]), Albano Maar, Colli Albani volcano, Italy (e.g., [13, 21]), Hule Maar, Costa Rica (e.g., [12]), or Barombi Mbo Maar, Cameroon [22, 59]. In the light of several previous discussions around the monogenetic volcanoes (e.g., [6, 58, 60]), we present some key similarities and dissimilarities within simple and complex small-volume volcanoes, especially maars, so that a better definition for those volcanic end members could be understood. We also discuss what processes might drive complex activity at maar volcanoes in order to propose a conceptual model that can summarize the origin and growth of this type of end-member volcano.

2. General features of maar-diatreme volcanoes

In many volcanic fields, it is usually common to see a low rim of bedded pyroclastic ejecta surrounding a dried or water-filled depression that cuts into the pre-eruptive ground (**Figure 1**). This structure is usually called maar. Maar is a German-derived word that means "crater lake," whose origin derived from the Latin word *mare* (sea). In 1819, in his book "Die erloschenen Vulkane in der Eifel und am Niederrheine," Johann Steininger was probably the first to coin the term maar to describe a volcanic feature while working in the Eifel volcanic area in Germany, in which the craters are usually occupied by lakes. The term was then widely used by Ollier [61] and Lorenz [62, 63] and others authors cited therein. These early papers put the term of maar into the scientific literature as an important volcanic landform formed through phreatomagmatic eruptions, which is now applied to similar craters (e.g., [12]). A maar stands for a volcanic crater that is cut into country rocks (a few meters or tens of meters above the preexisting ground surface) and surrounded by a low-height ejecta rim composed of pyroclastic deposits of few meters to up to 200 m thick above the syn-eruptive surface level (e.g., [64]). The term maar is sometimes also used only as a morphological term.

Figure 1. Photos showing the typical landform of some maar volcanoes. (**A**) Blue Lake Maar (Mount Gambier, South Australia), (**B**) Aci Golu Maar (Turkey), (**C**) Al Wahbah Maar (Harrat Khisb in western Saudi Arabia), (**D**) Cora Maar (Central Anatolia), and (**E**) Meke Gölü Maar (Turkey), (**F**) Alchichica Maar, (**G**) Aljojuca Maar, and (**H**) Atexcac Maar of the EMVB, Mexico.

Maar volcanoes are characterized by a relatively small crater size, hundred meters to up to 5 km in diameter [65], with few dozen of meters to up to 300 m deep [9]. The craters are mostly circular in shape, although in some cases an irregular morphology can be observed due to a formation through the injection of discrete dikes at several but closely spaced explosion craters/centers to the main crater (e.g., [55]). Examples of such simple maars are certainly the most widely spread in monogenetic volcanic fields, such as the Eifel volcanic field, western central Germany, where many maars are characterized by small crater diameters ranging from 83 to 1580 m [25] and low tephra deposit thickness, e.g., Ulmener Maar, 7.5 m thick, Pulvermaar, 27 m thick, or Meedelder Maar, 23 m thick[24]. The same feature is observed with some maars of the Trans-Mexican volcanic field (e.g., [66–68]) or those of the Quaternary Auckland Volcanic Field in New Zealand (e.g., [69]), as well as maars of the Sabatini Volcanic District in the Roman Province of Central Italy (e.g., [70]). The crater floor usually lies well below the surrounding ground level and frequently exhibits near-vertical crater wall escarpments (e.g., [6, 8, 71]). On the other hand, the ejecta rings of maars are characterized by sequences of successive alternation and contrastingly bedded pyroclastic deposits. Much of the bedding forms by dilute pyroclastic density currents (PDC), blast and fallout depositions after phreatomagmatic explosions. This produces a range of beds, typically changing from thick, structureless, and commonly block-rich near the vent to well-developed medial cross-bedding and dune-form and thin distal planar beds [12, 34, 42, 72, 73]. In many cases, occasionally interbedded Strombolian-derived spatter material or scoria fall units are observed. This exemplifies the changes in the eruptive styles during the formation of a maar volcano (e.g., [45, 49, 74, 75]). The deposit sequence also commonly contains large amounts of lithic material that is entrained from the country rock basement and in some cases accretionary lapilli that is an indicator of free moisture or water droplets in the moving two- or three-phase current [76–78]. Bedding sags are common sedimentary features and their abundance usually reflects the violent excavation of blocks of country rock or magmatic bombs during the formation of the diatreme and the ballistic nature of eruptions (e.g., [55–57]) (**Figure 2**).

Figure 2. Some features observed on maar deposits. (**A**) Contrasting surge and fall units of the Alberca de Guadalupe maar deposits (Zacapu basin, Michoacán, Mexico); (**B**) thinly stratified surge deposits of Nyos Maar, Cameroon; (**C**) soft structure deformation and impact sags in a sequence of Barombi Mbo Maar deposit, Cameroon.

The estimated volumes of bulk-ejected tephra and the corresponding dense rock equivalent (DRE) using different methods such as isopach and/or juvenile content of the bulk deposits or by applying interpolation techniques on digital elevation models (DEM) along with rock textural data collected from the field (e.g., [69]) are usually very small (≤1 km^3). This suggests that maars are very small-volume volcanoes compared to the middle-size shield volcanoes (1–10 km^3) and the large polygenetic volcanoes (10–10,000 km^3). The latter ones have a stable melt source over prolonged periods, where shallow magma storage systems are expected to develop and form a well-defined and stable vent zone over a long time, producing large volumes of materials and potentially chemically diverse eruptive products [60]. The duration of volcanic activity leading to their formation would therefore be probably short and even reduced to a single eruptive event (e.g., [6]).

Another feature that is usually associated to maars is a diatreme. Because of the occurrence in maar deposits of an important amount of accidental lithic fragments of the country rock, it was inferred that below a maar there is inevitably an extended subsurface inverted cone- or carrot-shaped structure called diatreme (e.g., [29, 79]). Many remnants of well-exposed diatremes have been identified in association with massifs of plutonic rocks of different compositions and also in ore deposits fields, where in the form of brecciated and pebbly pipes, they have frequently served as the most favorable ore- and diamond hosting structures. However, even if geophysical studies have demonstrated the presence of this structure beneath maars (e.g., [80, 81]) and that in rare occasions drill holes have reach the upper level of the diatreme facies of the maar-diatreme volcanoes (e.g., [62, 63]), the opportunity to examine a diatreme and an ejecta ring belonging to the same maar-diatreme volcano is rare, posing some difficulty to establish the direct relationships between the ejecta ring of a maar, the eruption processes, and the growth of its underlying diatreme. Nevertheless, the diatreme beneath maars might consist of deposits formed during eruptions that can be described collectively as "diatreme deposits," including bedded diatreme fill; un-bedded diatreme fill, including in zones that cut across bedded fill; as well as root zone deposits (e.g., [12], **Figure 3**).

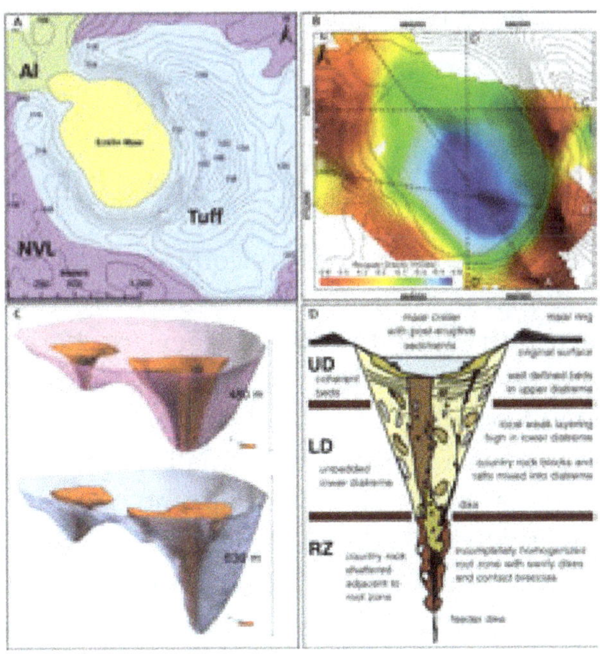

Figure 3. Example of geophysical response from the Ecklin Maar, Newer Volcanics Province, southeastern Australia, (images from Blaikie et al. [81]). (**A**) Simplified geology of the maar; (**B**) Bouguer anomaly with regional trend removed showing gravity low over the crater; (**C**) 3D model viewed from southwest (upper) and optimized geometry (lower) of the Ecklin diatreme. (**D**) Maar and diatreme structure with distinct parts of the diatremes defined by their typical lithofacies and structure (from White and Ross [82]).

These general features commonly characterize a simple maar volcano that is considered as monogenetic volcano sensu stricto. The latter corresponds to a volcano characterized by a single eruptive vent (single crater for maars and tuff rings and unique and regular cone shape for scoria cones) through which only a small and temporal magma supply of single or various compositions erupted once in a brief period of time. This implies that all the pathways of magma supply should have cooled down and ascending routes are no longer favored for the next magma batch (e.g., [60]). In reference to maars or tuff rings, the tephra ring for monogenetic maars would have a relatively regular shape that might follows the morphology of the crater. The stratigraphic sequence is as simple as possible in terms of tephra succession (e.g., no stratigraphic gap or discordance indicative of a potential break in the eruption progression). This simplicity does not only refer to the small thickness or the relatively homogeneous type of deposits (e.g., PDC) that can be observed at some maars, because some monogenetic maars can have complex deposit sequences including dilute PDC, tephra fall and spatter, and sometimes rootless lava flows. The Barombi Koto Maar (Cameroon volcanic line) is an example of this type of maar volcanoes. The deposit sequence of this maar indicates a volcanic evolution comprising an initial phreatomagmatic stage, followed by a late sustained Strombolian activity that formed a small scoria phase, then another phreatomagmatic phase, and a late sustained Strombolian-style explosive eruption that formed a small scoria cone constructing an islet in the middle of the crater lake, without any break in the preserved eruptive sequence [23]. Nyos Maar in Cameroon could also be a good example. Nyos Maar is characterized by a lower lava flow unit (8 m thick) and an upper dilute PDC unit (~70–80 m thick on the eastern lakeshore), indicating an initial fire-fountaining phase [83] and a series of phreatomagmatic explosions [83] without gap between the eruptive sequences indicating a continuous eruption [84].

3. Features of complex maar volcanoes

As discussed above, maar-diatreme volcanoes are commonly composed of a crater, an ejecta ring, and an underlying diatreme structure that is filled by various fragments from the ascending magma and the country rock. In addition, they are characterized by small eruptive volumes that usually result in the simplicity of their volcanic edifice. The small eruptive volume is also interpreted as a result of a short volcanic activity and even reduced to a single eruptive event. However, even characterized by a small eruptive volume, all maar-diatreme volcanoes are dissimilar in terms of volcanic edifice morphology. Like their "cousins" tuff rings and scoria cones, which are usually considered as monogenetic volcanoes, these volcanoes are very complex especially when their stratigraphic sequences, the morphology of their craters and/or their ejecta rings, or the chemical composition throughout the sequence are examined in detail. For instance, Németh et al. [85] and Németh and Kereszturi [60] following earlier definitions of monogenetic volcanoes (e.g., [4]) highlighted different types of small volcanoes that can be encountered in monogenetic volcanic fields. These included monogenetic volcanoes sensu stricto and complex monogenetic volcanoes with multiple eruptive episodes, which in some cases are characterized by a complex magmatic feeding system. In the literature there are numerous examples for such eruptive behavior: Crater Hill [32], the long-lived scoria cone and lava flow complex of Rangitoto Auckland volcanic field, New Zealand [86], and Motukorea tuff ring in Auckland volcanic field, New Zealand [87, 88]; the Kissomlyó in Hungary (e.g., [89]); the Udo, Songaksan, and Yangpory in South Korea (e.g., [8, 54, 90]); the Purrumbete Maar in Australia (e.g., [10]); Fekete-hegy [91], Bondoró [31], and Tihany [29] from the Bakony-Balaton Highland Volcanic Fields in Western Hungary; some maars in the Eifel volcanic field, Germany [92, 93]; the Cerro Negro scoria cone, Nicaragua [94, 95]; and El Volcancillo, Mexico [96]. All of these examples were likely constructed over a longer period of time (from Ky to My). This was inferred from the fact that those volcanoes, even having a small eruptive volume, have a complex stratigraphy and tephra ring architecture suggesting that multiple eruptive episodes contributed to the growth and destruction of the volcanic edifice (e.g., [85]). These volcano categories are revisited hereafter with an emphasis on maar-diatremes.

Figure 4. Textural unconformities in the Meke Gölü Maar deposit, Turkey (**A**), and laterally discontinuous thick cross-laminated beds in the Cora Maar deposit (**B**). (**C**) Complex well-marked discordant contacts and truncation surface between the deposit packages at Alchichica Maar, Easter Mexican Volcanic Belt (EMVB). Note the discordant contacts between the scoria cone, the lava flow, and the maar pyroclastic deposit sequence (PH). The left lower photo shows a sharp contact between the scoria cone and the pyroclastic level PH.

As with polygenetic volcanoes, multiple eruptive events have the capability to produce with time a large cumulative volume of tephra and/or lava products around a single or multiple volcanic vents. In the case of small volcanoes, this probably will result in the deposition of thick eruptive sequences. However, the volume or the thickness of deposits might not be a common feature to all small volcanoes where multiple eruptions or polycyclic activity is observed. This is mainly because these parameters depend directly on the volume of magma involved in each eruptive cycle and, in the cases of maars and tuff rings, to the depths at which explosions took place to excavate an important fraction of country rocks that compose up to 90% of ejected materials (e.g., [12]). Nevertheless, the consequence of the poly-activity within small volcanoes is the construction of complex stratigraphic sequences. These complex volcanoes usually display packages or depositional units made of erupted materials that in some cases can be directly apparent on the field by deposit textural differences, chaotic deposits separated by a lava flow horizon (e.g., [97]), and/or a dike cutting through the deposit units (e.g., [98]). Textural differences in pyroclastic sequences can also show altered or palagonized juvenile-rich deposits that underlie a fresh surge or fall unit within the same eruptive sequence (e.g., [33, 97, 99]) or the presence of centimeter- to decimeter-thick light brown to yellowish pedogenized ash horizons in some deposits [21]. Well-marked structural discordant contacts or truncation surface or erosional limits between the deposit packages (e.g., [26, 34, 66, 76, 92, 100]) are some of the main features observed within the stratigraphic sequence. These are characterized by high-angle, laterally discontinuous or thick cross-laminated levels and angular unconformities between pyroclastic deposits, ranging in outcrop scale from centimeter to decameter long (**Figure 4**). In many other cases, one of the features that separate the eruptive packages is a paleosoil (e.g., [12, 22, 42, 101, 102]).

Because the formation of a soil requires a minimum time ranging from hundreds to millions of years depending on the climatic conditions (e.g., [103]), this feature highlights how long was the period of the eruptive activity and is therefore commonly used to distinguish between simple monogenetic and complex polygenetic small volcanoes. However, multiple eruptive events might occur within a short timescale without the formation of paleosoils between eruptive packages, and the surrounding deposits can display the same stratigraphic and structural complexity [9]. Note that in historic times only a few maar-diatreme volcanoes erupted. In 1954 the Nilahue Maar erupted in Chile during almost half a year, but the main eruptive phase ended after 10 days producing a maar crater of 300 m in diameter. In contrast, in 1977 the Ukinrek West Maar erupted only for 3 days and generated 10 m-thick tephra ring, a 170 m wide (rim to rim) and 30 m deep maar crater (e.g., [100, 104]). This information is certainly not enough to generalize about the duration of a sequence of maar-forming eruption, making it difficult to easily distinguish between the complex maars. Fisher et al. [105] suggested that an eruptive pulse is a single explosion or detonation that may last a few seconds to minutes producing an eruption column from which particles will sediment to form a single well-defined tephra bed. On the other hand, an eruptive phase consists of series of strong explosions that can last a few hours to days generating pulsating eruptions columns and formation of several well-defined beds. Depending on the style of magma fragmentation, an eruptive phase may alternate between explosive and effusive eruptive phases [106]. It is also important to note that the eruption here is fed by a single magma batch or multiple magma batches that could be of the same or different compositions (e.g., [60]). The eruptive episode or single eruption is composed of several eruptive phases, which may last a few days to months and in some volcanoes for years [105 ,106].

Following these definitions, (1) a complex monogenetic volcano can be categorized as the one where multiple eruptive phases have been identified. This implies that magma batch or batches feeding the system erupted almost at the same time, with a very short break (days to years [9]) insufficient to allow any significant erosion or alteration (palagonization) at the top of each eruptive package (deposits of one eruptive phase) and especially the formation of a paleosoil. This type can experience vertical and lateral vent migration and dike arrests which are very common processes in the formation of maar-diatreme volcanoes (e.g., [28, 107, 108]).

In contrast, (2) complex monogenetic volcanoes with polygenetic inheritance are those in which at least two eruptive episodes have been identified, i.e., where a paleosoil or any indication for time gap from the eruptive sequence can be established (time obtained by conventional dating methods) that separates two sequences of deposits, each composed of multiple packages (e.g., [22, 101]). This also implies that the time gap between the eruptive episodes is significant, several thousands of years as observed with the Albano Maar (e.g., [13, 21]), the Barombi Mbo Maar (e.g., [59]), the Bondoró Volcanic Complex [31], the Hule Maar (e.g., [12]), or Ilchulbong tuff cone [9].

While the erosional limits or the presence of paleosoils within the stratigraphic sequence would mainly indicate a time gap between eruptive cycles, structural truncation surfaces or discordant contacts usually result in complex tephra ring architectures, especially when deposit packages have different dipping angles (e.g., [9, 20]). This suggests an influence of the variation in the eruptive vents or some tectonic activity with the progression of eruptions that have been attributed to the formation of complex craters morphologies (e.g., [17, 20, 26, 27, 90, 91]). Experimental studies have even demonstrated that the size and shape of maar craters might vary depending on the positions and numbers of the explosion loci during their formation (e.g., [109–112]). For instance, according to [111], final crater shapes tend to be roughly circular if subsurface explosion epicenters occur within each other's footprints (i.e., the plan view area of reference crater produced by a single explosion) and elongate if an epicenter lies somewhat beyond the footprint of the previous explosion, such that their footprints overlap. But if epicenters are too far apart, the footprints do not overlap and separate craters result (e.g., [29, 113]). This is likely the process that occurred at the Tihany volcanic complex in Western Hungary, where successive eruptions created three separated volcanic centers (e.g., [29]). **Figure 5** shows this complexity of the crater shape for some maar volcanoes of both monogenetic (e.g., Atexcac; **Figure 5c**) and polygenetic natures (e.g., Albano and Purrumbete Maars; **Figure 5B** and **D**). These maars are characterized by many small craters a minimum of three for the Purrumbete Maar to up to nine for the Atexcac maar [27] that coalesced to form relatively regular or irregular crater morphologies. Many other maars with such complex eruptive evolution and crater morphology have been identified in different volcanic fields. Crater Hill in Auckland Volcanic Field, New Zealand, is characterized by a nearly circular tuff ring of 900–1100 m wide and only 9–15 m thick, surrounding an elliptic irregular crater [115]. The crater resulted from the coalescence of at least four vents spaced along a NNE trending, 600 m-long fissure [115]. Tecuitlapa Maar located in the eastern Central Volcanic Belt of Mexico [28] is characterized by a 1 km-diameter irregular crater which is an alignment of scoria cones. It is thought that activity there began in the eastern part of the crater with phreatomagmatic eruption, where basaltic magma interacted with liquefied tuffaceous sediments. Then, the explosion locus gradually moved westward producing an elliptical crater. The eruptions then dried out and began to produce scoria/spatter cones with nested craters along the same alignment parallel to regional structural trends [28]. Chako Tchamabé et al. [11] demonstrated also that similar migration of explosion vent occurred at the polygenetic Barombi Mbo Maar, forming a very large, amalgamated maar crater with a total diameter of 2.5 km. A minimum of three and a maximum of five craters were suggested according to the three eruptive episodes identified and the potential implication of several dike injections within the progression of activity [55].

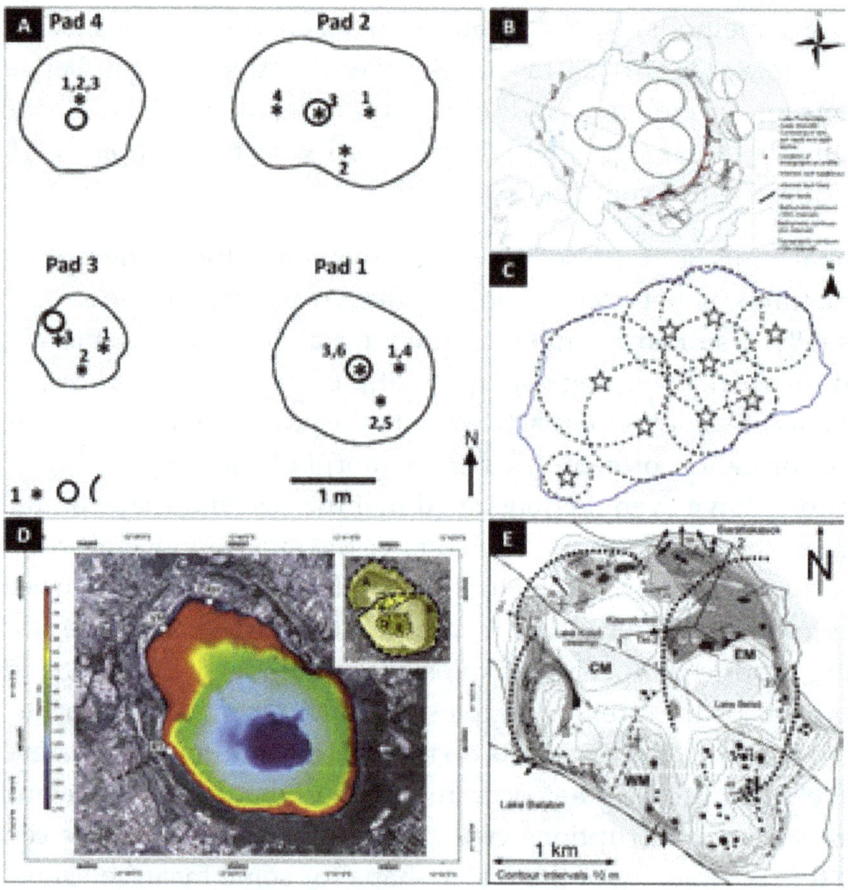

Figure 5. Complexity of crater morphology of maars. (**A**) Shapes of final crater rims resulting from experimental study showing the effect of vertical (Pad 4), lateral (Pads 2 and 3), and both (Pad 1) series of explosions (after Valentine et al. [111]). Bold circle, low point in each crater; 1, 2…, numbers of blast epicenters (asterisk). (**B, C and D**) Crater shapes and inferred number of vents (dash circles) for Purrumbete Maar (after Jordan et al. [20]), Atexcac Maar (after López-Rojas and Carrasco-Núñez [27]), and Albano Maar (after Anzidei et al. [114]) respectively. (**E**) Complex Tihany volcanic center where large distances between explosion vents have formed three distinct craters (after Németh et al. [29]).

Complex crater morphology (e.g., size and shape) could thus be considered as other useful features that characterize complex maars. However, distinguishing between simple maars and complex ones based on the morphology of the crater alone might be confusing. As noted earlier, simple monogenetic maars can present both regular (subcircular to circular) and irregular crater shapes, irrespective to their sizes. This is probably because multiple batches of magma might cause explosions simultaneously at several locations near the main center of the crater (e.g., [37, 103]), resulting to the formation of an irregular crater-shaped and a complex but simple deposit sequence in which discordances are scarce. Sill complexes are present in some monogenetic volcanic fields and suggested to fed some maar-diatreme-forming eruptions (e.g., [18, 34, 116–119]). In addition, investigations have shown that the crater morphology and even the architecture of pyroclastic deposits and evolution of maar-diatreme volcanoes can be highly affected by the type of environment—hard substrate (rocks) or a soft substrate (unconsolidated volcaniclastic or sedimentary deposits)—in which they are emplaced (e.g., [82, 91, 120, 121]). In soft substrates, maar-diatreme volcanoes tend to have large and bowl-shaped craters, with gently dipping inner walls [91]. Recent analog experiments as well as field observations from classical diatremes cut into "soft substrate" showed that the diatreme wall can be steep for such maars that cut through soft substrate (e.g., [121–125]). This might be valid for the geometry of the upper part of the maar-diatreme volcano, especially for its crater, given that the number of individual eruptions can also heavily affect the final crater-diatreme morphology, and as many explosive events take place hence as large and old as your maar, the role of the substrate physical conditions will be reduced (e.g., [125]). In contrast, maars formed in hard-rock environment tend to be irregular, small in size and characterized by funnel-shaped and vertical (e.g., Joya Honda, Mexico [126], Nyos Maar, Cameroon [127]) to steeply dipping crater walls. For instance, in the Calatrava volcanic field in Spain, [120] measured and compared the crater sizes and shapes of 60 maars formed in hard substrate and 66 maars formed in soft-substrate basin-filling sediments. While the average crater radius of maars in hard substrate setting is ~ 339 m, those in the other setting have an average of 556 m, indicating that in this volcanic field, the size of the craters for soft-substrate maars is 64% larger on average than that of hard-substrate maars, though the average crater shape in aerial view is quite similar [121]. Maar crater shapes can also be strongly controlled by the presence of any pre-volcanic lithological situations, including older cones that might have been dissected by the maar-forming eruption, or when explosions occur in a preexisting crater form by previous activity (e.g., [128]). The initial shape of the crater might even change with time due to erosion and slumping of the walls and tephra ring (e.g., [18, 79, 129–131]), shallowing the crater slope and reducing the relief. Older maar basins, for example, could have strong erosion modification along their margins and also could be filled with post-eruptive debris, enlarging the original size of the crater. Unusually large maar lake with irregular boundary might certainly results from complex and migrating explosion locus in the area of the crater floor resulting in complex collapse event and scalloped crater wall architecture. Therefore, it is possible to wrongly interpret a maar with complex crater outline as complex maar as its erosion progresses. Large and complex crater outlines can equally mean either a complex eruptive history or long-lasting erosion history; then one has to check the eruptive sequence carefully not only the morphology of the crater. Correlations should be done between the sequence of activity, the different eruptive packages to the number of craters/vents, and probably the distance between them before using the crater morphology to characterize complex maars, as the crater morphology reflects the complexity on the growth of the volcano (**Table 1**).

Maars	Stratigraphy				Crater morphology				Genetic nature
	Thickness (m) of deposit rim (maximum section)	**Number of eruptive units**	**Transition style between deposit packages**	**Geochemical composition of erupted materials**	**Size (km)**	**Number of vents or potential crater basins and relative distance between them**	**shape**	**Depth (m – under lake surface)**	
BMM	126	3	Paleosoils	Bimodal	2.5	~5 small craters closely distributed	Subcircular	110	Polygenetic
Tecuitlapa	50–70	4	Facies transition	–	1.3 × 1	Undetermined but at least six cinder vents have been highlighted [28]. The positions of the explosion locus are unknown but are distributed laterally following a structural fault	Elliptical	–	Monogenetic
Purrumbete	40	4	Three major Structural discordances	Bimodal (polymagmatic)	3	Three craters closely distributed [10]	Subcircular	45	Polycyclic
Albano		3	Paleosoils	Complex		Five craters closely distributed	Elliptical		Polygenetic
Hule		3	Paleosoils and facies transition	Bimodal [12]	2.3 × 1.8	Three lakes separated by two intra-maar pyroclastic cones and lava flows [12]. Assuming	Sub circular for the whole basin,	Vary for each of the three lakes (Hule, 26.5; Congo, 14.6;	Polygenetic

Maars	Stratigraphy				Crater morphology				Genetic nature
	Thickness (m) of deposit rim (maximum section)	Number of eruptive units	Transition style between deposit packages	Geochemical composition of erupted materials	Size (km)	Number of vents or potential crater basins and relative distance between them	shape	Depth (m – under lake surface)	
						the lakes are lying in resulting craters, we may have three small craters among which the main Hule lake shows two vents for [116] located ~500 m from a basin to another (see **Figure 2** in [117])	elongated for the main Hule lake	unnamed, 4 m) [12]	
Atexcac	61	4	Facies transition	–	1.15 × 0.85	Nine potential craters randomly distributed [27]	Elliptical	120	Polycyclic
Crater Hill	9–15	7	Truncation surface and facies transition	Compositional variation due to clino pyroxene ± spinel fractionation	0.656	4 Aligned along a fissure	Elliptical	100–120	Polycyclic

Table 1. Some characteristics of complex maar volcanoes formed from multiple eruptive events. The number of eruptive units is based here on the number and style of transitions identified and in some case corresponds to the number of eruptive events.

4. Growth of complex monogenetic volcanoes

The eruptive mechanism associated with the formation and growth of monogenetic volcanoes is neither well known nor uniform actually and is somehow attributed to a wide range of magmatic and magma-water interaction-driven explosions at both shallow and deep levels vertically and laterally within the substrate [22, 23, 34, 88, 98, 99, 132, 133]. However, the eruptive timespan for the development of complex monogenetic volcanoes makes a big difference compared to the monogenetic sensu stricto end members. The time in this context is certainly related the timescale of magmatic process in the mantle beneath the volcano. In fact, it is much longer and it is sometimes comparable with large polygenetic volcanoes that are characterized by subsequent production of significant volumes of magma with time. Recent studies have shown that at monogenetic volcanoes, small volumes of melt can segregate from the mantle and readily ascend to the surface through dike or crack propagations (e.g., [8, 54, 55, 134–137]). The segregated melts can rise and erupt simultaneously. In such cases, a polymagmatic monogenetic volcano would form, assuming that the magma batches are of different chemical compositions, such as the Udo volcano in Korea (e.g., [8]). On the other hand, the melts can form with time (e.g., yrs to My), rise, and erupt sporadically. In this latter case, successive vents can be constructed and, depending on the distance between the feeder conduits in the system, can produce the nested or separated vents that characterize these relatively complex volcanoes (**Figure 5**). This process can occur in a typical intra-plate volcanic field such as Saudi Arabia [138] or at basaltic-andesitic polygenetic volcanoes such as Tongariro volcano in New Zealand [139] or at complex maars such as Albano Maar (e.g., [21]) or Barombi Mbo Maar [55] and is broadly accompanied by polymagmatic activity. At Tongariro volcanic Complex in New Zealand, for example, diverse lava flows and pyroclastic units with contrasting chemical and isotopic composition were deposited in a period of 275 Ky, constructing 17 small (>0.3 km^3) to large (>12 km^3) nested and overlapping volcanic cones in a non-systematic orderly progression in space for cone-building events and without any systematic distribution of the vents as well [139]. Freda et al. [21] demonstrated based on $^{40}Ar/^{39}Ar$ ages dating that volcanic activity at Albano Maar (Italy) was strongly discontinuous in time, with a first eruptive cycle at 69±1 ka producing at least two eruptive phases and a second cycle with two peaks at 39±1 and 36±1 ka producing at least four eruptive phases. All these cycles occurred in a narrow surface area centered from each other within only hundreds of meters away, forming a compound volcanic edifice. Using geochemical constraints, they also could demonstrate that each eruptive phase was fed by magmas with different compositions. The complexity in chemical composition was attributed either to the arrival of a new batch of magma during the different eruptive cycles, or to the feeding of the system by the same magma that continuously differentiated and erupted during the whole life of the activity. The eruptive activity at Barombi Mbo Maar in Cameroon follows also such complex volcanic and petrogenetic evolution [55]. In this case, three distinct eruptive events occurred subsequently at 0.5 Ma, 0.2 Ma, and 0.08 Ma [59], fed by magmas with different compositions (**Figure 6**). Petrogenetic constrains there also highlighted the segregation and rise of distinct magma batches with time. During the first eruptive event at Barombi Mbo, successive magma batches of same composition created a first crater, and after a significant reposed period of about 0.3 My, other magma batches some with the same composition with the former one and other with distinct composition were involved. This indicates that during this second eruptive episode, at least two dikes contributed to the formation of another crater close to the first one. The same process occurred during the third episode after another repose period of about 0.1 My.

It can be observed that the production of magmas within these volcanoes is distributed in a longer timescale, covering a 500 ka range for the Barombi Mbo Maar, less than 300 ka at Tongariro, and only 30 ka at Albano Maar. These observations suggest that one of the main factors that might favor polygenetic activity at monogenetic volcanoes is certainly the time necessary for the segregation of small volumes of melt, mantle fertility, available melt, melting and discharge rates, and the quick potential of magma batches to rise to the surface through regional tectonic setting and stress distribution in the crust. It is important to note that beneath such polygenetic volcanoes, there could be several pockets of melting in the mantle.

Because the degree of partial melting may also vary in each pocket of melting depending on various factors (e.g., the P-T condition, mineral phases present and volume of volatile phases in the mantle zone, or the geotectonic context where the volcano is located), the melts can segregate simultaneously or individually in the different melting spots in the mantle and erupt with time. Still, it is not excluded that the same melting point can produce, with time, small but sufficient volumes of melt that can erupt at different locations near the previous vents due to the tectonic control in the volcanic area or following cracks produced during precedent eruptions. This also allows us to suggest that, if beneath a monogenetic volcanic field, there are conditions that can favor in a local mantle zone the existence of multiple melting spots; the melts might raise with time as they are produced to develop complex small volcanoes with multiple eruptions. If the rising magma batches encountered a wet zone near the surface, a complex maar-diatreme will develop (**Figure 6**).

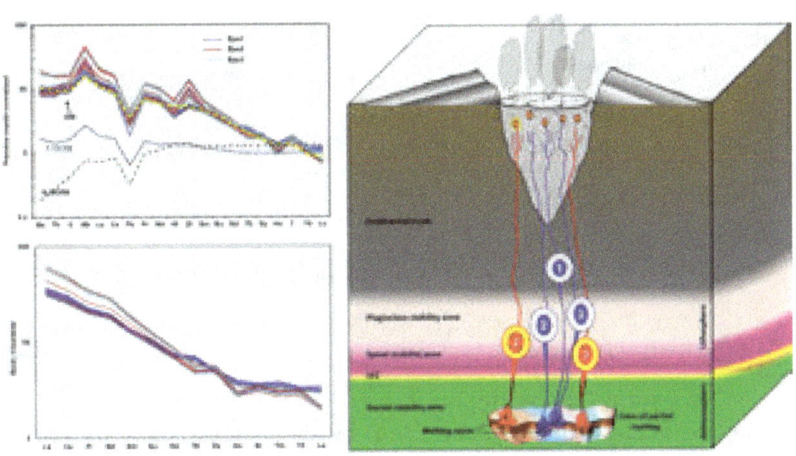

Figure 6. Example of complex compositional variation highlighting complex evolution at Barombi Mbo Maar in Cameroon. The schematic diagram (not to scale) presents the expected feeding system beneath the BMM complex. (1), (2), and (3) correspond to different magma batches feeding the system during the 1st, 2nd, and 3rd eruptive events, respectively, after Chako Tchamabé et al. [59]. The blue and red colors are used here to highlight the different magmas and not the melting loci. Note the involvement of at least two distinct dikes during the second and the third eruptive episodes (details in Chako Tchamabé [55]).

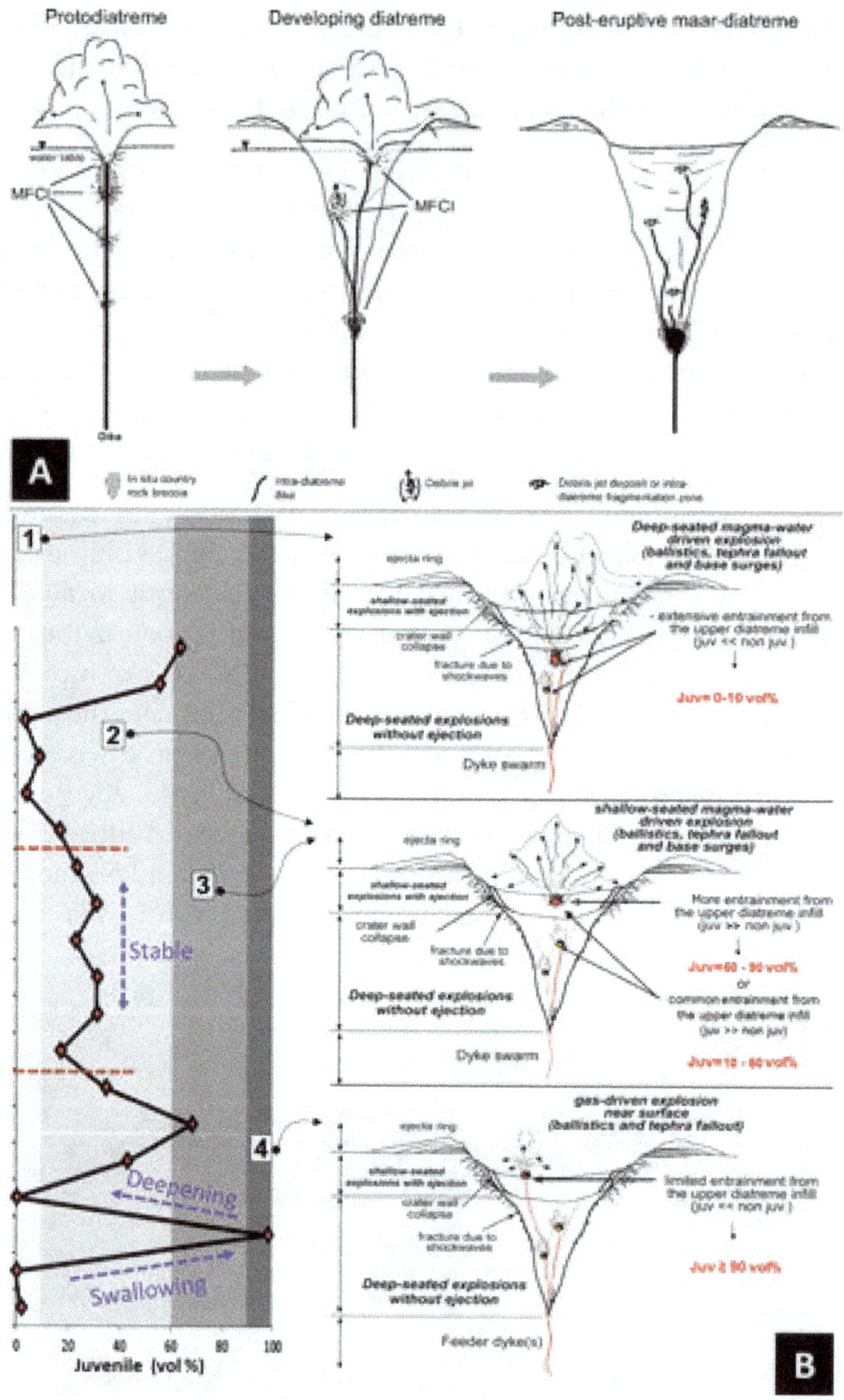

Figure 7. (**A**) Conceptual model after Valentine and White [148] showing explosive molten-fuel-coolant interactions (MFCI) that might take place over a range of depths, brecciating country rock where the explosions take place, but being most effective at shallow depths. (**B**) Comparative schematic model for interpreting the evolution and potential explosion sites (shallow or deep) at maars based on stratigraphic distribution of juvenile components in the BMM ejecta ring. From left to right we have the variation of juvenile populations with a delimitation of domains of juvenile proportion that might reflect a potential model of explosions during maar-diatreme formation. Dashed red lines represent the volcanic hiatus (paleosoils in the deposit) separating the different episodes. The different domains: 1 (juvenile ≤ 10 vol.%), 2 (juvenile = 10–60 vol.%), 3 (juvenile = 60–90 vol.%), and 4 (juvenile ≥ 90) are described in the text and more details in Chako Tchamabé et al. [11].

In the context of growth of such complex monogenetic volcanoes, these observations have an important consequence. The classical growth model of maar-diatremes has long been interpreted following the conceptual model of Lorenz [8], who suggested that the locus of subsurface phreatomagmatic explosions propagates downward with the deepening of a groundwater drawdown cone, as water is used and ejected by explosions [8]. This model implies that the diatremes widen due to slumping and subsidence of host material as their explosion loci deepen [140, 141]. As a result, near-surface occurring lithics would dominate the base of the ejecta rings, while lithics originating from deep-seated explosions location will be deposited on the upper parts of the ejecta ring. Many authors, however, have interpreted the variations in grain size and component distributions in tephra deposits of maars to reflect variations in the intensity of fragmentation during the phreatomagmatic explosions and/or intervening magmatic volatile-driven phases (e.g., [11, 20, 74, 142]) which in turn are often inferred to be related to magma-water ratios (e.g., [143]). It has thus been observed that some maars record intermediate and/or closing phases of magmatic volatile-driven activity in the form of lavas and/or scoria accumulations (e.g., Barombi Koto Maar [23], Tecuitlapa Maar [28]) which are interpreted to result from the absence of groundwater according to [8]. But, the presence of magmatic fragmentation with the evolution of a maar may certainly indicate shallow explosions (e.g., [74]). For instance, Valentine and White [29] propose an alternative model that allows multiple levels of country rock disruption and fragmentation, based on effective mixing by debris jets, an important subsurface transport phenomenon in phreatomagmatic vent complexes that is defined as an upward-moving stream of volcaniclastic debris, magmatic gases, and water vapor ± liquid water droplets, occurring on multiple vertical levels within a growing subsurface diatreme (e.g., [144]). This conceptual model is in accordance with the observed irregular distribution of accidental lithics in ejecta rings (e.g., [145]), field examples on diatreme geometry (e.g., [79]), but also on experimental cratering studies (e.g., [109, 124, 146]) and geophysical modeling (e.g., [80, 81, 147]). Chako Tchamabé et al. [11] also suggested that the variation of juvenile populations within the stratigraphic sequence of maars might reflect a potential mode of explosions during maar-diatreme formation (**Figure 7**). They proposed four domains varying from 0 to 100 vol.% of juvenile contain with the corresponding mode of explosion. For example, a juvenile content of ≤10 vol.% (domain 1) might suggest deep-seated explosions with limited ejection of juveniles and extensive entrainment of fragmented lithics. For 10–60 vol.% juvenile contents, deep- and shallow-seated explosions might occur, with a common entrainment of juveniles and more fragmented lithics, whereas juvenile contents of 60–90 vol.% would suggest shallow-seated explosions with more ejection of juvenile and limited entrainment of fragmented lithics. Up to 90 vol.% of juvenile indicates very shallow (near-surface) gas-driven explosions with ejection of more juveniles. This observation, supported by the conceptual model of [148] for the growth of maars and their diatremes (**Figure 7**), makes clear that explosions may occur at multiple levels, laterally and vertically, contributing to fragmentation and mixing of debris through a combination of upward-directed jets and downward subsidence (e.g., [109, 110, 124, 128, 149]).

However, while those models allow for understanding the diverse eruption scenarios within the formation of simple maars, it might be difficult to determine the growth of complex monogenetic volcanoes, especially complex maars that formed from multiple eruption

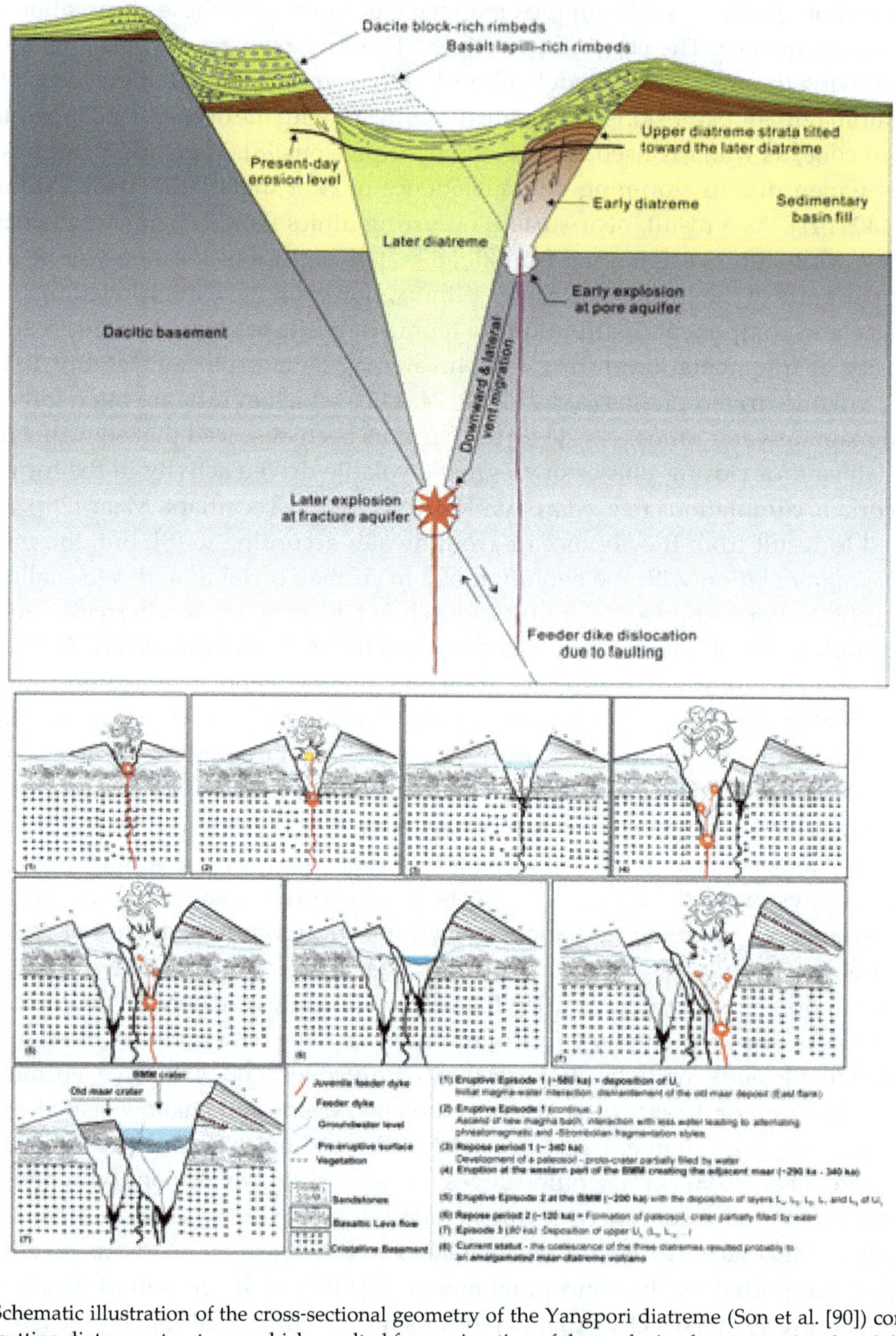

Figure 8. Schematic illustration of the cross-sectional geometry of the Yangpori diatreme (Son et al. [90]) consisting of two cross-cutting diatreme structures, which resulted from migration of the explosion locus associated with basin-margin fault movement (left). Sketch of the temporal evolution and growth of the BMM and its diatreme; here, the explosions started at shallow depth. Afterward, a vertical shift of explosion locus in the substrate followed, producing a scoria-rich layer through alternating phreatomagmatic- and Strombolian-type explosions. Explosions started again after a quiescent period of ~0.3 Ma and magma-water interactions occurring at deeper and at various lateral positions within the diatreme. These explosions widened the crater and deposited more tephra onto the ejecta ring. The explosion pattern may have been the same during the third eruptive episode, continuing to widen the crater and the diatreme (details in Chako Tchamabé et al. [11]).

episodes. Such volcanoes can have dramatic change in the eruption processes given to the overlapping nature of the eruptive products. These can also create truncation and bias in the sedimentary and stratigraphic record as a response of lateral and vertical variation of subsurface explosive loci. The formation of the Yangpori diatreme (South Korea), for example, occurred in two distinct eruption phases, punctuated by sudden lowering of the explosion locus [90]. The first phase of eruption was initiated and maintained at a relatively shallow level within the water-logged basin fills, whereas the second eruptive phase was generated by explosions within a fracture-controlled or joint aquifer within the dacitic basement. This generated two cross-cutting diatreme structures, which resulted from migration of the explosion locus associated with basin-margin fault movement (**Figure 8**). Similar processes were suggested for the Barombi Mbo Maar in Cameroon, but in contrast to the Yangpori diatreme where a tectonically controlled migration was highlighted, new diatremes grew close to the first one at Barombi Mbo due to the discrete injection of new dikes. This implies that the growth process of these complex volcanoes cannot be "predicted" using such growth models, because they are way too complicated in terms of eruptive evolution. A generalized model may not apply for these volcanoes. Each complex monogenetic or polygenetic small volcano should be treated independently, and the growth model for its formation should be done taking into consideration the number of vents identified, the discontinuities observed within the stratigraphy, the eruptive timespan, and probably the geochemistry of the erupted materials.

5. Conclusions

1. Maar-diatreme volcanoes are small volcanic landforms formed as a result of strong MFCI explosive eruptions and usually following a single evolution with a succession of eruptive phases all related to a single eruption, that is closely related in time, and therefore they are usually considered as simple monogenetic volcanoes. However, recent examples of maar volcanoes show a more complex evolution, involving important timescale and breaks in the eruptive activity, changes in the eruptive style, and variations in the magma composition, suggesting the injection of different magma batches during long periods of time. Such complex volcanoes can be grouped into two end members:

    - Complex monogenetic volcanoes that are characterized by multiple eruptive phases but which evolved in a single eruptive episode. Here magma batches feeding the system erupt almost at the same time, with a very short break (months to years) insufficient to allow any significant erosion or alteration (e.g., palagonization) at the top of each eruptive package (deposits of one eruptive phase) and especially the formation of a paleosoil. These are polycyclic monogenetic volcanoes.

    - If the volcano formed during a very large timescale (e.g., Ky to My) and if at least two eruptive episodes are identified with significant time gaps that can be measured by radiogenic dating methods, the volcano surely is a polygenetic volcano. In such cases, paleosoil layers or highly eroded or altered surfaces may separate the eruptive units.

It is also important to note that for such polygenetic volcanoes, all the eruptions should take place in very close vents that will form a final compound volcanic edifice with overlapping deposits. If the vents are distant ones from others, distinct, but very closely distributed monogenetic volcanic edifices might form.

2. Maars are characterized by composite stratigraphic sequences that are dominated by PDCs and minor fall beds and in some case spatter lava flows. However, for complex maars, sedimentological evidences to establish time gap during the growth of the edifice are crucial to establish the polygenetic nature of the volcano. Maars are also characterized by complex craters morphologies that reflect the complex eruptive evolution and the influence of numerous other factors such as the geologic and tectonic settings, the presence of any pre-volcanic lithological situations including older cones that might have been dissected by the maar-forming eruption or preexisting crater. Because the complexity of the crater morphology applies for both simple and complex maars, observed crater margin needs to be evaluated in respect to establish if the size and shape of the crater reflect the structural boundary of the maar or if this results from an erosion enlarged and/or lake overfilled boundary. In both cases, however, the structural boundary of the maar crater commonly results from the complex explosive excavation history, which is linked to multiple concomitant or timely spaced dike injections, and vent migration in the crater floor that can either be randomly distributed or followed by some structural element such as fissures.

3. The magmatic plumbing system also plays an important role on the growth of complex monogenetic volcanoes, especially maar volcanoes in which diatremes are present. Geochemical variations are sometimes noted at many simple and complex volcanoes. This either means that multiple but near-simultaneous magma batch rise took place or the chemical variations reflect magmatic differentiation en route or both. Thus, if no time gap can be established between the eruptive units, a polymagmatic monogenetic volcano will develop. In contrast, if the complex magmatic activity is correlated with many eruptive episodes, the volcano will be presented as a complex polymagmatic monogenetic volcano with polygenetic inheritance.

4. Though a significant number of large and complex maar volcanoes are known, many of them might really be a reflection of short-lived volcanic events taking place nearly in the same place over longer time (ka range). This chapter clearly demonstrates the detailed complexity of maar eruptions that also emerged from other recent studies on other small-volume volcanoes. Even if the low levels of magmatic differentiation within some of these volcanoes do not allow observation of contrasting magmas in any single volcanic construct, systematic stratigraphically constrained analysis of sample sets might bring significant information on the formation and growth of maars. A complex combination of controlled factors includes the nature of the magmatic plumbing system, the substrate and the influence of local tectonic settings, the melting and ascent rates, groundwater availability, and the multiple injections of magmas successively or, concomitantly during a single eruption, vent migration and establishment of multiple sequential or even possibly concurrent eruption sites. Such detailed investigation would be necessary to

understand each volcanic system and it is only at the end that the volcano may be declare monogenetic or polygenetic.

5. These complex monogenetic volcanoes occur more often than it was previously thought, which is perhaps the reflection of the source region complexity and ascent mechanism. This line of research should be systematically examined in the future because it might hold important clue to understand the geological evolution and volcanic hazard associated with these small-volume magmatic systems located usually far from tectonic boundaries.

Acknowledgements

We thank the book editor for inviting this contribution. The main idea of the work originated from CTB's PhD results, conducted in the framework of the SATREPS-Ny-Mo project entitled "Magmatic Fluid Supply into Lakes Nyos and Monoun, and Mitigation of Natural Disasters in Cameroon." The project organizers and the funding institutions, Japan Science and Technology (JST) and Japan International Cooperation Agency (JICA), are greatly thanked here. Postdoc scholarship supports from National University Autonomous of Mexico (UNAM) and funding from *Consejo Nacional de Ciencia y Tecnología* (CONACyT) through the CONACyT-0150900 project, led by G. Carrasco-Núñez, have given the opportunity to CTB to work on maars of the Eastern Mexican Volcanic Belt (EMVB). Review by B. van Wyk de Vries and language editing by D. Miggins (Oregon State University, USA) significantly increased the readability of the text.

Author details

Boris Chako Tchamabé[1*], Gabor Kereszturi[2], Karoly Németh[2] and Gerardo Carrasco-Núñez[1]

*Address all correspondence to: boris.chako@yahoo.fr

1 Centre for Geosciences, National University Autonomous of Mexico (UNAM), Campus UNAM Juriquilla, Querétaro, Mexico

2 Volcanic Risk Solutions, Massey University, Palmerston North, New Zealand

References

[1] Lorenz, V. (2007) Syn- and posteruptive hazards of maar–diatreme volcanoes. *J Volcanol Geotherm Res*, 159, 285–312. Doi:10.1016/j.jvolgeores.2006.02.015.

[2] Wilson, L. (2009) Volcanism in the Solar System. Nature Geoscience, vol. 2, p. 389–97. Doi:10.1038/ngeo529.

[3] Shoemaker, E.M., Robinson, M.S., Eliason, E.M. (1994) The south pole region of the moon as seen by clementine. *Science*, 266, 1851–54. Doi:10.1126/science.266.5192.1851.

[4] Takada, A. (1994) The influence of regional stress and magmatic input on styles of monogenetic and polygenetic volcanism. *J Geophys Res*, 99, 13563–73. Doi: 10.1029/94JB00494.

[5] Walker, G.P.L. (2000) Basaltic volcanoes and volcanic systems. In: Sigurdsson, H., Houghton, B., Rymer, H., Stix, J., and McNutt, S., editors. Encyclopedia of Volcanoes. Academic Press, p. 283–9.

[6] Nemeth, K. (2010) Monogenetic volcanic fields; origin, sedimentary record, and relationship with polygenetic volcanism. *Spec Paper Geol Soc Am*, 470, 43–66. Doi: 10.1130/2010.2470(04).

[7] Johnson, P.J., Valentine, G.A., Cortés, J.A., Tadini, A. (2014) Basaltic tephra from monogenetic Marcath Volcano, central Nevada. *J Volcanol Geotherm Res*, 281, 27–33. Doi: 10.1016/j.jvolgeores.2014.05.007.

[8] Brenna, M., Cronin, S.J., Smith, I.E.M., Sohn, Y.K., Németh, K. (2010) Mechanisms driving polymagmatic activity at a monogenetic volcano, Udo, Jeju Island, South Korea. *Contr Mineral Petrol*, 160, 931–50. Doi:10.1007/s00410-010-0515-1.

[9] Sohn, Y.K., Cronin, S.J., Brenna, M., Smith, I.E.M., Németh, K., White, J.D.L., et al. (2012) Ilchulbong tuff cone, Jeju Island, Korea, revisited: a compound monogenetic volcano involving multiple magma pulses, shifting vents, and discrete eruptive phases. *Bull Geol Soc Am*, 124, 259–74. Doi:10.1130/B30447.1.

[10] Jordan, S.C., Jowitt, S.M., Cas, R.A.F. (2015) Origin of temporal–compositional variations during the eruption of Lake Purrumbete Maar, Newer Volcanics Province, southeastern Australia. *Bull Volcanol*, 77, 883. Doi:10.1007/s00445-014-0883-x.

[11] Chako Tchamabé, B., Ohba, T., Kereszturi, G., Németh, K., Aka, F.T., Youmen, D., et al. (2015) Towards the reconstruction of the shallow plumbing system of the Barombi Mbo Maar (Cameroon) implications for diatreme growth processes of a polygenetic maar volcano. *J Volcanol Geotherm Res*, 301, 293–313. Doi:10.1016/j.jvolgeores.2015.06.004.

[12] Alvarado, G.E., Soto, G.J., Salani, F.M., Ruiz, P., de Mendoza, L.H. (2011) The formation and evolution of Hule and Róo Cuarto maars, Costa Rica. *J Volcanol Geotherm Res*, 201, 342–56. Doi:10.1016/j.jvolgeores.2010.12.017.

[13] Giaccio, B., Marra, F., Hajdas, I., Karner, D.B., Renne, P.R., Sposato, A. (2009) $^{40}Ar/^{39}Ar$ and ^{14}C geochronology of the Albano maar deposits: implications for defining the age and eruptive style of the most recent explosive activity at Colli Albani Volcanic District, Central Italy. *J Volcanol Geotherm Res*, 185, 203–13. Doi:10.1016/j.jvolgeores.2009.05.011.

[14] Wohletz, K.H., McQueen, R.G. (1984) Volcanic and stratospheric dustlike particles produced by experimental water-melt interactions. *Geology*, 12, 591–4. Doi: 10.1130/0091-7613(1984)12<591:VASDPP>2.0.CO.

[15] Zimanowski, B., Fröhlich, G., Lorenz, V. (1995) Experiments on steam explosion by interaction of water with silicate melts. *Nucl Eng Des*, 155, 335–43. Doi:10.1016/0029-5493(94)00880-8.

[16] Zimanowski, B., Fröhlich, G., Lorenz, V. (1991) Quantitative experiments on phreatomagmatic explosions. *J Volcanol Geotherm Res*, 48, 341–58. Doi:10.1016/0377-0273(91)90050-A.

[17] Autin-Erickson, A., Büttner, R., Dellino, P., Ort, M.H., Zimanowski, B. (2008) Phreatomagmatic explosions of rhyolitic magma: experimental and field evidence. *J Geophys Res: Solid Earth*, 113, 1–12. Doi:10.1029/2008JB005731.

[18] Lorenz, V. (2003) Maar–diatreme volcanoes, their formation, and their setting in hard-rock or soft-rock environments. *Geolines*, 15, 72–83.

[19] White, J.D.L. (1996) Impure coolants and interaction dynamics of phreatomagmatic eruptions. *J Volcanol Geotherm Res*, 74, 155–70. Doi:10.1016/S0377-0273(96)00061-3.

[20] Jordan, S.C., Cas, R.A.F., Hayman, P.C. (2013) The origin of a large (>3 km) maar volcano by coalescence of multiple shallow craters: Lake Purrumbete maar, southeastern Australia. *J Volcanol Geotherm Res*, 254, 5–22. Doi:10.1016/j.jvolgeores.2012.12.019.

[21] Freda, C., Gaeta, M., Karner, D.B., Marra, F., Renne, P.R., Taddeucci, J., et al. (2006) Eruptive history and petrologic evolution of the Albano multiple maar (Alban Hills, Central Italy). *Bull Volcanol*, 68, 567–91. Doi:10.1007/s00445-005-0033-6.

[22] Chako Tchamabé, B., Youmen, D., Owona, S., Issa, Ohba, T., Nemeth, K., et al. (2013) Eruptive history of the Barombi Mbo Maar, Cameroon Volcanic Line, Central Africa: constraints from volcanic facies analysis. *Central Eur J Geosci*, 5, 480–96. Doi:10.2478/s13533-012-0147-2.

[23] Tamen, J., Nkoumbou, C., Mouafo, L., Reusser, E., Tchoua, F.M. (2007) Petrology and geochemistry of monogenetic volcanoes of the Barombi Koto volcanic field (Kumba graben, Cameroon volcanic line): implications for mantle source characteristics. *C R Geosci*, 339, 799–809. Doi:10.1016/j.crte.2007.09.007.

[24] Rausch, J., Grobéty, B., Vonlanthen, P. (2015) Eifel maars: quantitative shape characterization of juvenile ash particles (Eifel Volcanic Field, Germany). *J Volcanol Geotherm Res*, 291, 86–100. Doi:10.1016/j.jvolgeores.2014.11.008.

[25] Seib, N., Kley, J., Büchel, G. (2013) Identification of maars and similar volcanic landforms in the West Eifel Volcanic Field through image processing of DTM data: efficiency of different methods depending on preservation state. *Int J Earth Sci*, 102, 875–901. Doi: 10.1007/s00531-012-0829-5.

[26] Carrasco-Núñez, G., Ort, M.H., Romero, C. (2007) Evolution and hydrological conditions of a maar volcano (Atexcac crater, Eastern Mexico). *J Volcanol Geotherm Res*, 159, 179–97. Doi:10.1016/j.jvolgeores.2006.07.001.

[27] López-Rojas, M., Carrasco-Núñez, G. (2015) Depositional facies and migration of the eruptive loci for Atexcac axalapazco (central Mexico): implications for the morphology of the crater. Revista Mexicana de Ciencias Geológicas 377–94.

[28] Ort, M.H., Carrasco-Núñez, G. (2009) Lateral vent migration during phreatomagmatic and magmatic eruptions at Tecuitlapa Maar, east-central Mexico. *J Volcanol Geotherm Res*, 181, 67–77. Doi:10.1016/j.jvolgeores.2009.01.003.

[29] Németh, K., Martin, U., Harangi, S. (2001) Miocene phreatomagmatic volcanism at Tihany (Pannonian Basin, Hungary). *J Volcanol Geotherm Res*, 111, 111–35. Doi:10.1016/S0377-0273(01)00223-2.

[30] Németh, K., Martin, U., Harangi, S. (1999) Miocene maar/diatreme volcanism at the Tihany Peninsula (Pannonian Basin); the Tihany Volcano. *Acta Geol Hung*, 42, 349–77.

[31] Kereszturi, G., Csillag, G., Németh, K., Sebe, K., Balogh, K., Jáger, V. (2010) Volcanic architecture, eruption mechanism and landform evolution of a Plio/Pleistocene intracontinental basaltic polycyclic monogenetic volcano from the Bakony–Balaton Highland Volcanic Field, Hungary. *Central Eur J Geosci*, 2, 362–84. Doi:10.2478/v10085-010-0019-2.

[32] Houghton, B.F., Wilson, C.J.N., Rosenberg, M.R., Smith, I.E.M., Parker, R.J. (2000) Mixed deposits of complex magmatic and phreatomagmatic volcanism: an example from Crater Hill, Auckland, New Zealand. *Bull Volcanol*, 58, 59–66. Doi:10.1007/s004450050126.

[33] Németh, K., White, J.D.L. (2003) Reconstructing eruption processes of a Miocene monogenetic volcanic field from vent remnants: Waipiata Volcanic Field, South Island, New Zealand. *J Volcanol Geotherm Res*, 124, 1–21. Doi:10.1016/S0377-0273(03)00042-8.

[34] Agustín-Flores, J., Németh, K., Cronin, S.J., Lindsay, J.M., Kereszturi, G., Brand, B.D., et al. (2014) Phreatomagmatic eruptions through unconsolidated coastal plain sequences, Maungataketake, Auckland Volcanic Field (New Zealand). *J Volcanol Geotherm Res*, 276, 46–63. Doi:10.1016/j.jvolgeores.2014.02.021.

[35] Wohletz, K.H., Sheridan, M.F. (1983) Martian rampart crater ejecta: experiments and analysis of melt–water interaction. *Icarus*, 56, 15–37. Doi:10.1016/0019-1035(83)90125-2.

[36] Kokelaar, P. (1986) Magma–water interactions in subaqueous and emergent basaltic volcanism. *Bull Volcanol*, 48, 275–89.

[37] Büttner, R., Zimanowski, B. (1998) Physics of thermohydraulic explosions. *Phys Rev E*, 57, 5726–9. Doi:10.1103/PhysRevE.57.5726.

[38] Wohletz, K.H. (1986) Explosive magma–water interactions: thermodynamics, explosion mechanisms, and field studies. *Bull Volcanol*, 48, 245–64. Doi:10.1007/BF01081754.

[39] Büttner, R., Dellino, P., La Volpe, L., Lorenz, V., Zimanowski, B. (2002) Thermohydraulic explosions in phreatomagmatic eruptions as evidenced by the comparison between pyroclasts and products from Molten Fuel Coolant Interaction experiments. *J Geophys Res: Solid Earth*, 107, ECV 5–1–ECV 5–14. Doi:10.1029/2001JB000511.

[40] Sheridan, M.F., Wohletz, K.H. (1981) Hydrovolcanic explosions: the systematics of water–pyroclast equilibration. *Science*, 212, 1387–9. Doi:10.1126/science.212.4501.1387.

[41] Valentine, G.A. (1987) Stratified flow in pyroclastic surges. *Bull Volcanol*, 49, 616–30 Doi: 10.1007/BF01079967.

[42] Sohn, Y.K., Chough, S.K. (1989) Depositional processes of the Suwolbong tuff ring, Cheju Island (Korea). *Sedimentology*, 36, 837–55. Doi:10.1111/j.1365-3091.1989.tb01749.x.

[43] White, J.D.L., Schmincke, H.U. (1999) Phreatomagmatic eruptive and depositional processes during the 1949 eruption on La Palma (Canary Islands). *J Volcanol Geotherm Res*, 94, 283–304. Doi:10.1016/S0377-0273(99)00108-0.

[44] Sulpizio, R., Mele, D., Dellino, P., La Volpe, L. (2007) Deposits and physical properties of pyroclastic density currents during complex Subplinian eruptions: the AD 472 (Pollena) eruption of Somma-Vesuvius, Italy. *Sedimentology*, 54, 607–35. Doi:10.1111/j.1365-3091.2006.00852.x.

[45] Brand, B.D., Clarke, A.B., Semken, S. (2009) Eruptive conditions and depositional processes of Narbona Pass Maar volcano, Navajo volcanic field, Navajo Nation, New Mexico (USA). *Bull Volcanol*, 71, 49–77. Doi:10.1007/s00445-008-0209-y.

[46] Ngwa, C.N., Suh, C.E., Devey, C.W. (2010) Phreatomagmatic deposits and stratigraphic reconstruction at Debunscha Maar (Mt Cameroon volcano). *J Volcanol Geotherm Res*, 192, 201–11. Doi:10.1016/j.jvolgeores.2010.02.012.

[47] Gernon, T.M., Upton, B.G.J., Hincks, T.K. (2013) Eruptive history of an alkali basaltic diatreme from Elie Ness, Fife, Scotland. *Bull Volcanol*, 75, 1–20. Doi:10.1007/s00445-013-0704-7.

[48] Houghton, B.F., Hackett, W.R. (1984) Strombolian and phreatomagmatic deposits of ohakune craters, Ruapehu, New Zealand: a complex interaction between external water and rising basaltic magma. J Volcanol Geotherm Res. 21, 207–31.

[49] Clarke, H., Troll, V. (2005) Changing eruptive styles and textural features from phreatomagmatic to strombolian activity of basaltic littoral cones: Los Erales cinder cone, Tenerife, Canary Islands. *Estud Geol*, 134, 121–34.

[50] Clarke, H., Troll, V.R., Carracedo, J.C. (2009) Phreatomagmatic to Strombolian eruptive activity of basaltic cinder cones: Montaña Los Erales, Tenerife, Canary Islands. *J Volcanol Geotherm Res*, 180, 225–45. Doi:10.1016/j.jvolgeores.2008.11.014.

[51] D'Oriano, C., Poggianti, E., Bertagnini, A., Cioni, R., Landi, P., Polacci, M., et al. (2005) Changes in eruptive style during the A.D. 1538 Monte Nuovo eruption (Phlegrean Fields, Italy): the role of syn-eruptive crystallization. *Bull Volcanol*, 67, 601–21. Doi: 10.1007/s00445-004-0397-z.

[52] Nicholson, R.S., Gardner, J.E., Neal, C.A. (2011) Variations in eruption style during the 1931 A.D. eruption of Aniakchak volcano, Alaska. *J Volcanol Geotherm Res*, 207, 69–82. Doi:10.1016/j.jvolgeores.2011.08.002.

[53] Van Otterloo, J., Raveggi, M., Cas, R.A.F., Maas, R. (2014) Polymagmatic activity at the monogenetic Mt Gambier volcanic complex in the Newer Volcanics Province, SE Australia: new insights into the occurrence of intraplate volcanic activity in Australia. *J Petrol*, 55, 1317–51. Doi:10.1093/petrology/egu026.

[54] Brenna, M., Cronin, S.J., Németh, K., Smith, I.E.M., Sohn, Y.K. (2011) The influence of magma plumbing complexity on monogenetic eruptions, Jeju Island, Korea. *Terra Nova*, 23, 70–5. Doi:10.1111/j.1365-3121.2010.00985.x.

[55] Chako Tchamabé, B. (2015) Volcano-Stratigraphy and Geochemistry of Tephra Deposits and its Relevance for Understanding the Polygenetic Inheritance and Plumbing System of Maar–Diatreme Volcanoes: Clues for Hazards Prospective, A Case Study for the Barombi Mbo Maar, Cameroon, Central Africa. Tokai University. Doi:ci.nii.ac.jp/naid/500000935695.

[56] Keating, G.N., Valentine, G.A., Krier, D.J. Perry, F.V. (2008) Shallow plumbing systems for small-volume basaltic volcanoes. *Bull Volcanol*, 70, 563–82. Doi:10.1007/s00445-007-0154-1.

[57] Hintz, A.R., Valentine, G.A. (2012) Complex plumbing of monogenetic scoria cones: new insights from the Lunar Crater Volcanic Field (Nevada, USA). *J Volcanol Geotherm Res*, 239–240, 19–32. Doi:10.1016/j.jvolgeores.2012.06.008.

[58] Kereszturi, G., Nemeth, K. (2013) Monogenetic basaltic volcanoes: genetic classification, growth, geomorphology and degradation. *Update Volcanol New Adv Understand Volcan System*, 3–88. Doi:10.5772/51387.

[59] Chako Tchamabé, B., Ohba, T., Ooki, S., Youmen, D., Owona, S., et al. (2014) Temporal evolution of the Barombi Mbo Maar, a polygenetic maar–diatreme volcano of the Cameroon Volcanic Line. *Int J Geosci*, 5, 1315–23. Doi:10.4236/ijg.2014.511108.

[60] Németh, K., Kereszturi, G. (2015) Monogenetic volcanism: personal views and discussion. *Int J Earth Sci*, 104, 2131–46. Doi:10.1007/s00531-015-1243-6.

[61] Ollier, C.D. (1967) Maars their characteristics, varieties and definition. *Bull Volcanol*, 31, 45–73. Doi:10.1007/BF02597005.

[62] Lorenz, V. (1973) On the formation of maars. *Bull Volcanol*, 37, 183–204. Doi:10.1007/BF02597130.

[63] Lorenz, V. (1986) On the growth of maars and diatremes and its relevance to the formation of tuff rings. *Bull Volcanol*, 48, 265–74. Doi:10.1007/BF01081755.

[64] Németh, K. (2001) Deltaic density currents and turbidity deposits related to maar crater rims and their importance for palaegeographic reconstruction of the Bakony–Balaton Highland volcanic field, Hungary. *Spec Publs. Int Ass Sediment*, 31, 261–77. Doi: 10.1002/9781444304275.ch19.

[65] Begét, J.E., Hopkins, D.M., Charron, S.D. (1996) The largest known maars on earth, Seward Peninsula, Northwest Alaska. *Arctic*, 49, 62–9.

[66] Siebe, C., Salinas, S. (2014) Distribution of monogentic phreatomagmatic volcanoes (maars, tuff-cones and tuff-rings) in the Mexican Volcanic Belt and their tectonic and hydrogeologic environment. In: Carrasco-Núñez, G., Aranda-Gómez, J.J., Ort, M.H., Silva-Corona J.J., editor. *IAVCEI-5IMC-Meeting, Querétaro November 2014 (Mexico)*, p. 183–4.

[67] Kshirsagar, P., Siebe, C., Guilbaud, M.N., Salinas, S., Layer, P.W. (2015) Late Pleistocene Alberca de Guadalupe maar volcano (Zacapu basin, Michoacán): stratigraphy, tectonic setting, and paleo-hydrogeological environment. *J Volcanol Geotherm Res*, 304, 214–36. Doi:10.1016/j.jvolgeores.2015.09.003.

[68] Siebe, C., Macías, J.L., Abrams, M., Rodríguez, S., Castro, R., Delgado, H. (1995) Quaternary explosive volcanism and pyroclastic deposits in east central Mexico: implications for future hazards. *Guidebook of Geological Excursions: In Conjunction with the Annual Meeting of the Geol Soc Am*, New Orleans, Louisiana, November 6–9, 1995, p. 1–48.

[69] Kereszturi, G., Németh, K., Cronin, S.J., Agustín-Flores, J., Smith, I.E.M., Lindsay, J. (2013) A model for calculating eruptive volumes for monogenetic volcanoes—implication for the Quaternary Auckland Volcanic Field, New Zealand. *J Volcanol Geotherm Res*, 266, 16–33. Doi:10.1016/j.jvolgeores.2013.09.003.

[70] Sottili, G., Palladino, D.M., Gaeta, M., Masotta, M. (2012) Origins and energetics of maar volcanoes: examples from the ultrapotassic Sabatini Volcanic District (Roman Province, Central Italy). *Bull Volcanol*, 74, 163–86. Doi:10.1007/s00445-011-0506-8.

[71] Lorenz, V. (1970) Some aspects of the eruption mechanism of the big hole maar, central oregon. *Bull Geol Soc Am*, 81, 1823–30. Doi:10.1130/0016-7606(1970)81[1823:SAOTEM]2.0.CO;2.

[72] White, J.D. (1991) Maar–diatreme phreatomagmatism at Hopi Buttes, Navajo Nation (Arizona), USA. *Bull Volcanol*, 53, 239–58. Doi:10.1007/BF00414522.

[73] Vazquez, J.A., Ort, M.H. (2006) Facies variation of eruption units produced by the passage of single pyroclastic surge currents, Hopi Buttes volcanic field, USA. *J Volcanol Geotherm Res*, 154, 222–36. Doi:10.1016/j.jvolgeores.2006.01.003.

[74] Sottili, G., Taddeucci, J., Palladino, D.M., Gaeta, M., Scarlato, P., Ventura, G. (2009) Subsurface dynamics and eruptive styles of maars in the Colli Albani Volcanic District, Central Italy. *J Volcanol Geotherm Res*, 180, 189–202. Doi:10.1016/j.jvolgeores.2008.07.022.

[75] Kereszturi, G., Németh, K., Csillag, G., Balogh, K., Kovács, J. (2011) The role of external environmental factors in changing eruption styles of monogenetic volcanoes in a Mio/Pleistocene continental volcanic field in western Hungary. *J Volcanol Geotherm Res*, 201, 227–40. Doi:10.1016/j.jvolgeores.2010.08.018.

[76] Heiken, G.H. (1971) Tuff rings: examples from the Fort Rock-Christmas Lake Valley basin, south-central Oregon. *J Geophys Res*, 76, 1626–5615. Doi:10.1029/JB076i023p05615.

[77] Miyabuchi, Y., Ikebe, S., Watanabe, K. (2008) Geological constraints on the 2003–2005 ash emissions from the Nakadake crater lake, Aso Volcano, Japan. *J Volcanol Geotherm Res*, 178, 169–83. Doi:10.1016/j.jvolgeores.2008.06.025.

[78] Chough, S.K., Sohn, Y.K. (1990) Depositional mechanics and sequences of base surges, Songaksan tuff ring, Cheju Island, Korea. *Sedimentology*, 37, 1115–1135. Doi:10.1111/j.1365-3091.1990.tb01849.x.

[79] Kurszlaukis, S., Fulop, A. (2013) Factors controlling the internal facies architecture of maar–diatreme volcanoes. *Bull Volcanol*, 75, 1–12. Doi:10.1007/s00445-013-0761-y.

[80] Blaikie, T.N., Ailleres, L., Cas, R.A.F., Betts, P.G. (2012) Three-dimensional potential field modelling of a multi-vent maar–diatreme—the Lake Coragulac maar, Newer Volcanics Province, south-eastern Australia. *J Volcanol Geotherm Res*, 235–236, 70–83. Doi:10.1016/j.jvolgeores.2012.05.002.

[81] Blaikie, T.N., Ailleres, L., Betts, P.G., Cas, R.A.F. (2014) A geophysical comparison of the diatremes of simple and complex maar volcanoes, Newer Volcanics Province, south-eastern Australia. *J Volcanol Geotherm Res*, 276, 64–81. Doi:10.1016/j.jvolgeores.2014.03.001.

[82] White, J.D.L., Ross, P.S. (2011) Maar–diatreme volcanoes: a review. *J Volcanol Geotherm Res*, 201, 1–29. Doi:10.1016/j.jvolgeores.2011.01.010.

[83] Lockwood, J.P., Rubin, M. (1989) Origin and age of the Lake Nyos maar, Cameroon. *J Volcanol Geotherm Res*, 39, 117–24. Doi:10.1016/0377-0273(89)90052-8.

[84] Aka, F.T., Yokoyama, T., Kusakabe, M., Nakamura, E., Tanyileke, G., Ateba, B., et al. (2008) U-series dating of Lake Nyos maar basalts, Cameroon (West Africa): implications for potential hazards on the Lake Nyos dam. *J Volcanol Geotherm Res*, 176, 212–24. Doi:10.1016/j.jvolgeores.2008.04.009.

[85] Németh, K., Cronin, S.J., Haller, M.J., Brenna, M., Csillag, G. (2010) Modern analogues for Miocene to Pleistocene alkali basaltic phreatomagmatic fields in the Pannonian Basin: "soft-substrate" to "combined" aquifer controlled phreatomagmatism in intraplate volcanic fields. *Cent Eur J Geosci*, 2, 339–61 (Research Article). Doi:10.2478/v10085-010-0013-8.

[86] Shane, P., Gehrels, M., Zawalna-Geer, A., Augustinus, P., Lindsay, J., Chaillou, I. (2013) Longevity of a small shield volcano revealed by crypto-tephra studies (Rangitoto volcano, New Zealand): change in eruptive behavior of a basaltic field. *J Volcanol Geotherm Res*, 257, 174–83. Doi:10.1016/j.jvolgeores.2013.03.026.

[87] McGee, L.E., Millet, M.A., Smith, I.E.M., Németh, K., Lindsay, J.M. (2012) The inception and progression of melting in a monogenetic eruption: Motukorea Volcano, the Auckland Volcanic Field, New Zealand. *Lithos*, 155, 360–74. Doi:10.1016/j.lithos.2012.09.012.

[88] Agustín-Flores, J., Németh, K., Cronin, S.J., Lindsay, J.M., Kereszturi, G. (2015) Construction of the North Head (Maungauika) tuff cone: a product of Surtseyan volcanism, rare in the Auckland Volcanic Field, New Zealand. *Bull Volcanol*, 77, 97–120.. Doi:10.1007/s00445-014-0892-9.

[89] Jankovics, M.É., Harangi, S., Németh, K., Kiss, B., Ntaflos, T. (2015) A complex magmatic system beneath the Kissomlyó monogenetic volcano (western Pannonian Basin): evidence from mineral textures, zoning and chemistry. *J Volcanol Geotherm Res*, 301, 38–55. Doi:10.1016/j.jvolgeores.2015.04.010.

[90] Son, M., Kim, J.S., Jung, S., Ki, J.S., Kim, M.C., Sohn, Y.K. (2012) Tectonically controlled vent migration during maar–diatreme formation: an example from a Miocene half-graben basin in SE Korea. *J Volcanol Geotherm Res*, 223–224, 29–46. Doi:10.1016/j.jvolgeores.2012.02.002.

[91] Auer, A., Martin, U., Németh, K. (2007) The Fekete-hegy (Balaton Highland Hungary) "soft-substrate" and "hard-substrate" maar volcanoes in an aligned volcanic complex —implications for vent geometry, subsurface stratigraphy and the palaeoenvironmental setting. *J Volcanol Geotherm Res*, 159, 225–45. Doi:10.1016/j.jvolgeores.2006.06.008.

[92] Shaw, C.S.J., Woodland, A.B., Hopp, J., Trenholm, N.D. (2010) Structure and evolution of the Rockeskyllerkopf Volcanic Complex, West Eifel Volcanic Field, Germany. *Bull Volcanol*, 72, 971–90. Doi:10.1007/s00445-010-0380-9.

[93] Pirrung, M., Fischer, C., Büchel, G., Gaupp, R., Lutz, H, Neuffer, F.O. (2003) Lithofacies succession of maar crater deposits in the Eifel area (Germany). *Terra Nova*, 15, 125–32. Doi:10.1046/j.1365-3121.2003.00473.x.

[94] McKnight, S.B., Williams, S.N. (1997) Old cinder cone or young composite volcano? The nature of Cerro Negro, Nicaragua. *Geology*, 25, 339–42. Doi:10.1130/0091-7613(1997)025<0339:OCCOYC>2.3.CO;2.

[95] Courtland, L.M., Kruse, S.E., Connor, C.B., Connor, L.J., Savov, I.P., Martin, K.T. (2012) GPR investigation of tephra fallout, Cerro Negro volcano, Nicaragua: a method for constraining parameters used in tephra sedimentation models. *Bull Volcanol*, 74, 1409–24. Doi:10.1007/s00445-012-0603-3.

[96] Siebert, L., Carrasco-Núñez, G. (2002) Late-Pleistocene to precolumbian behind-the-arc mafic volcanism in the eastern Mexican Volcanic Belt; implications for future hazards. *J Volcanol Geotherm Res*, 115, 179–205. Doi:10.1016/S0377-0273(01)00316-X.

[97] Befus, K.S., Hanson, R.E., Lehman, T.M., Griffin, W.R. (2008) Cretaceous basaltic phreatomagmatic volcanism in West Texas: Maar complex at Peña Mountain, Big Bend National Park. *J Volcanol Geotherm Res*, 173, 245–64. Doi:10.1016/j.jvolgeores.2008.01.021.

[98] Muirhead, J.D., Van Eaton, A.R., Re, G., White, J.D.L., Ort, M.H. (2016) Monogenetic volcanoes fed by interconnected dikes and sills in the Hopi Buttes volcanic field, Navajo Nation, USA. *Bull Volcanol*, 78, 11. Doi:10.1007/s00445-016-1005-8.

[99] Valentine, G.A., Perry, F.V., WoldeGabriel, G. (2000) Field characteristics of deposits from spatter-rich pyroclastic density currents at Summer Coon volcano, Colorado. *J Volcanol Geotherm Res*, 104, 187–99. Doi:10.1016/S0377-0273(00)00206-7.

[100] Kienle, J., Kyle, P.R., Self, S., Motyka, R.J., Lorenz, V. (1980) Ukinrek Maars, Alaska, I. April 1977 eruption sequence, petrology and tectonic setting. *J Volcanol Geotherm Res*, 7, 11–37. Doi:10.1016/0377-0273(80)90018-9.

[101] Giordano, G. (1998) Facies characteristics and magma–water interaction of the White Trachytic Tuffs (Roccamonfina Volcano, southern Italy). *Bull Volcanol*, 60, 10–26. Doi: 10.1007/s004450050213.

[102] Valentine, G.A., Sottili, G., Palladino, D.M., Taddeucci, J. (2015) Tephra ring interpretation in light of evolving maar–diatreme concepts: Stracciacappa maar (central Italy). *J Volcanol Geotherm Res*, 308, 19–29. Doi:10.1016/j.jvolgeores.2015.10.010.

[103] Dosseto, A., Buss, H.L., Suresh, P.O. (2012) Rapid regolith formation over volcanic bedrock and implications for landscape evolution. *Earth Planet Sci Lett*, 337–338, 47–55. Doi:10.1016/j.epsl.2012.05.008.

[104] Self, S., Kienle, J., Huot, J.P. (1980) Ukinrek Maars, Alaska, II. Deposits and formation of the 1977 craters. *J Volcanol Geotherm Res*, 7, 39–65. Doi:10.1016/0377-0273(80)90019-0.

[105] Fisher, R.V., Heiken, G., Hulen, J.B. (1998) Volcanoes: Crucibles of Change. Princeton: Nature Publishing Group.

[106] Fisher, R.V., Schmincke, H.-U. (1984) Pyroclastic Rocks. Berlin, Heidelberg: Springer-Verlag. Doi:10.1007/978-3-642-74864-6.

[107] Lefebvre, N.S., White, J.D.L., Kjarsgaard, B.A. (2016) Arrested diatreme development: standing Rocks East, Hopi Buttes, Navajo Nation, USA. *J Volcanol Geotherm Res*, 310, 186–208. Doi:10.1016/j.jvolgeores.2015.12.007.

[108] Kereszturi, G., Németh, K. (2011) Shallow-seated controls on the evolution of the Upper Pliocene Kopasz-hegy nested monogenetic volcanic chain in the Western Pannonian Basin (Hungary). *Geol Carpath*, 62, 535–46. Doi:10.2478/v10096-011-0038-3.

[109] Graettinger, A.H., Valentine, G.A., Sonder, I., Ross, P.S., White, J.D.L., Taddeucci, J. (2014) Maar–diatreme geometry and deposits: subsurface blast experiments with variable explosion depth. *Geochem Geophys Geosyst*, 15, 740–64. Doi: 10.1002/2013GC005198.

[110] Graettinger, A.H., Valentine, G.A., Sonder, I. (2015) Circum-crater variability of deposits from discrete, laterally and vertically migrating volcanic explosions: experimental evidence and field implications. *J Volcanol Geotherm Res*, 308, 61–9. Doi:10.1016/j.jvolgeores.2015.10.019.

[111] Valentine, G.A., Graettinger, A.H., Macorps, É., Ross, P.S., White, J.D.L., Döhring, E., et al. (2015) Experiments with vertically and laterally migrating subsurface explosions with applications to the geology of phreatomagmatic and hydrothermal explosion craters and diatremes. *Bull Volcanol*, 77, 1–17. Doi:10.1007/s00445-015-0901-7.

[112] Sonder, I., Graettinger, A. H. Valentine, G. A. (2015) Scaling multiblast craters: general approach and application to volcanic craters." *Journal of Geophys Res: Solid Earth*, 120, 6141–6158. Doi: 15 10.1002/2015JB012018

[113] Abrams, M.J., Siebe, C. (1994) Cerro Xalapaxco: an unusual tuff cone with multiple explosion craters, in central Mexico (Puebla). *J Volcanol Geotherm Res*, 63, 183–99. Doi: 10.1016/0377-0273(94)90073-6.

[114] Anzidei, M., Carapezza, M.L., Esposito, A., Giordano, G., Lelli, M., Tarchini, L. (2008) The Albano Maar Lake high resolution bathymetry and dissolved CO_2 budget (Colli Albani volcano, Italy): constrains to hazard evaluation. *J Volcanol Geotherm Res*, 171, 258–68. Doi:10.1016/j.jvolgeores.2007.11.024.

[115] Houghton, B.F., Wilson, C.J.N., Smith, I.E.M. (1999) Shallow-seated controls on styles of explosive basaltic volcanism: a case study from New Zealand. *J Volcanol Geotherm Res*, 91, 97–120. Doi:10.1016/S0377-0273(99)00058-X.

[116] Francis, E.H. (1968) Effect of sedimentation on volcanic processes, including neck-sill relationships, in the British Carboniferous. *PROC 23RD INT GEOL CONGR PRAGUE 2*, Czech Republic, 1968, p. 163–74.

[117] Stollhofen, H. (1997) Regional European Meeting of Sedimentology. Heidelberg.

[118] Lorenz, V., Haneke, J. (2004) Relationship between diatremes, dykes, sills, laccoliths, intrusive-extrusive domes, lava flows, and tephra deposits with unconsolidated water-saturated sediments in the late Variscan intermontane Saar-Nahe Basin, SW Germany. *Geol Soc Lond Special Publ*, 234, 75–124. Doi:10.1144/gsl.sp.2004.234.01.07.

[119] Németh, K., Martin, U. (2007) Shallow sill and dyke complex in western Hungary as a possible feeding system of phreatomagmatic volcanoes in "soft-rock" environment. *J Volcanol Geotherm Res*, 159, 138–52. Doi:10.1016/j.jvolgeores.2006.06.014.

[120] Martín-Serrano, A., Vegas, J., García-Cortés, A., Galán, L., Gallardo-Millán, J.L., Martín-Alfageme, S., et al. (2009) Morphotectonic setting of maar lakes in the Campo de

Calatrava Volcanic Field (Central Spain, SW Europe). *Sediment Geol*, 222, 52–63. Doi: 10.1016/j.sedgeo.2009.07.005.

[121] Ross, P.S., Delpit, S., Haller, M.J., Németh, K., Corbella, H. (2011) Influence of the substrate on maar–diatreme volcanoes—an example of a mixed setting from the Pali Aike volcanic field, Argentina. *J Volcanol Geotherm Res*, 201, 253–71. Doi:10.1016/j.jvolgeores.2010.07.018.

[122] Delpit, S., Ross, P.S., Hearn, B.C. (2014) Deep-bedded ultramafic diatremes in the Missouri River Breaks volcanic field, Montana, USA: 1 km of syn-eruptive subsidence. *Bull Volcanol*, 76, 1–22. Doi:10.1007/s00445-014-0832-8.

[123] Valentine, G.A., van Wyk de Vries, B. (2014) Unconventional maar diatreme and associated intrusions in the soft sediment-hosted Mardoux structure (Gergovie, France). *Bull Volcanol*, 76, 1–16. Doi:10.1007/s00445-014-0807-9.

[124] Ross, P.S., White, J.D.L., Valentine, G.A., Taddeucci, J., Sonder, I., Andrews, R.G. (2013) Experimental birth of a maar–diatreme volcano. *J Volcanol Geotherm Res*, 260, 1–12. Doi: 10.1016/j.jvolgeores.2013.05.005.

[125] Macorps, É., Graettinger, A.H., Valentine, G.A., Ross, P.-S., White, J.D.L., Sonder, I. (2016) The effects of the host–substrate properties on maar–diatreme volcanoes: experimental evidence. *Bull Volcanol*, 78, 26. Doi:10.1007/s00445-016-1013-8.

[126] Aranda-Gómez, J.J., Luhr, J.F. (1996) Origin of the Joya Honda maar, San Luis Potosí, México. *J Volcanol Geotherm Res*, 74, 1–18. Doi:10.1016/S0377-0273(96)00044-3.

[127] McCord, S., Schladow, S.G. (1998) Numerical simulations of degassing scenarios for CO_2-rich Lake Nyos, Cameroon. *J Geophys Res*, 103, 12355–64.

[128] Taddeucci, J., Valentine, G.A., Sonder, I., White, J.D.L., Ross, P.S., Scarlato, P. (2013) The effect of pre-existing craters on the initial development of explosive volcanic eruptions: an experimental investigation. *Geophys Res Lett*, 40, 507–10. Doi:10.1002/grl.50176.

[129] Carn, S.A. (2000) The Lamongan volcanic field, East Java, Indonesia: physical volcanology, historic activity and hazards. *J Volcanol Geotherm Res*, 95, 81–108. Doi:10.1016/S0377-0273(99)00114-6.

[130] Gençalioğlu-Kuscu, G., Atilla, C., Cas, R.A.F., Kuscu, I. (2007) Base surge deposits, eruption history, and depositional processes of a wet phreatomagmatic volcano in Central Anatolia (Cora Maar). *J Volcanol Geotherm Res*, 159, 198–209. Doi:10.1016/j.jvolgeores.2006.06.013.

[131] Németh, K., Cronin, S.J., Smith, I.E.M., Agustín-Flores, J. (2012) Amplified hazard of small-volume monogenetic eruptions due to environmental controls, Orakei Basin, Auckland Volcanic Field, New Zealand. *Bull Volcanol*, 74, 2121–37. Doi:10.1007/s00445-012-0653-6.

[132] Smith, I.E.M., Blake, S., Wilson, C.J.N., Houghton, B.F. (2008) Deep-seated fractionation during the rise of a small-volume basalt magma batch: Crater Hill, Auckland, New Zealand. *Contr Miner Petrol*, 155, 511–27. Doi:10.1007/s00410-007-0255-z.

[133] Martin, U., Németh, K. (2005) Eruptive and depositional history of a Pliocene tuff ring that developed in a fluvio-lacustrine basin: Kissomlyó volcano (western Hungary). *J Volcanol Geotherm Res*, 147, 342–56. Doi:10.1016/j.jvolgeores.2005.04.019.

[134] Valentine, G.A., Graettinger, A.H., Sonder, I. (2014) Explosion depths for phreatomagmatic eruptions. *Geophys Res Lett*, 41, 3045–51. Doi:10.1002/2014GL060096.

[135] Valentine, G.A., Perry, F. V. (2007) Tectonically controlled, time-predictable basaltic volcanism from a lithospheric mantle source (central Basin and Range Province, USA). *Earth Planet Sci Lett*, 261, 201–16. Doi:10.1016/j.epsl.2007.06.029.

[136] Rasoazanamparany, C., Widom, E., Valentine, G.A., Smith, E.I., Cortés, J.A., Kuentz, D. et al. (2015) Origin of chemical and isotopic heterogeneity in a mafic, monogenetic volcanic field: a case study of the Lunar Crater Volcanic Field, Nevada. *Chem Geol*, 397, 76–93. Doi:10.1016/j.chemgeo.2015.01.004.

[137] Tadini, A., Bonali, F.L., Corazzato, C., Cortés, J.A., Tibaldi, A., Valentine, G.A. (2014) Spatial distribution and structural analysis of vents in the lunar crater volcanic field (Nevada, USA). *Bull Volcanol*, 76, 1–15. Doi:10.1007/s00445-014-0877-8.

[138] Camp, V.E., Roobol, M.J., Hooper, P.R. (1991) The Arabian continental alkali basalt province: part II. Evolution of Harrats Khaybar, Ithnayn, and Kura, Kingdom of Saudi Arabia. *Geol Soc Am Bull*, 363–91. Doi:10.1130/0016-7606(1991)103<0363:TACABP>2.3.CO;2.

[139] Hobden, B.J., Houghton, B.F., Davidson, J.P., Weaver, S.D. (1999) Small and short-lived magma batches at composite volcanoes: time windows at Tongariro volcano, New Zealand. *J Geol Soc*, 156, 865–8. Doi:10.1144/gsjgs.156.5.0865.

[140] Valentine, G.A., Shufelt, N.L., Hintz, A.R.L. (2011) Models of maar volcanoes, Lunar Crater (Nevada, USA). *Bull Volcanol*, 73, 753–65. Doi:10.1007/s00445-011-0451-6.

[141] Lorenz, V., Kurszlaukis, S. (2007) Root zone processes in the phreatomagmatic pipe emplacement model and consequences for the evolution of maar–diatreme volcanoes. *J Volcanol Geotherm Res*, 159, 4–32. Doi:10.1016/j.jvolgeores.2006.06.019.

[142] van Otterloo, J., Cas, R.A.F., Sheard, M.J. (2013) Eruption processes and deposit characteristics at the monogenetic Mt. Gambier Volcanic Complex, SE Australia: implications for alternating magmatic and phreatomagmatic activity. *Bull Volcanol*, 75, 1–21. Doi:10.1007/s00445-013-0737-y.

[143] Wohletz, K.H., Sheridan, M.F. (1983) Hydrovolcanic explosions. II. Evolution of basaltic tuff rings and tuff cones. *Am J Sci*, 385–413. Doi:10.2475/ajs.283.5.385.

[144] Ross, P.S., White, J.D.L. (2006) Debris jets in continental phreatomagmatic volcanoes: a field study of their subterranean deposits in the Coombs Hills vent complex, Antarctica. *J Volcanol Geotherm Res*, 149, 62–84. Doi:10.1016/j.jvolgeores.2005.06.007.

[145] Valentine, G.A. (2012) Shallow plumbing systems for small-volume basaltic volcanoes, 2: evidence from crustal xenoliths at scoria cones and maars. *J Volcanol Geotherm Res*, 223–224, 47–63. Doi:10.1016/j.jvolgeores.2012.01.012.

[146] Valentine, G.A., White, J.D.L., Ross, P.S., Amin, J., Taddeucci, J., Sonder, I., et al. (2012) Experimental craters formed by single and multiple buried explosions and implications for volcanic craters with emphasis on maars. *Geophys Res Lett*, 39. L20301. Doi: 10.1029/2012GL053716.

[147] Blaikie, T.N., Ailleres, L., Betts, P.G., Cas, R.A.F. (2014) Interpreting subsurface volcanic structures using geologically constrained 3-D gravity inversions: examples of maar–diatremes, Newer Volcanics Province, southeastern Australia. *J Geophys Res: Solid Earth*, 119, 3857–78. Doi:10.1002/2013JB010751.

[148] Valentine, G.A., White, J.D.L. (2012) Revised conceptual model for maar–diatremes: subsurface processes, energetics, and eruptive products. *Geology*, 40, 1111–4. Doi: 10.1130/G33411.1.

[149] McClintock, M., White, J.D.L., Houghton, B.F., Skilling, I.P. (2008) Physical volcanology of a large crater-complex formed during the initial stages of Karoo flood basalt volcanism, Sterkspruit, Eastern Cape, South Africa. *J Volcanol Geotherm Res*, 172, 93–111. Doi:10.1016/j.jvolgeores.2005.11.012.

8

Updates in Volcanology – From Volcano Modeling to Volcano Geology

Károly Németh

The book "*Updates in Volcanology: From volcano modelling to volcano geology*" is composed of 13 book chapters provided by authors from a great variety of disciplines. Each of the book chapter genuinely reflects the diversity of volcanological researches in recent years and documents new look at geological problems associated with volcanism and volcanic hazard research. The chapters from this book represent perfectly the current trends in volcanology as a merging research directions from geophysical aspects of volcanology and its traditional field-based methods. The book chapters have been grouped into three sections.

Section 1 is titled "Understanding the volcano system from petrology, geophysics to large-scale experiments" and provides a total of five chapters covering geophysical aspects of volcanic researches including their geochemical perspectives. The section starts with a comprehensive summary on the volcanic plumbing systems we know today and their relevance to understand the volcanic behavior from the magmatic source to a magma fragmentation that provides pyroclasts to be transported and deposited away from their source. Volcanic plumbing systems commonly defined as a network of various magmatic intrusive bodies (sheet- or dyke-like) and diverse size and shape of magmatic storage places (chambers) that located between the primary source and the surface anywhere the geological conditions allow to stall magma migration toward the surface [1–7]. The magmatic plumbing system of a volcano is a complex array of injected melts where various chemical processes take place that are strongly or loosely linked to the primary melt source and/or interact with the wall rocks. This book chapter provides a detailed summary of the methods recently applied to harvest information about these complex system feeding volcanoes on the surface. This chapter provides a summary on the potentials and the limitations of each applied methodology commonly used in magmatic plumbing system studies and highlight the fact that magmatic plumbing systems are complex geo-environments where physical and geochemical processes interact and create a complex 3D array of solidified melts that can then be traced in the geological record as igneous rocks.

Following the detailed summary of the melt generation and migration toward the surface, the next chapter takes the reader to the "surface" and provides a methodological summary on the power of photogrammetric surveys in volcanology. This chapter not only provides an exciting and new read on a traditional method to define volcanic morphology and their temporal changes during volcanic eruptions but it also takes the reader to a very active volcanic zone—Kamchatka—that produced several of the most spectacular volcanic eruptions in the last centuries including the Bezymianny 1955 and other [8, 9], Great Tolbachik 1975 [10–13] and recent 2012 events [14, 15] (**Figure 1**), Karymsky-lake Surtseyan eruption 1996 [16] or the Klyuchevskoy [17], Shiveluch [18], Ksudach [19], Mutnovsky, or Karymsky (**Figure 2**) volcanoes ongoing eruptions. Many of these eruptions not only provided new insight to volcanic eruption processes but also gave new conceptual models for gold recycling beneath volcanoes [20] or let us discover new minerals such as the tolbachite ($CuCl_2$) [21] and other unique minerals [22] or special rock names such as avachites a high-magnesian basalt group documented from the Avachinsky volcano [23]. This chapter looks back 100 years of research history and provides some forgotten images and documents from an area largely unknown for the majority of volcanologists. The chapter through its critical review and global comparison also provides a way forward for the development of the photogrammetric method in volcanology.

Figure 1. Strombolian style explosive eruption through a fissure vent at the Tolbachik volcano in Kamchatka (2012) [Photograph by Dmitry Melnikov, Institute of volcanology and seismology FEB RAS, Petropavlovsk-Kamchatsky, Russia].

Linked to the previous chapter, the third chapter provides a critical review with some original data and research about the role of volcano gravimetry. Volcano gravimetry has developed dramatically in recent years and became a fundamental tool to understand activity changes inside a volcano and hence use collected in situ or near real-time data to estimate volcanic alert status in active volcanoes [24–29]. Gravimetry studies also been used to detect various volcanic structures such as maar-diatreme volcanoes and establish the geological origin of sediment field terrestrial basins, which indeed not as simple process as we might think [30–33]. Similarly, gravimetry has been applied successfully to describe the internal architecture of large volcanoes [34] as well as relatively small-scale features as cavities in lava flows [35]. This chapter provides an easy to understand review of the science behind gravimetry studies then provides an evaluation for the future of gravimetry research on volcanoes.

Figure 2. Summit activity on the Karymsky strato-volcano in Kamchatka [Photo by Dmitry Melnikov, Institute of volcanology and seismology FEB RAS, Petropavlovsk-Kamchatsky, Russia].

The fourth chapter takes the reader to Kyushu in Japan and provides a state-of-art research report with a high-value review of the methods applied to understand the gravity gradient tensor in active volcanic regions in southern Japan. Southern Japan is a home of several post-Pliocene calderas that define the landscape today and the style of expected volcanism in the region [36]. The Kagoshima Bay and its surrounding is in particular one of the region where the highest frequency of silicic caldera-forming eruptions is known on Earth that produced multiple large (10 km<) across calderas and associated ignimbrite and silicic pyroclastic tephra sheets [37]. The large number of volcanoes in these regions alongside with the thick deposits made difficult to see the internal structure of many of these complex volcanoes and their structural relationship to the regional and volcano-tectonic structures; hence, geophysical methods are useful to provide data. This book chapter provides an introduction to the applicability of special gravity gradiometry survey with a new technique to define gravity gradient tensors. The chapter provides clear example to demonstrate that this method capable to locate structural elements such as caldera walls as it has been tested across the Aso caldera [38] hence offer some new avenue to explore for others in other complex volcanic terrains such as the caldera-pitted North Island in New Zealand [39, 40].

The fifth and closing chapter of this section of the book provides a summary of large-scale experiments to understand the source conditions and controlling parameters of explosive eruptions. This chapter is a valuable summary of the methods applied for large-scale experiments written by a group of researchers pioneered experimental volcanology through large-scale experiments; hence, the reader can get a very genuine view of such researches limitations and extent [41, 42]. In addition, the author group is also strongly linked to field-based research on pyroclastic deposits of pyroclastic density currents; therefore, this chapter provides well-established ideas and concepts [43]. In general, large-scale experiment to model various types of explosive processes, magma fragmentations, crater formation, and their numerical modeling is probably one of the most dynamic parts of volcanology in recent years [44–56]. In this chapter, different methodologies for investigating eruptive source conditions and the subsequent evolution of the eruptive plumes are presented. The methodologies range from observational techniques to large-scale experiments and numerical models. The chapter also poses fundamental research questions that can only be answered if we follow the proposed techniques to define the effect of unsteady flow conditions at the source on the eruptive column dynamics and the interaction between a convective plume and wind.

The second section of the book "volcanic eruptions and their impact to the environment" consists of five chapters. Each of these chapters provides various aspects of volcanism from their hazard and consequence perspective. The opening chapter of this section sum up the volcanological information associated with one of the less known volcano that produced one of the largest set of eruptions in the past 10 ka. Baekdusan volcano that is located in the frontier between China and North Korea is a volcano that has a basaltic shield capped by a trachyte-dominated sequence and a large caldera [57]. This chapter highlights the fact that Baekdusan volcano is an active and very complex volcano as its volcanic seismicity, ground deformation, and volcanic gas geochemistry yields indicate that a magmatic unrest takes place in the period between 2002 and 2006. The geological record and the recent volcano monitoring data suggest that the Mt. Baekdusan is an active volcano [58]; hence, proper volcanic hazard studies are needed to estimate the potential volcanic eruption scenarios a new eruption would follow.

The second chapter in this section takes the reader to Japan where an analysis of the magnitude-frequency distribution of slope failures was studied. The statistical data provided in this chapter are based on many years of research, and this chapter gives a very useful summary to the readers how to apply similar method to their fields. Interestingly, this chapter provides a relatively simple equation for the magnitude-frequency distribution of slope failures for larger than 10^7 m^3 volumes. The chapter concludes that this magnitude-frequency may applicable over several thousands of years of record making possible for future large-scale volcanic failure predictions. It is an alarming conclusion however, that larger than 10^9 m^3 mega collapses can occur in every 1000–2000 years period; hence, such mega events cannot be looked at as a rare events in Japan. Volcanic debris avalanches are common volcanic processes that generate a specific volcanic sedimentary successions and geomorphological disturbances; hence, their study is important [59, 60]. Such large-scale failures of volcanic edifices need to be studied in similar way as this chapter demonstrated to be able to provide magnitude-frequency distribution relationships in volcanic terrains prone to collapse.

The third chapter of this section deals with a research area of volcanic aerosol behavior developing dynamically in recent years to provide functioning models to volcanic aerosol effect to radiation in the Arctic. This chapter claims that despite the suspected perturbance of volcanic aerosols to the Arctic radiation balance, we know very little about their radiative impacts in the Arctic. It is also poorly understood partially due to the limited attention, this process has received what the effects of other aerosol types that are often present in the region [61], both natural and anthropogenic [62–64]. The link between volcanic events and their potential climatic impact is one of the subjects reach beyond the limits of volcano sciences and commonly connected to social studies, history, and art [65–69]. Research activity to understand volcanic aerosol behavior over the Arctic has also increased in recent years [70–72]; hence, this chapter provides a very useful summary of the current state of knowledge. The chapter points out that the Arctic environment is both unique and complicated, and the perturbations caused by volcanic aerosol need to be examined in a regional context. Due to the harsh environment, more data are derived from remote sensing; hence, this chapter provides a comprehensive summary of remotely sensed data collection techniques. The numerous models presented in this chapter show the strengths and shortcomings of volcanic ash transport and dispersion models in general [73–75] and calling future targeted research specifically designed for Arctic conditions. Among many, the authors claim that the effect of ash aggregation such as accretionary lapilli formation [76, 77] not many cases considered in ash transport and deposition models in spite their huge effect on the potential distribution pattern their deposition can show [78].

The fourth chapter in this section takes a reader to a fairly new research area that studies the link between corrosion effects on various alloys in a volcano-polluted seawater. Corrosion damage in human-built environment caused by volcanic processes can cause significant economic loss especially in marine steel infrastructures such as bridges, wharfs, platforms, and pipeline systems. While understanding the corrosion in general relatively well documented in subaerial conditions such as volcanic ash and acid rain effect on man-made structures [79–81]. Similar studies in a subaqueous environment are relatively rare [82]. This chapter provides an interesting overview of the world of corrosion including the description of techniques that can measure their rates. The results of the study of corrosion in volcanic-polluted waters can contribute to understanding of the volcanic hazards and associated risks of such processes that can act on human-built environments.

The final chapter in this section deals with the mineral assemblages recorded in various fumarolic systems across the most famous fumarole fields in Europe. This chapter provides a descriptive overview of the mineral phases recorded from various fumarolic systems focusing on examples from Italy and Greece. Fumarolic systems viewed as important environments where rare elements can concentrate even in economically significant amount [20]. The mineral variations associated with fumarolic systems are great, and such systems often act as harvesting ground to identify previously unknown minerals [10, 83]. Fumarolic systems are also viewed as a window to the magmatic system; hence, they commonly studied with an aim to understand the magma behavior provides the heat beneath such systems [84–86]. This chapter takes examples from well-known fumarolic systems such as those at Vulcano Island in Italy to raise some awareness among researchers on the significance of fumarolic systems in volcanic regions.

The final, third section of the book consists of three chapters dealing with volcanic geology problems that can be applied to older volcanic successions. The section starts with an exhaustive summary of the recent results on the study of subaqueous explosive and effusive volcanism in the Cabo de Gata region in SE Spain. Subaqueous volcanism in recent years became the center of volcanic researches [87] partially due to the racing to identify reliable evidences of presence of water or ice on extra-terrestrial regions commonly associated with volcanism [88, 89] and understand ore-forming processes in volcanic regions located or closely linked to subaqueous environments [90, 91]. It is just an addition to the increase of research activity that in the geological record, volcanic rocks are commonly associated with subaqueous sedimentary rocks; hence, there is a need to understand well the limitations of identification of various eruptive environment hence using volcanic rocks for paleoenvironmental (eruptive environment) reconstructions. Identification of peperites [92] for instance became a trademark to contribute to the eruptive environment discussions such as establishing subaqueous versus subaerial conditions [93]. Similarly, new researches on distinguishing explosive versus effusive non-explosive fragmentation of magma to feed large volume of hyaloclastite piles provided useful tools to use these deposits for paleoenvironmental reconstruction of rock units in the geological record [94, 95]. As a result of these research activities, a new commission called Commission of Subaqueous Volcanism of the International Association of Volcanology and Chemistry of the Earth's Interior was established in 2016. Cabo de Gata is probably one of the most perfectly exposed regions where the eruptive products (effusive and explosive) of subaqueous volcanism can be accessed along coastal sections [96]. This chapter provides a very detailed summary for the research results of the study of the Cabo de Gata volcanic rocks based on many previous research works [97].

Figure 3. The AD 1256 fissure eruption site near the city of Al Madinah in Saudi Arabia, a potential geosite in the proposed Al Madinah Volcanic Geopark.

The second chapter in this section addresses a question: how polygenetic are monogenetic volcanoes? This question sounds strange as the two word, monogenetic versus polygenetic are opposing and negating each other. The authors take the reader to a journey to demonstrate the difficulty to define monogenetic volcanism especially if we are looking at volcanic successions preserved in a geological record. Such problem has been addressed recently in many places arguing that in old but partially preserved and eroded successions, the distinction between volcanic rock units formed in a single and short volcanic process can look very similar to those that formed over longer time without significant evidence of erosional surfaces in the rock units. This chapter draws again the attention to view small-volume volcanoes as a system that functions as a source to surface model [98]. *In such model, the timing, longevity, and the total eruptive volume of individual eruptive episodes and the total volcanic activity can produce complex volcanic facies architecture composed of multiple eruptive units commonly forming such small-volume volcanoes* [99, 100]. This chapter explores the volcanic facies architecture of small-volume volcanoes that carry obvious signs of complex eruptive history that commonly can be translated to relatively long eruptive history and an elevated volume of magma involvement [101–103]. This chapter is useful to understand the gradual transition of small to large volcanoes, hence to see the link between monogenetic and polygenetic volcanism [104].

The final chapter of this section and the entire book takes the reader to a volcanic region in Inner Mongolia, China, where volcanic rocks help to understand the evolution of the Central Asian Orogenic Belt [105–108]. While this region is not directly considered as a locality of volcanological studies, its geological history documents significant magmatological processes associated with terrane accretion and associated volcanic processes [109]. The region is also a home of advanced studies of adakite magmas, and hence, it is inferred that magmatism resulted from intermediate to felsic magmas that carry geochemical characteristics indicative for the melt to be sourced by partial melting of the altered basalt that is subducted below volcanic arcs [110]. This chapter also highlights the methods need to be applied to recognize various stages of terrain accretion and petrogenetic processes associated with a complex plate margin process.

The book overall provides a great diversity of subjects relevant to volcanic researches. While the book purely a result of "blue sky" attempt to collect new research outputs reflecting the current trends in volcanic researches, the selection of chapters reflects well the dynamic nature of volcanological researches. The book naturally cannot provide a balanced summary of the current advances in volcanology. Large and dynamically developing segments of current volcanological research are not covered in this book. Reports on the advances of crater formation based on analog experiments [47, 50], experimental volcanological studies [76, 111–113], lava flow dynamic modeling through experiments and numerical codes [114, 115], developing new methods for eruption forecasting especially the eruptions' economic impact [116–119], experimental and field-based studies of maar-diatreme volcanism [48] or new advances in understanding magma fragmentation, vesiculation, and their internal and external controlling parameters are among many new research fields this book has not provided overviews. Also, recently, IAVCEI has established a new commission called Commission on Volcano Geoheritage and Protected Volcanic Landscape, which is a clear sign that geoeducation, geoconservation, and geotouristic researches and programs associated with volcanism claiming their own place among volcanic sciences [120] (**Figure 3**). It is hoped that in a future "volcanology" book, reports on these field's research results will be incorporated. Until that please enjoy this book!

Author details

Károly Németh

Address all correspondence to: k.nemeth@massey.ac.nz

Volcanic Risk Solutions, Massey University, Palmerston North, New Zealand

References

[1] Allard, P., et al., Mount Etna 1993–2005: Anatomy of an evolving eruptive cycle. Earth-Science Reviews, 2006. 78(1–2): p. 85–114.

[2] Cartwright, J. and D.M. Hansen, Magma transport through the crust via interconnected sill complexes. Geology, 2006. 34(11): p. 929–932.

[3] Holt, S.J., S.P. Holford, and J. Foden, New insights into the magmatic plumbing system of the South Australian Quaternary Basalt province from 3D seismic and geochemical data. Australian Journal of Earth Sciences, 2013. 60(8): p. 797–816.

[4] Kavanagh, J.L., D. Boutelier, and A.R. Cruden, The mechanics of sill inception, propagation and growth: Experimental evidence for rapid reduction in magmatic overpressure. Earth and Planetary Science Letters, 2015. 421: p. 117–128.

[5] Gudmundsson, A., Strengths and strain energies of volcanic edifices: Implications for eruptions, collapse calderas, and landslides. Natural Hazards and Earth System Sciences, 2012. 12(7): p. 2241–2258.

[6] Geshi, N., S. Kusumoto, and A. Gudmundsson, Effects of mechanical layering of host rocks on dike growth and arrest. Journal of Volcanology and Geothermal Research, 2012. 223: p. 74–82.

[7] Geshi, N., K. Németh, and T. Oikawa, Growth of phreatomagmatic explosion craters: A model inferred from Suoana crater in Miyakejima Volcano, Japan. Journal of Volcanology and Geothermal Research, 2011. 201(1–4): p. 30–38.

[8] Belousov, A., Deposits of the 30 March 1956 directed blast at Bezymianny volcano, Kamchatka, Russia. Bulletin of Volcanology, 1996. 57(8): p. 649–662.

[9] Belousov, A., et al., Pyroclastic surges and flows from the 8-10 May 1997 explosive eruption of Bezymianny volcano, Kamchatka, Russia. Bulletin of Volcanology, 2002. 64(7): p. 455–471.

[10] Chaplygin, I., et al., Native gold from volcanic gases at Tolbachik 1975–76 and 2012–13 Fissure Eruptions, Kamchatka. Journal of Volcanology and Geothermal Research, 2015. 307: p. 200–209.

[11] Fedotov, S.A., A.P. Khrenov, and A.M. Chirkov, Tolbachik Great Fissure Eruption of 1975, kamchatka. Doklady Akademii Nauk Sssr, 1976. 228(5): p. 1193–1196.

[12] Fedotov, S.A., et al., Major Tolbachik Fissure Eruption in Kamchatka, South Break of 1975–1976. Doklady Akademii Nauk Sssr, 1977. 237(5): p. 1155–1158.

[13] Budmikov, V.A., Y.K. Markhinin, and A.A. Ovsyannikov, The quantity, distribution and petrochemical features of pyroclastics of the Great Tolbachik Fissure Eruption., in The Great Tolbachik Fissure Eruption, S.A. Fedotov and Y.K. Markhinin, Editors. 1983, Cambridge University Press: Cambridge. p. 41–56.

[14] Lundgren, P., et al., Dike model for the 2012–2013 Tolbachik eruption constrained by satellite radar interferometry observations. Journal of Volcanology and Geothermal Research, 2015. 307: p. 79–88.

[15] Volynets, A.O., et al., Monitoring of the volcanic rock compositions during the 2012–2013 fissure eruption at Tolbachik volcano, Kamchatka. Journal of Volcanology and Geothermal Research, 2015. 307: p. 120–132.

[16] Belousov, A. and M. Belousova, Eruptive process, effects and deposits of the 1996 and the ancient basaltic phreatomagmatic eruptions in Karymskoye Lake, Kamchatka, Russia. Special Publication of the International Association of Sedimentologists, 2001. 30: p. 35–60.

[17] Koulakov, I.Y., et al., Magma sources in the mantle wedge beneath the volcanoes of the Klyuchevskoy group and Kizimen based on seismic tomography modeling. Russian Geology and Geophysics, 2016. 57(1): p. 82–94.

[18] Belousov, A.B., The Shiveluch volcanic-eruption of 12 November 1964—Explosive eruption provoked by failure of the edifice. Journal of Volcanology and Geothermal Research, 1995. 66(1–4): p. 357–365.

[19] Braitseva, O.A., et al., The caldera-forming eruption of Ksudach volcano about cal AD 240: The greatest explosive event of our era in Kamchatka, Russia. Journal of Volcanology and Geothermal Research, 1996. 70(1–2): p. 49–65.

[20] Zelenski, M., V.S. Kamenetsky, and J. Hedenquist, Gold recycling and enrichment beneath volcanoes: A case study of Tolbachik, Kamchatka. Earth and Planetary Science Letters, 2016. 437: p. 35–46.

[21] Vergasova, L.P. and S.K. Filatov, A new mineral Tolbachite, CuCl2. Doklady Akademii Nauk Sssr, 1983. 270(2): p. 415–417.

[22] Pekov, I.V., et al., New zinc and potassium chlorides from fumaroles of the Tolbachik volcano, Kamchatka, Russia: Mineral data and crystal chemistry. III. Cryobostryxite, $KZnCl_3 \bullet 2H_2O$. European Journal of Mineralogy, 2015. 27(6): p. 805–812.

[23] Portnyagin, M.V., et al., Petrology of avachites, high-magnesian basalts of Avachinsky volcano, Kamchatka: I. General characteristics and composition of rocks and minerals. Petrology, 2005. 13(2): p. 99–121.

[24] Delgado, F. and A. Pavez, New insights into La Pacana caldera inner structure based on a gravimetric study (central Andes, Chile). Andean Geology, 2015. 42(3): p. 313–328.

[25] Crossley, D., J. Hinderer, and U. Riccardi, The measurement of surface gravity. Reports on Progress in Physics, 2013. 76(4):046101. doi: 10.1088/0034-4885/76/4/046101.

[26] Battaglia, M., et al., 4D volcano gravimetry. Geophysics, 2008. 73(6): p. WA3–WA18.

[27] Behncke, B., et al., Ground deformation and gravity changes on the island of Pantelleria in the geodynamic framework of the Sicily Channel. Journal of Volcanology and Geothermal Research, 2006. 150(1–3): p. 146–162.

[28] Pingue, F., et al., Ground deformation and gravimetric monitoring at Somma-Vesuvius and in the Campanian volcanic area (Italy). Physics and Chemistry of the Earth Part A: Solid Earth and Geodesy, 2000. 25(9–11): p. 747–754.

[29] Saibi, H., J. Gottsmann, and S. Ehara, Post-eruptive gravity changes from 1999 to 2004 at Unzen volcano (Japan): A window into shallow aquifer and hydrothermal dynamics. Journal of Volcanology and Geothermal Research, 2010. 191(1–2): p. 137–147.

[30] Sesulka, V., et al., Identification of a buried Late Cenozoic maar-diatreme structure (North Moravia, Czech Republic). Geologica Carpathica, 2014. 65(6): p. 471–479.

[31] Nickschick, T., H. Kaempf, and T. Jahr, The "Triasscholle" near Greiz, Germany-a volcanic origin? Bulletin of Volcanology, 2014. 76:806. DOI 10.1007/s00445-014-0806-x.

[32] Valenta, J.J., et al., Problems and challenges in detection of pre-Mesozoic maar volcanoes: Example from the Principalek Volcano in the Permian Krkonose Piedmont Basin. Journal of Geosciences, 2014. 59(3): p. 169–181.

[33] Flechsig, C., et al., Integrated geophysical and geological methods to investigate the inner and outer structures of the Quaternary Mytina maar (W-Bohemia, Czech Republic). International Journal of Earth Sciences, 2015. 104(8): p. 2087–2105.

[34] La Delfa, S., G. Patane, and J.C. Tanguy, Kilometer-scale heterogeneities inside volcanoes revealed by using a set of geophysical methods: Variable stress field at Mount Etna, Sicily. Physics of the Earth and Planetary Interiors, 2000. 121(1–2): p. 157–173.

[35] Deroussi, S., et al., Localization of cavities in a thick lava flow by microgravimetry. Journal of Volcanology and Geothermal Research, 2009. 184(1–2): p. 193–198.

[36] Suzukikamata, K. and H. Kamata, The proximal facies of the Tosu Pyroclastic-flow deposit erupted from Aso Caldera, Japan. Bulletin of Volcanology, 1990. 52(5): p. 325–333.

[37] Aramaki, S., Formation of the Aira caldera, southern Kyushu, approximately 22,000 years ago. Journal of Geophysical Research, 1984. 89(NB10): p. 8485–8501.

[38] Miyabuchi, Y., Post-caldera explosive activity inferred from improved 67–30 ka tephrostratigraphy at Aso Volcano, Japan. Journal of Volcanology and Geothermal Research, 2011. 205(3–4): p. 94–113.

[39] Wilson, C.J.N. and J.V. Rowland, The volcanic, magmatic and tectonic setting of the Taupo Volcanic Zone, New Zealand, reviewed from a geothermal perspective. Geothermics, 2016. 59: p. 168–187.

[40] Cole, J.W., et al., Okataina Volcanic Centre, Taupo Volcanic Zone, New Zealand: A review of volcanism and synchronous pluton development in an active, dominantly silicic caldera system. Earth-Science Reviews, 2014. 128: p. 1–17.

[41] Dioguardi, F. and P. Dellino, PYFLOW: A computer code for the calculation of the impact parameters of Dilute Pyroclastic Density Currents (DPDC) based on field data. Computers and Geosciences, 2014. 66: p. 200–210.

[42] Duerig, T., et al., A new method for the determination of the specific kinetic energy (SKE) released to pyroclastic particles at magmatic fragmentation: Theory and first experimental results. Bulletin of Volcanology, 2012. 74(4): p. 895–902.

[43] Mele, D., et al., Hazard of pyroclastic density currents at the Campi Flegrei Caldera (Southern Italy) as deduced from the combined use of facies architecture, physical modeling and statistics of the impact parameters. Journal of Volcanology and Geothermal Research, 2015. 299: p. 35–53.

[44] Ross, P.S., et al., Experimental birth of a maar-diatreme volcano. Journal of Volcanology and Geothermal Research, 2013. 260: p. 1–12.

[45] Mader, H.M., et al., Dynamics of explosive degassing of magma: Observations of fragmenting two-phase flows. Journal of Geophysical Research-Solid Earth, 1996. 101(B3): p. 5547–5560.

[46] Roche, O., D.C. Buesch, and G.A. Valentine, Slow-moving and far-travelled dense pyroclastic flows during the Peach Spring super-eruption. Nature Communications, 2016. 7.

[47] Macorps, E., et al., The effects of the host-substrate properties on maar-diatreme volcanoes: Experimental evidence. Bulletin of Volcanology, 2016. 78:26. DOI 10.1007/s00445-016-1013-8.

[48] Valentine, G.A., et al., Experiments with vertically and laterally migrating subsurface explosions with applications to the geology of phreatomagmatic and hydrothermal explosion craters and diatremes. Bulletin of Volcanology, 2015. 76:15. DOI 10.1007/s00445-015-0901-7.

[49] Sweeney, M.R. and G.A. Valentine, Transport and mixing dynamics from explosions in debris-filled volcanic conduits: Numerical results and implications for maar-diatreme volcanoes. Earth and Planetary Science Letters, 2015. 425: p. 64–76.

[50] Sonder, I., A.H. Graettinger, and G.A. Valentine, Scaling multiblast craters: General approach and application to volcanic craters. Journal of Geophysical Research-Solid Earth, 2015. 120(9): p. 6141–6158.

[51] Graettinger, A.H., et al., Facies distribution of ejecta in analog tephra rings from experiments with single and multiple subsurface explosions. Bulletin of Volcanology, 2015. 77:8. DOI 10.1007/s00445-015-0951-x.

[52] Valentine, G.A., et al., Experimental craters formed by single and multiple buried explosions and implications for volcanic craters with emphasis on maars. Geophysical Research Letters, 2012. Vol. 39, L20301, doi:10.1029/2012GL053716.

[53] Doronzo, D.M., et al., Numerical analysis of the effect of topography on deposition from dilute pyroclastic density currents. Earth and Planetary Science Letters, 2010. 300(1–2): p. 164–173.

[54] Doronzo, D.M., et al., Dust storms, volcanic ash hurricanes, and turbidity currents: Physical similarities and differences with emphasis on flow temperature. Arabian Journal of Geosciences, 2016. 9:290. DOI 10.1007/s12517-016-2351-8.

[55] Roche, O., Nature and velocity of pyroclastic density currents inferred from models of entrainment of substrate lithic clasts. Earth and Planetary Science Letters, 2015. 418: p. 115–125.

[56] Lube, G., et al., Synthesizing large-scale pyroclastic flows: Experimental design, scaling, and first results from PELE. Journal of Geophysical Research-Solid Earth, 2015. 120(3): p. 1487–1502.

[57] Yun, S.-H. and J. Koh, Petrochemical characteristics of volcanic rocks of historic era at Mt. Baekdusan. Journal of the Geological Society of Korea, 2014. 50(6): p. 753–769.

[58] Park, K., Research Trends on Mt. Baekdu Volcano - Is She Dormant or Active? Journal of The Korean Geomorphological Association, 2013. 20(4): p. 117–131.

[59] Tost, M., S.J. Cronin, and J.N. Procter, Transport and emplacement mechanisms of channelised long-runout debris avalanches, Ruapehu volcano, New Zealand. Bulletin of Volcanology, 2014. 76(12).

[60] Procter, J.N., S.J. Cronin, and A.V. Zernack, Landscape and sedimentary response to catastrophic debris avalanches, western Taranaki, New Zealand. Sedimentary Geology, 2009. 220(3–4): p. 271–287.

[61] Iwi, A.M., et al., Mechanisms linking volcanic aerosols to the atlantic meridional overturning circulation. Journal of Climate, 2012. 25(8): p. 3039–3051.

[62] VanCuren, R.A., et al., Aerosols and their sources at Summit Greenland – First results of continuous size- and time-resolved sampling. Atmospheric Environment, 2012. 52: p. 82–97.

[63] Scalabrin, E., et al., Amino acids in Arctic aerosols. Atmospheric Chemistry and Physics, 2012. 12(21): p. 10453–10463.

[64] Allen, R.J. and S.C. Sherwood, The impact of natural versus anthropogenic aerosols on atmospheric circulation in the Community Atmosphere Model. Climate Dynamics, 2011. 36(9–10): p. 1959–1978.

[65] Toohey, M., et al., Climatic and societal impacts of a volcanic double event at the dawn of the Middle Ages. Climatic Change, 2016. 136(3–4): p. 401–412.

[66] Miller, G.H., et al., Abrupt onset of the Little Ice Age triggered by volcanism and sustained by sea-ice/ocean feedbacks. Geophysical Research Letters, 2012. Vol. 39, L02708,doi:10.1029/2011GL050168.

[67] Timmreck, C., Modeling the climatic effects of large explosive volcanic eruptions. Wiley Interdisciplinary Reviews-Climate Change, 2012. 3(6): p. 545–564.

[68] Meronen, H., et al., Climate effects of northern hemisphere volcanic eruptions in an Earth System Model. Atmospheric Research, 2012. 114: p. 107–118.

[69] Zhong, Y., et al., Centennial-scale climate change from decadally-paced explosive volcanism: A coupled sea ice-ocean mechanism. Climate Dynamics, 2011. 37(11–12): p. 2373–2387.

[70] Kravitz, B. and A. Robock, Climate effects of high-latitude volcanic eruptions: Role of the time of year. Journal of Geophysical Research-Atmospheres, 2011. Vol. 116, D01105, doi:10.1029/2010JD014448.

[71] Harris, B.M. and E.J. Highwood, A simple relationship between volcanic sulfate aerosol optical depth and surface temperature change simulated in an atmosphere-ocean general circulation model. Journal of Geophysical Research-Atmospheres, 2011. Vol. 116, D05109, doi:10.1029/2010JD014581.

[72] Robock, A., L. Oman, and G.L. Stenchikov, Regional climate responses to geoengineering with tropical and Arctic SO_2 injections. Journal of Geophysical Research-Atmospheres, 2008. Vol. 113, D16101, doi:10.1029/2008JD010050.

[73] Wilkins, K.L., et al., Using data insertion with the NAME model to simulate the 8 May 2010 Eyjafjallajokull volcanic ash cloud. Journal of Geophysical Research-Atmospheres, 2016. 121(1): p. 306–323.

[74] Mulena, G.C., et al., Examining the influence of meteorological simulations forced by different initial and boundary conditions in volcanic ash dispersion modelling. Atmospheric Research, 2016. 176: p. 29–42.

[75] Stevenson, J.A., et al., Big grains go far: Understanding the discrepancy between tephrochronology and satellite infrared measurements of volcanic ash. Atmospheric Measurement Techniques, 2015. 8(5): p. 2069–2091.

[76] Mueller, S.B., et al., Experimental volcanic ash aggregation: Internal structuring of accretionary lapilli and the role of liquid bonding. Earth and Planetary Science Letters, 2016. 433: p. 232–240.

[77] Van Eaton, A.R., et al., Hail formation triggers rapid ash aggregation in volcanic plumes. Nature Communications, 2015. 6:7860. DOI: 10.1038/ncomms8860.

[78] Folch, A., A. Costa, and G. Macedonio, FPLUME-1.0: An integral volcanic plume model accounting for ash aggregation. Geoscientific Model Development, 2016. 9(1): p. 431–450.

[79] Wilson, G., et al., Vulnerability of laptop computers to volcanic ash and gas. Natural Hazards, 2012. 63(2): p. 711–736.

[80] Watanabe, M., et al., Corrosion of copper and silver plates by volcanic gases. Corrosion Science, 2006. 48(11): p. 3759–3766.

[81] Kelestemur, O. and B. Demirel, Corrosion behavior of reinforcing steel embedded in concrete produced with finely ground pumice and silica fume. Construction and Building Materials, 2010. 24(10): p. 1898–1905.

[82] Morinaga, Y., et al., Comparative Study of Corrosion Resistance and Corrosion Products in Hot Spring, Seaside, and Other Environments Between Zn and Zn-7Al Alloy-Coated Steel. Corrosion, 2008. 64(12): p. 929–938.

[83] Jacobsen, M.J., et al., Oskarssonite, AlF3, a new fumarolic mineral from Eldfell volcano, Heimaey, Iceland. Mineralogical Magazine, 2014. 78(1): p. 215–222.

[84] Caliro, S., G. Chiodini, and A. Paonita, Geochemical evidences of magma dynamics at Campi Flegrei (Italy). Geochimica Et Cosmochimica Acta, 2014. 132: p. 1–15.

[85] Tassi, F., et al., Geochemical and isotopic changes in the fumarolic and submerged gas discharges during the 2011-2012 unrest at Santorini caldera (Greece). Bulletin of Volcanology, 2013. 75:711. DOI: 10.1007/s00445-013-0711-8.

[86] Paonita, A., et al., The episodic and abrupt geochemical changes at La Fossa fumaroles (Vulcano Island, Italy) and related constraints on the dynamics, structure, and compositions of the magmatic system. Geochimica Et Cosmochimica Acta, 2013. 120: p. 158–178.

[87] Cas, R.A.F. and G. Giordano, Submarine Volcanism: A review of the constraints, processes and products, and relevance to the Cabo de Gata volcanic succession. Italian Journal of Geosciences, 2014. 133(3): p. 362–377.

[88] Bernhardt, H., et al., Photogeologic mapping and the geologic history of the Hellas basin floor, Mars. Icarus, 2016. 264: p. 407–442.

[89] Zealey, W.J., Glacial, periglacial and glacio-volcanic structures on the Echus Plateau, upper Kasei Valles. Planetary and Space Science, 2009. 57(5–6): p. 699–710.

[90] Large, R.R., Australian volcanic-hosted massive sulfide deposits—features, styles, and genetic models. Economic Geology and the Bulletin of the Society of Economic Geologists, 1992. 87(3): p. 471–510.

[91] McPhie, J., et al., Origin of the supergiant Olympic Dam Cu-U-Au-Ag deposit, South Australia: Was a sedimentary basin involved? Geology, 2011. 39(8): p. 795–798.

[92] Skilling, I.P., J.D.L. White, and J. McPhie, Peperite: A review of magma-sediment mingling. Journal of Volcanology and Geothermal Research, 2002. 114(1–2): p. 1–17.

[93] Rosa, C.J.P., J. McPhie, and J.M.R.S. Relvas, Distinguishing peperite from other sediment-matrix igneous breccias: Lessons from the Iberian Pyrite Belt. Journal of Volcanology and Geothermal Research, 2016. 315: p. 28–39.

[94] Jutzeler, M., J. McPhie, and S.R. Allen, Explosive destruction of a Pliocene hot lava dome underwater: Dogashima (Japan). Journal of Volcanology and Geothermal Research, 2015. 304: p. 75–81.

[95] Allen, S.R. and J. McPhie, Products of neptunian eruptions. Geology, 2009. 37(7): p. 639–642.

[96] Biber, J.L. and T. McCann, Peperite development in a Miocene-age carbonate succession, Cabo de Gata volcanic area (Agua Amarga/Carboneras basins), SE Spain - The Breche Rouge revisited. Zeitschrift Der Deutschen Gesellschaft Fur Geowissenschaften, 2014. 165(2): p. 229–245.

[97] Soriano, C., et al., Facies architecture, emplacement mechanisms and eruption style of the submarine andesite El Barronal complex, Cabo de Gata, SE Spain. Journal of Volcanology and Geothermal Research, 2013. 264: p. 210–222.

[98] Canon-Tapia, E., Reappraisal of the significance of volcanic fields. Journal of Volcanology and Geothermal Research, 2016. 310: p. 26–38.

[99] Muirhead, J.D., et al., Monogenetic volcanoes fed by interconnected dikes and sills in the Hopi Buttes volcanic field, Navajo Nation, USA. Bulletin of Volcanology, 2016. 78:11. DOI: 10.1007/s00445-016-1005-8.

[100] Lefebvre, N.S., J.D.L. White, and B.A. Kjarsgaard, Arrested diatreme development: Standing Rocks East, Hopi Buttes, Navajo Nation, USA. Journal of Volcanology and Geothermal Research, 2016. 310: p. 186–208.

[101] Sheth, H.C. and E. Canon-Tapia, Are flood basalt eruptions monogenetic or polygenetic? International Journal of Earth Sciences, 2015. 104(8): p. 2147–2162.

[102] Jankovics, M.E., et al., A complex magmatic system beneath the Kissomlyo monogenetic volcano (western Pannonian Basin): Evidence from mineral textures, zoning and chemistry. Journal of Volcanology and Geothermal Research, 2015. 301: p. 38–55.

[103] Brenna, M., et al., Co-located monogenetic eruptions similar to 200 kyr apart driven by tapping vertically separated mantle source regions, Chagwido, Jeju Island, Republic of Korea. Bulletin of Volcanology, 2015. 77:43. DOI: 10.1007/s00445-015-0928-9.

[104] Nemeth, K. and G. Kereszturi, Monogenetic volcanism: Personal views and discussion. International Journal of Earth Sciences, 2015. 104(8): p. 2131–2146.

[105] Zhao, Y., et al., The Dunhuang block is a Paleozoic orogenic belt and part of the Central Asian Orogenic Belt (CAOB), NW China. Gondwana Research, 2016. 30: p. 207–223.

[106] Shi, Y., et al., Zircon ages and Hf isotopic compositions of Ordovician and Carboniferous granitoids from central Inner Mongolia and their significance for early and late Paleozoic evolution of the Central Asian Orogenic Belt. Journal of Asian Earth Sciences, 2016. 117: p. 153–169.

[107] Li, H., et al., Ordovician intrusive rocks from the eastern Central Asian Orogenic Belt in Northeast China: Chronology and implications for bidirectional subduction of the early Palaeozoic Palaeo-Asian Ocean. International Geology Review, 2016. 58(10): p. 1175–1195.

[108] Xiao, W. and M. Santosh, The western Central Asian Orogenic Belt: A window to accretionary orogenesis and continental growth. Gondwana Research, 2014. 25(4): p. 1429–1444.

[109] Liu, J., et al., The tectonic setting of early Permian bimodal volcanism in central Inner Mongolia: Continental rift, post-collisional extension, or active continental margin? International Geology Review, 2016. 58(6): p. 737–755.

[110] Castillo, P.R., Adakite petrogenesis. Lithos, 2012. 134: p. 304–316.

[111] Kueppers, U., et al., Fragmentation efficiency of explosive volcanic eruptions: A study of experimentally generated pyroclasts. Journal of Volcanology and Geothermal Research, 2006. 153(1–2): p. 125–135.

[112] Daniels, K.A. and T. Menand, An experimental investigation of dyke injection under regional extensional stress. Journal of Geophysical Research-Solid Earth, 2015. 120(3): p. 2014–2035.

[113] Dioguardi, F., P. Dellino, and S. de Lorenzo, Integration of large-scale experiments and numerical simulations for the calibration of friction laws in volcanic conduit flows. Journal of Volcanology and Geothermal Research, 2013. 250: p. 75–90.

[114] Kereszturi, G., et al., Emplacement conditions of the 1256 AD Al-Madinah lava flow field in Harrat Rahat, Kingdom of Saudi Arabia - Insights from surface morphology and lava flow simulations. Journal of Volcanology and Geothermal Research, 2016. 309: p. 14–30.

[115] Del Negro, C., A. Cappello, and G. Ganci, Quantifying lava flow hazards in response to effusive eruption. Geological Society of America Bulletin, 2016. 128(5–6): p. 752–763.

[116] Bebbington, M. and R. Zitikis, Dynamic Uncertainty in Cost-Benefit Analysis of Evacuation Prior to a Volcanic Eruption. Mathematical Geosciences, 2016. 48(2): p. 123–148.

[117] Sobradelo, R., et al., Probabilistic approach to decision-making under uncertainty during volcanic crises: Retrospective application to the El Hierro (Spain) 2011 volcanic crisis. Natural Hazards, 2015. 76(2): p. 979–998.

[118] Kereszturi, G., et al., Influences on the variability of eruption sequences and style transitions in the Auckland Volcanic Field, New Zealand. Journal of Volcanology and Geothermal Research, 2014. 286: p. 101–115.

[119] Runge, M.G., et al., Integrating geological and geophysical data to improve probabilistic hazard forecasting of Arabian Shield volcanism. Journal of Volcanology and Geothermal Research, 2016. 311: p. 41–59.

[120] Moufti, M.R. and K. Nemeth, The intra-continental Al Madinah volcanic field, Western Saudi Arabia: A proposal to establish Harrat Al Madinah as the first volcanic Geopark in the Kingdom of Saudi Arabia. Geoheritage, 2013. 5(3): p. 185–206.

PERMISSIONS

All chapters in this book were first published in UVFVMVG, by InTech Open; hereby published with permission under the Creative Commons Attribution License or equivalent. Every chapter published in this book has been scrutinized by our experts. Their significance has been extensively debated. The topics covered herein carry significant findings which will fuel the growth of the discipline. They may even be implemented as practical applications or may be referred to as a beginning point for another development.

The contributors of this book come from diverse backgrounds, making this book a truly international effort. This book will bring forth new frontiers with its revolutionizing research information and detailed analysis of the nascent developments around the world.

We would like to thank all the contributing authors for lending their expertise to make the book truly unique. They have played a crucial role in the development of this book. Without their invaluable contributions this book wouldn't have been possible. They have made vital efforts to compile up to date information on the varied aspects of this subject to make this book a valuable addition to the collection of many professionals and students.

This book was conceptualized with the vision of imparting up-to-date information and advanced data in this field. To ensure the same, a matchless editorial board was set up. Every individual on the board went through rigorous rounds of assessment to prove their worth. After which they invested a large part of their time researching and compiling the most relevant data for our readers.

The editorial board has been involved in producing this book since its inception. They have spent rigorous hours researching and exploring the diverse topics which have resulted in the successful publishing of this book. They have passed on their knowledge of decades through this book. To expedite this challenging task, the publisher supported the team at every step. A small team of assistant editors was also appointed to further simplify the editing procedure and attain best results for the readers.

Apart from the editorial board, the designing team has also invested a significant amount of their time in understanding the subject and creating the most relevant covers. They scrutinized every image to scout for the most suitable representation of the subject and create an appropriate cover for the book.

The publishing team has been an ardent support to the editorial, designing and production team. Their endless efforts to recruit the best for this project, has resulted in the accomplishment of this book. They are a veteran in the field of academics and their pool of knowledge is as vast as their experience in printing. Their expertise and guidance has proved useful at every step. Their uncompromising quality standards have made this book an exceptional effort. Their encouragement from time to time has been an inspiration for everyone.

The publisher and the editorial board hope that this book will prove to be a valuable piece of knowledge for researchers, students, practitioners and scholars across the globe.

LIST OF CONTRIBUTORS

Angelo Paone
Pusan National University, Department of Earth Science Education, Busan, Republic of Korea
Liceo Scientifico Meucci, Aprilia, Latina, Italy

Sung-Hyo Yun
Pusan National University, Department of Earth Science Education, Busan, Republic of Korea

Tonči Balić-Žunić
University of Copenhagen, Copenhagen, Denmark

Anna Garavelli and Pasquale Acquafredda
University of Bari, Bari, Italy

Sveinn Peter Jakobsson and Kristjan Jonasson
Icelandic Institute of Natural History, Gardabaer, Iceland

Athanasios Katerinopoulos and Konstantinos Kyriakopoulos
University of Athens, Athens, Greece

Cindy L. Young
School of Earth and Atmospheric Sciences, Georgia Institute of Technology, Atlanta, GA, USA

Jennifer W. Telling
National Center for Airborne Laser Mapping (NCALM), University of Houston, Houston, TX, USA

Carles Soriano
Institute of Earth Sciences "Jaume Almera", Superior Council for Scientific Research c/ Lluís Solé Sabarís s/n, Barcelona, Spain

Ray A.F. Cas
School of Earth, Atmosphere and Environment, Monash University, Monash University, Calyton Campus, Australia

Nancy R. Riggs
Geology Program, School of Earth Sciences and Environmental Sustainability, Northern Arizona University, Flagstaff, USA

Guido Giordano
Department of Science, Geological Sciences Section, Roma Tre University, Largo S. Leonardo Murialdo, Roma, Italy

Yuruo Shi
Beijing SHRIMP Center, Institute of Geology, Chinese Academy of Geological Sciences, Beijing, China

Rita Sánchez-Tovar, Ramón Fernández-Domene, Clara Escrivà-Cerdán, Bianca Lucas-Granados and José García-Antón
Chemical and Nuclear Engineering Department, Polytechnic University of Valencia, Electrochemical Engineering and Corrosion (IEC), Valencia, Spain

Rafael Leiva-García
School of Materials, University of Manchester, Manchester, UK

Boris Chako Tchamabé and Gerardo Carrasco-Núñez
Centre for Geosciences, National University Autonomous of Mexico (UNAM), Campus UNAM Juriquilla, Querétaro, Mexico

Gabor Kereszturi and Karoly Németh
Volcanic Risk Solutions, Massey University, Palmerston North, New Zealand

Index

S

T

V

www.ingramcontent.com/pod-product-compliance
Lightning Source LLC
LaVergne TN
LVHW060516210926
848485LV00005B/1086